Structural Materials
for Harbor
and Coastal
Construction

Structural Materials for Harbor and Coastal Construction

Lawrence L. Whiteneck, P.E.
Harbor Engineer

Lester A. Hockney, P.E.
Port Planner

McGraw-Hill Book Company
New York   St. Louis   San Francisco   Auckland
Bogotá   Hamburg   London   Madrid   Mexico
Milan   Montreal   New Delhi   Panama
Paris   São Paulo   Singapore
Sydney   Tokyo   Toronto

Library of Congress Cataloging-in-Publication Data

Whiteneck, Lawrence L.
    Structural materials for harbor and coastal
construction.

    Bibliography: p.
    Includes index.
    1. Shore protection.   2. Building materials.
3. Hydraulic structures.   I. Hockney, Lester A.
II. Title.
TC373.W45   1989   627'.2   87-27865
ISBN 0-07-068153-8

The material in this book was originally prepared for the U.S.
Army Engineer Waterways Experiment Station's Coastal
Engineering Research Center, report #10, dated February 1983.
This material is now in the public domain, and the U.S.
Department of the Navy has given McGraw-Hill permission to
publish it.

1234567890   DOC/DOC   89321098

ISBN 0-07-068153-8

The editors for this book were Joel Stein and Galen H. Fleck, the
designer was Naomi Auerbach, and the production supervisor was
Richard A. Ausburn. This book was set in Century Schoolbook. It was
composed by the McGraw-Hill Book Company Professional &
Reference Division composition unit.

Printed and bound by R. R. Donnelley & Sons Company.

Contents

Preface

This book is designed to provide coastal engineers with specific guidelines for selecting materials suitable for construction in the marine environment. It also condenses the subject of materials adequacy and suitability into a single reference. The coastal engineer can more easily select suitable materials without referring to disparate and often voluminous treatises on the properties of materials. The discussion is limited to the properties and treatments of materials that have proved most effective and durable in coastal structures. Emphasis has been put on giving full coverage to those materials.

Preparation of this treatise required the contributions of many authors. Chapters 2 and 3 were written by Ross A. Morrison and Louis J. Lee of Woodward-Clyde, geological consultants. Carrol M. Wakeman contributed much to Chapter 4. Dr. Albert L. Roebuck wrote all of Chapters 6 and 10. Robert J. Barrett contributed all of the geotextile data of Chapter 8; William J. Herron wrote all of Chapter 9. The project director wrote all of the remaining chapters with assistance from L. A. Hockney in writing Chapter 7. L. A. Hockney and S. H. Anderson were invaluable in reviewing and clarifying much of the manuscript.

The authors know that much remains to be done in the field of material use in coastal structures. In Section 11.3 some of the on-going investigations into the uses and protection of materials in coastal structures are indicated. It is the authors' belief that the compilation of these data will be helpful to all coastal engineers, teachers, and investigators.

Lawrence L. Whiteneck
Lester A. Hockney

Introduction

Construction materials for coastal structures can be classified as stone and earth, concrete, metals, wood, and synthetics. Some of those classes are treated in more than one section in the book in order to better clarify their use and performance in different structures. For example, stone and earth are discussed in separate sections, as are portland cement and other types of concrete and grout. The material requirements are discussed in sufficient detail to permit the coastal engineer and structural designer to evaluate materials on the basis of physical properties and past performance in coastal structures. The coastal structures generally considered are breakwaters, groins, seawalls, bulkheads, revetments, jetties, piers, wharves, piles, and navigation aids, as well as other less common structures.

Background

A number of excellent coastal engineering manuals and guides incorporate the best principles and criteria for design of coastal structures that have evolved through several decades of experience and research. However, most of these publications give only light treatment to the subject of materials adequacy. In-depth coverage of the pros and cons of each material used would, of course, make the publications too unwieldy for efficient use. Thus materials adequacy is usually covered by references to a number of disparate and often voluminous treatises on properties of materials, which places a heavy burden of literature research on the design professional. Many treatises cover aspects of materials that have little relevance to coastal work or information of value to the coastal engineer. In some cases, on the other hand, a reference may ignore unique effects of the coastal environment on a material.

1

Objective

This book aims to condense the subject of materials adequacy and suitability into a single document to support the coastal engineering profession and eliminate the superfluous coverage. The study is confined to the properties of materials, and treatments or variations thereof, that are applicable to coastal engineering structures. Emphasis is placed on full coverage of materials that have proved most effective and long-lasting in coastal structures. Materials useful in coastal structures can be equally applied to structures in inland waterways, rivers, streams, lakes, and reservoirs.

Organization

Experience has demonstrated the successes and failures of many materials used in the past to create various types of coastal structures. This book sets forth the principal physical properties of the materials and their importance in the material selection process. New synthetic materials and protective systems, including coatings and cathodic protection of metals, are included so that an evaluation of the long-term use of metals in the coastal zone can be considered in the material selection process. Problems associated with the uses of different materials are discussed when the physical properties which may impact upon and establish limited uses for the materials are considered.

Chapter 11 briefly discusses the significant uses of each material as well as some investigations and research that may improve the performance of the material in the coastal zone. The specific material chapters and the summary provide the coastal engineer with the fundamental physical properties information with which he or she must be concerned. Specific problems brought about by unusual or nonrecurring local conditions may require further investigation or research to ascertain the potential performance of a given material in a specific environment.

This book aims to assist the coastal engineer in the selection of materials appropriate for use in the coastal zone. Information on the placement, repair, and treatment of materials to improve durability is included, but it should be considered to be only a guide and not definitive instructions for field use of materials. Professional personnel experienced in such areas as wood treatment, protective systems, and cathodic protection should be consulted for implementation of design parameters presented here.

Material Requirements
for Coastal Structures

The primary considerations in the selection of a material for use in a coastal structure are availability, strength, durability, material life as compared to the desired life of the structure, costs, ease of maintenance, and maintenance costs. Also, consideration must be given to structural stability and flexibility. Depending on location, the impact of the environment on the material may be a determining factor.

1.1 Structural Properties

Consideration of six structural properties is a key part of the material selection process.

1.1.1 Specific gravity

Specific gravity must be considered whether the material is to be placed in the water, on the ocean bottom, or on dry land and whether it is to be used as a floating structure. High-specific-gravity materials such as rock, earth, and concrete are essential for submerged structures, whereas floating structures must be of low-specific-gravity materials such as wood and most synthetics. Heavy materials are used for floating structures only if the design provides for adequate buoyancy.

1.1.2 Material strength

Material strength in tension, compression, and flexure may determine the size and stability of a structure. Most metals are high in tensile

---

This chapter was written by Lawrence L. Whiteneck.

strength, whereas unreinforced concrete is low in tensile strength but high in compressive strength. Rock and soils may have high strength as individual pieces or particles, but they may require the addition of high-strength adhesives to create acceptable coastal structures.

1.1.3 Resistance to cyclical and impact loading

Resistance to cyclical and impact loading, such as that caused by waves or coastal storm conditions, may require that consideration be given to material flexibility within the elastic limit or the flexible limits of an adhesive used in the structure. For example, cement used in making concrete has virtually no flexibility, whereas asphalt used as a binder can be quite flexible.

1.1.4 Resistance to seismic forces

Seismic forces, which can be both horizontal and vertical, may result in excessive structural stresses. They may require a structure to be stiff and rigid like one made of concrete or steel or one constructed of nonhomogeneous materials such that the seismic forces can be relieved in planned isolated areas of the structure.

1.1.5 Material flexibility

The material flexibility property includes the ability to bend without breaking or to be adaptable to change in configuration. Wood has some degree of flexibility; rubber and certain synthetic materials have high degrees of flexibility. Structural steel is usually considered a stiff material, but steel shapes such as cable, wire rope, and rods are highly flexible.

1.1.6 Structural size

Structural size may determine the structural material. Large structures such as breakwaters are usually composed of many pieces of one or more materials that may not be bound together to create a homogeneous mass, or they may be composed of sections of the same material bound together to create a single structure. Small structures such as sand fences may be a series of independent pieces or sections of a material acting independently.

Structures built of stone, earth, and asphalt generally cannot resist tensile stress. They can take loads in compression, shear, and impact only, and they must be designed accordingly. Concrete and wood may or may not be subjected to tensile stress or bending moments. If concrete is subjected to such a stress, reinforcing steel or prestressed cable must be employed to carry the tensile load. Steel, when properly designed, is capable of withstanding all types of stress. The sections flex or deflect when subjected to bending loads, and that movement should be considered in the design phase. Synthetics, particularly the sandbag and filter cloth materials, are mainly subjected to tension, impacts, flexing, and fatigue. They are seldom required to accept compressive loads.

1.2 Nonstructural Properties

Nonstructural properties may also be important, but the three that follow are generally considered to be the most important ones.

1.2.1 Durability

Durability is the ability of a material to withstand the effects of the service conditions to which the material is exposed. Many laboratory tests are available to measure the durability of materials, but it is extremely difficult to obtain direct correlations between laboratory tests and field use. Because of the severity of the coastal environment, it is important that field experience be carefully assessed when materials for coastal structures are selected. If a coastal structure is to function properly, the planned structural life must be known; a structural life projected for a short term (e.g., less than 10 years) may have a major impact on material selection and cost. The location of the structure with respect to local resources and materials will also have an impact on material selection.

Durability is generally related to the desired lifespan of the structure and the relation between first costs and maintenance. Igneous rock, among the stones, is usually the most durable variety. Depending on makeup, it may be extremely durable or, after a few years, it may fracture and partially disintegrate. Sedimentary rock should be examined very carefully because it is usually stratified, may not be well consolidated, and is subject to failure through shear, impact, fracture due to changes of water content, and/or chemical deterioration.

Earth is generally considered durable unless changes in water content or chemistry reduce grain size to the silt and clay range, which will result in plastic flow. Concrete is considered durable and will generally last the planned life of the structure so long as it is not exposed to adverse chemical reaction or excessive abrasion. Asphalt is not generally considered a durable material. It has low strength in compression or tension; it is subject to chemical reactions; and it is not resistant to impact or abrasion.

Steel is considered durable if it is properly maintained, but it is subject to rapid deterioration through corrosion and abrasion. Particularly in a wet-dry tidal area, abrasion can be severe at the sandline, where steel may deteriorate rapidly.

Wood is considered less durable than concrete, but its lifespan depends to a great extent on its characteristics, its usage, and the quality of its maintenance. It is an organic material and is therefore subject to attack by both plants and animals. Obviously, it is more subject to damage by fire than other materials are.

Synthetic materials are generally quite durable to chemical attack, but many of them will deteriorate rapidly when exposed to sunlight. They require very little maintenance; but because of their short history, the service life of most synthetics relative to the life of the structure is yet to be determined.

1.2.2 Adaptability

There is usually more than one material or combination of compatible materials that can satisfy the performance requirements of any given coastal structure. Selecting the proper materials, as well as a design adaptable to the structure site, is important. The size of the structure and the accessibility of construction materials must also be considered.

Stone and earth structures can assume a wide variety of shapes, and the materials are generally available. When formed into nonrigid structures, they can, for example, accommodate changes in foundation elevations or slope adjustments without losing structural integrity or the capacity to perform their functions in the structure. Stone can be used under most weather conditions, because it will accept major and rapid changes in temperature and moisture without major failures.

Concrete, with or without steel reinforcement, is very adaptable. Its use is similar to that of stone except for cost and the ability to resist abrasion. Concrete can compete in both strength and durability with steel or wood as piles or as sheet piles.

Because of its cost, steel normally is used only for piles, sheet piles, and beams, but it is extremely adaptable for use in complex structures. That is particularly true if the steel parts can be hot-dipped zinc-coated or be otherwise protected.

Wood is very adaptable except for limitations on the capacity to function well against large wave forces or in greater than moderate depths of water. It resists impact and abrasion well; it can resist tension, compression, or shear; and it is easily handled in construction. Synthetic and protective coating materials are usually special-use items; they are not considered to have a wide range of uses.

1.2.3 Fire resistance

When materials are to be used around water, fire resistance may not at first thought seem to be important; but in the case of cargo wharves, it can be a critical consideration. Stone and earth are generally regarded as very fire-resistant, especially those that are of an igneous or metamorphic nature. Sedimentary stone, because of stratification, is less fire-resistant.

Concrete is generally fire-resistant unless it is exposed to very high temperatures. Reinforced concrete, when exposed to extreme temperatures for an extended period of time, may fail because of excessive expansion of the steel, which will result in spalling and cracking of the concrete. Steel, of course, is not combustible; but if it is exposed to high temperatures, it will tend to warp and lose strength. A rigid steel structure may also tend to warp and buckle because of excessive expansion of its members.

Asphalt is vulnerable to relatively low fire-induced temperature rises and is not considered to be fire-resistant. Wood, the least fire-resistant of all coastal construction materials, is vulnerable to fire. Synthetics also are vulnerable to fire, and they may generate dangerous toxic fumes as well.

1.3 Availability

The availability of both the materials and the equipment necessary to build a structure may limit the selection of materials. Lack of availability of the ingredients required for mixing concrete onsite may necessitate the use of more readily available wood; lack of availability of utilities during construction may limit emplacement methods and thus limit the materials used. Site access with respect to local resources and materials must also be considered. For example, transporting steel and concrete into remote areas may be difficult, whereas timber may be readily available. In that instance, the durability and other properties of wood must be carefully considered to determine whether wood is a suitable material. Availability is one of the most important factors influencing materials selection, especially when considered in conjunction with transportation costs.

1.3.1 Abundance

Stone. Stone is generally abundant in the continental United States and most outlying areas. Along the coasts of the Gulf of Mexico and the South Atlantic, however, sources of stone are 240 kilometers (km) [150 miles (m)] or more from projects, so handling costs can become a major factor. In many areas, particularly the volcanic areas of the Pacific Ocean, the stone may be of such low density or will be so badly fractured in use as not to be suitable as armor stone. The mere presence of large stone sources is not a guarantee of either suitability or availability. Only for a very large project is it feasible to develop a new quarry from virgin rock. Even when a quarry exists, it may not be equipped to produce stone of the type and size needed for a particular coastal project. The handling equipment available at the quarry may be a critical factor. The cost of quarrying and transporting the stone will affect the choice of construction material.

Earth.   In most parts of the world an adequate quantity of earth material is available for fills, dikes, and beaches. There are two exceptions: In some delta areas, the immediately adjacent earths may be predominantly silts and clays; in some rocky coast areas, beach sands may not be available.

Concrete.   The cement, sand, and stone required to make concrete are available in all parts of the United States. Some of the smaller Pacific islands may require the importation of cement.

Wood.   Wood used to be one of the most available construction materials in the United States. It is generally produced within reasonable shipping range of a coastal project. Because certain types and sizes of hardwoods are becoming more difficult to obtain, a designer now has to compare the costs and advantages of wood with those of other construction materials in cases in which wood would once have been an almost automatic selection.

Asphalt.   Asphalt is generally available in the United States, but it may not be available for projects in other areas for reason of either lack of the material or lack of handling equipment.

Synthetics.   Synthetics are manufactured materials, and the plants may not be located near the construction site. Synthetics are, however, easily and economically shipped. There may be a timing problem because some lead time may be required for delivery, particularly of large orders.

1.3.2 Transportability

If a material cannot be brought to the construction site or costs too much to haul, obviously it cannot be used. The usual restriction on transportability is that the material is too heavy or too large to move or the equipment necessary to move it is unavailable.

Transport mode.   Most construction materials for coastal projects can be transported by conventional freight haulers, i.e., rail, truck, barge, or ship. Armor stone for breakwaters and jetties may present transport problems because of its large dimensions and extreme weight. The design size of armor stone is frequently from 9 to 27 metric tons (t) (10 to 30 tons) per stone. Most state highway departments have a load limit of 178 to 214 kilonewtons (178 to 214 kN) (20 to 24 tons) per truck. That is not a problem with rail or barge haul, but most coastal projects require some use of public highways. The load limitation not only dictates the design size of the armor rock but also imposes a requirement of careful load scheduling to maximize the use of either trucks or railway cars.

Handling limitations.   Coastal projects in isolated locations must be carefully analyzed so that materials selected can be handled by available equipment. This involves not only placement equipment but also transport and processing equipment.

Stone.   The primary problem with stone is handling armor stone. Quarry processing and loading equipment usually has the capability to handle larger armor stone than public highways will permit. Not only does the placement of armor stone on breakwaters and jetties require a certain tonnage lift capability, but the equipment must be able to reach outward a sufficient distance to place the toe rock accurately.

Earth.   Earth can generally be handled with conventional construction equipment, but the availability of compaction equipment may control the method by which an earthfill is compacted.

Concrete.   Many special designs of concrete structures may require the use of highly specialized handling equipment. The cost or availability of such equipment may influence selection of the concrete structure. Underwater placement, the shaping of concrete armor units such as tribars and dolos, and the fabrication of reinforced or prestressed concrete piles all require specially designed handling equipment. The equipment may or may not have a reuse capability.

Steel.   Most conventional steel shapes can be handled by conventional construction equipment. However, specially designed units, very heavy units, and some types of underwater placement may require handling equipment specifically designed for that particular job.

Wood. Conventional construction equipment can usually transport and handle wood members, but special equipment may be required to penetrate the cells effectively with some types of chemical treatment.

Asphalt. Special heating or hot-asphalt-handling equipment may be required for transport and use of asphalt. If asphalt is placed underwater, specially designed handling equipment will be required.

Synthetics. Special handling equipment may be required for placement, particularly underwater placement, of synthetic materials. However, great weights are not often involved, and the equipment can usually be made or adapted in the field.

1.4   Compatibility with Other Materials

Compatibility problems may be physical, chemical, or esthetic. When structures are composed of more than one material, such as reinforced concrete, the compatibility of the constituent material properties must be considered. Steel of high tensile strength and concrete of high compressive strength, when used together properly, can result in a highly efficient structure, and asphalt with high adhesion to aggregate properties can create high wear resistance and a cohesive structure.

Materials may not be compatible because of the effects of abrasion between different or even the same materials (e.g., two stones of different hardnesses may not be compatible in a structure). Flexible and inflexible structural units may be incompatible. Major differences in shrink-swell or expansion-contraction coefficients may induce physical stresses. The weight of heavy structural units on a fragile substructure may cause failure. All these are generally stress, fatigue, or abrasion problems.

Chemical incompatibility is particularly critical in the choice of cement and aggregates, selection of synthetics, use of asphalt, electrolysis or corrosion of steel members, and corrosive interaction of dissimilar metals. The effects of incompatibility of materials may take a long time to appear; and if chemical action is allowed to continue, it can result in structural failure. Compatibility must be carefully considered when traditional methods or materials are to be used with new and sometimes untried construction techniques or materials.

1.5   Maintenance and Preservation Requirements

Maintenance and preservation requirements and their annual costs are generally influenced by the initial costs of construction. All materials require some maintenance and preservation. The maintenance problem may be physical or chemical, and it may vary not only between different materials but within the same material.

Stone. It is unusual, but possible, that stone will deteriorate chemically. The main problems are reduction in size through abrasion, reduction in size through splitting or breaking (particularly of armor stone), and loss of stone due to the power of waves or currents or the undermining of the structure. Preservation of stone as a material is not generally feasible, and maintenance is normally a matter of replacing damaged or missing stones. Generally, damage to a breakwater, jetty, or groin does not present a severe immediate hazard, and mobilization costs to do maintenance work are high. Therefore, maintenance is frequently deferred until a great deal of damage has occurred.

Earth. Little can be done to preserve an earth structure except to protect it from erosion. As with stone, maintenance is a matter of replacing lost material. Ease of access to the earth part of an installation determines the maintenance cost.

Concrete. The quality and life of concrete are largely controlled by the methods of mixing and placing. Coatings are available to improve the set period of the concrete and protect the concrete surface from flaking or dusting. In salt water, and to a lesser extent in fresh water, the reinforcing steel will corrode if it is exposed to oxygen. The corroded surface of the steel expands greatly, and that results in cracks in the concrete that admit more water and accelerate the process. The consequent physical spalling, cracking, or splitting of the concrete leads to total failure. Therefore, cracks must be kept sealed to slow the corrosion process. Because cement has a calcium base, it may also be necessary to protect concrete from chemical change due to pollutants or biological attack.

The primary need for maintenance or preservation is to prevent deterioration, which may be from abrasion by harder and sharper substances such as quartz sands or from the force of storm waves overstressing the structure. Impact by rocks, barges, ships, or debris may overload concrete, as in the case of dock structures. Seismic damage may occur. Maintenance may consist of sealing cracks, patching abraded or worn areas before the reinforcing steel is exposed, or actually replacing individual concrete units within the structure.

Steel. In contrast to stone and concrete, the primary purpose of maintenance or preservation of steel structures is to prevent chemical or galvanic deterioration. Unless it is made of special and expensive alloys, exposed steel is subject to rapid deterioration through oxidation or rusting, especially in the wet-dry tidal area and at the sandline. The latter can be very severe in the surf zone, where the corrosive process is accelerated by the abrasive action of the sand continually removing the rust and exposing new steel. Paint or some of the new protective coatings can greatly increase the life of steel members. The galvanic process can be greatly reduced or eliminated by the installation and maintenance of "cathodic protection systems."

Physical failure will not normally occur from wave or current forces if steel structures are properly designed. The primary cause of failure will be severe damage by ships, barges, or debris or, in the case of a dock, overloading. Prompt replacement of buckled members is mandatory to prevent further damage to adjacent members.

Wood.   The greatest cause of deterioration of wooden structures is biological attack by plant or animal life. Damage may occur completely above or below the waterline but is most likely to occur in the wet-dry tidal area. Most biological species require some sunlight, but some are active in total darkness. Only below the mud line are wood members safe from such attack. Two methods of preventive maintenance are available: complete impregnation of the cells of the wood by chemicals or application of a surface coating that prevents entry of borers into the timber. The surface coating may be an antifouling paint or a material resistant to borer penetration, such as a 0.5-millimeter (mm) [0.020-inch (in)] epoxy coating, a synthetic film wrapped around the wood members, or a membrane of concrete [usually about 50 m (2 in) thick] completely surrounding the wood members. Without such protection, a wood structure will deteriorate rapidly and make a sustained maintenance program of inspection and prompt replacement necessary to a long service life of the stucture.

Physical damage generally consists of members broken by ships, barges, or debris or by the force of storm waves. Prompt replacement of broken members is necessary to avoid deterioration of adjacent members. Damage through chemical reaction is largely confined to unusual events in industrial harbor basins.

Asphalt.   Asphalt is primarily used as a surfacing material for harbor roads, parking areas, and storage zones. It is subject to chemical deterioration, abrasion, plastic flow under heavy loads, and vehicular impact, particularly in areas subjected to high temperatures. Continuous maintenance is required; otherwise, damage will be extensive not only to the asphalt structure but also to vehicles and equipment using it.

When asphalt is used on jetties, frequent inspection and replacement of broken or dislodged asphalt is required.

1.6 Environmental Considerations

The physical properties and performance experience of each of the coastal structural materials are discussed in detail in the following sections. The environmental impacts on each of the materials are the corrosive and pollutant attacks on exposed surfaces: the effects of sunlight, water penetration, wave and currents, severe temperature, ice, marine organisms, periodic wetting and drying, wind erosion, burrowing animals, flora, fire, abrasion, seismic effects, and human activity. Although not all of the above impacts affect all structural materials (e.g., fire does not change the physical properties of earth), the environmental impacts that may affect a specific structural material to be considered for use in coastal construction are discussed in the appropriate material section.

2

Stone

2.1 Types and General Characteristics of Stone

"Stone" refers to individual blocks, masses, or fragments that have been broken or quarried from bedrock exposures or obtained from boulders and cobbles in alluvium and are intended for commercial use. Stone is used for many purposes, which are generally divided into two main classes:

1. Physical uses, in which the stone is broken, crushed, pulverized, shaped, or polished but its physical and chemical characteristics remain essentially unchanged

2. Chemical uses, in which the stone is changed physically or chemically to yield an end product that differs from the raw stone in composition

The use in coastal structures is primarily of a physical nature.

Crushed and broken stone. Includes all stone in which the shape is not specified, such as that used as aggregate and riprap. Riprap is well graded within wide size limits.

Quarrystone armor consists of comparatively large broken stone that is typically a specified size and is used without a binder principally for breakwaters, jetties, groins, and revetments, which are intended primarily to resist the physical action of water.

Stone for coastal structures should be sound, durable, and hard. It

---

This chapter was written by Ross A. Morrison and Louis J. Lee.

should be free from laminations and weak cleavages, and it should be of such character that it will not disintegrate from the action of air, seawater, and undesirable weathering or from handling and placing. In general, stone with a high specific gravity should be used to decrease the volume of material required in the structure and to increase the resistance to movement by the action of waves or currents.

Characteristics that affect the durability of stone are texture, structure, mineral composition, hardness, toughness, and resistance to disintegration on exposure to wetting and drying and to freezing and thawing. Ordinarily, the most durable stone is one that is dense or fine-textured, hard, and tough, but exceptions to this general rule occur. The character of the stone for any project depends on what is available, and often the choice of material involves weighing the relative economy of using a local stone of lower quality against using a better-quality stone from a distance. When the local stone is markedly inferior, the greater cost of transporting durable, high-quality stone from outside the immediate area may be justified and advisable. Because of the wide range in climatic conditions, and thus of the severity of exposure in different regions of the United States, acceptable standards of durability for the various regions will vary.

The stone industry recognizes the following stone classification (Gay, 1957) based mainly on composition and texture: (1) granite, (2) basalt and related rocks, (3) carbonate stone including limestone and marble, (4) sandstone, and (5) miscellaneous stone (including chert, conglomerate, greenstone, serpentine, shale, slate, tufaceous volcanic rocks, and coral).

2.1.1 Granite

The term "granite" is commonly applied to medium- to coarse-grained igneous stones that consist mainly of feldspar and quartz, and ordinarily contain subordinate proportions of ferromagnesian minerals. Mica may also be present. In small quantities mica is not particularly harmful, but in larger quantities it sets up planes of structural weakness and provides a starting point for disintegration. Granite occurs mainly in large bodies, known as batholiths, which are exposed over many square miles. Batholiths commonly consist of numerous individual bodies of various granitic rock types with contrasting colors, textures, and mineral compositions.

Although granites vary widely in texture and appearance, most are dense and have a porosity of less than 1 percent. Granite spalls badly under the combined effects of fire and water, so it is not particularly resistant to fire. Most unweathered granitic stones are hard, strong, tough, and resistant to abrasion, impact, and chemical attack. The average unit weights range from approximately 24.3 to 27.5 kilograms per cubic meter (kg/m^3) [155 to 175 pounds per cubic foot (lb/ft^3)]. These properties make granitic stones well suited for use as riprap and quarrystone armor units.

2.1.2 Basalt and related rocks

In commercial usage, the term "basalt" is applied to any of the dense, fine-grained, dark-gray or black volcanic stones. The term ordinarily includes stone types that geologists classify as dacite, andesite, basalt, trachyte, or latite. Basaltic rock has solidified by the cooling of lava either as flows on the earth's surface or as shallow intrusive bodies beneath the surface. It is composed primarily of feldspar and ferromagnesian minerals in crystals that range in size from submicroscopic to clearly visible. Commonly, an appreciable percentage of glassy material is present. Some effusive basalt is vesicular, and the vesicles may have become filled with potentially reactive substances such as opaline, silica, or zeolites which render the rock unfit for use as aggregate.

Basalts are among the heaviest of stones with an average specific gravity of 2.9 to 3.2 and average unit weights of 28.3 to 31.4 kg/m^3 [180 to 200 pounds per cubic foot (lb/ft^3] but in certain areas they may contain many small cavities (vesicles) which result in stone with low densities. Basaltic stones are characteristically hard, tough, and durable, so they are well suited for use as aggregate, riprap, and quarrystone armor units.

2.1.3 Carbonate stone

Carbonate stones are broadly divided by geologists into (1) limestone, which consists almost entirely of calcite ($CaCO_3$), (2) dolomite, which consists mainly of the mineral dolomite ($CaCO_3 \cdot MgCO_3$), and (3) marble, which is the metamorphosed crystalline equivalent of either type. There are all gradations between limestone and dolomite and between very fine-grained and very coarse-grained material.

In the stone industry the term "limestone" is applied to many types of rock that contain a high percentage of calcium carbonate, although large proportions of other substances also may be present. Such substances include siderite ($FeCO_3$), magnesite ($MgCO_3$), and rhodochrosite ($MnCO_3$). They also commonly contain clay, silt, and sand grains. A high percentage of clay commonly weakens carbonate rock, making it unfit for use as stone. A high content of sand grains or silica may harden carbonate rock. Marble is similar to limestone chemically, but it has been subjected to a metamorphic process which has made it more crystalline in structure, harder, and better able to hold a polish.

For use as stone, carbonate rock should be physically sound, dense, and relatively pure. The approximate range of the porosity of limestone is from 1 to 15 percent. Limestones have an average unit weight of approximately 22.0 to 25.9 kg/m^3 (140 to 165 lb/ft^3). Marble has an average unit weight of 25.1 to 26.7 kg/m^3 (160 to 170 lb/ft^3). Carbonate stone that is tough, strong, and durable is well suited for use as concrete aggregate, riprap, and quarrystone armor units.

2.1.4 Sandstone

Sandstone is clastic sedimentary rock composed of particles mainly in the size range of about 0.25 to 6.4 mm (0.01 to 0.25 in) in diameter. Although some sandstones consist almost wholly of quartz grains, most sandstones are feldspathic; some contain a high proportion of ferromagnesian minerals. The strength and durability of sandstone are mainly determined by the type of material that cements the grains together. Only well-indurated sandstone cemented with silica or calcite (rather than with the weaker cements, clay, or iron oxide) is well suited to use as crushed and broken stone. The porosity of sandstone is typically high, ranging from 5 to 25 percent. The average unit weight ranges from approximately 21.2 to 25.1 kg/m^3 (135 to 160 lb/ft^3).

2.1.5 Miscellaneous types of stone

Other types of stone may be suitable for some applications depending upon tests for durability and less severe environmental conditions.

Chert. Chert is a sedimentary rock composed almost entirely of silica, in the form of opal, chalcedony, or microgranular quartz. It commonly occurs in thin-bedded deposits. The most desirable form of chert is hard and dense, which is well suited for use as crushed and broken stone and riprap. Some chert, however, is too laminated or contains too much silt or shale for such use. It is generally not used for large quarrystone armor units.

Conglomerate. Conglomerate is clastic sedimentary rock containing abundant fragments of pebble size or larger in a matrix of sand and finer-grained materials. Conglomerates show various degrees of induration which depend largely on the nature and amount of cementing material—clay, calcium carbonate, iron oxides or silica—in the matrix. Conglomerate is not abundantly used for riprap or quarrystone armor because relatively few deposits are sufficiently well indurated for this use.

Greenstone. "Greenstone" is a general term applied by geologists to basic or intermediate volcanic rocks that contain abundant green secondary minerals. In the stone industry the term is also applied to a variety of fine-grained green rocks, including arkosic sandstone, graywacke, impure quartzite, and various pyroclastic rocks. Physically sound greenstone may be used for aggregate, riprap, or quarrystone armor if it is available economically.

Serpentine. Serpentine is an ultrabasic igneous rock composed mainly of the mineral serpentine, a hydrous magnesium silicate. Serpentine rock is moderately soft but commonly massive and dense in structure, and it is very resistant to chemical and physical weathering. These properties make it desirable as crushed and broken stone for riprap but generally not for quarrystone armor units.

Shale. Shale is a very fine-grained, thinly bedded sedimentary rock composed mostly of clay- and silt-size particles. PreMesozoic shales are commonly well indurated, if not metamorphosed. Most Mesozoic and Tertiary shales are moderately to poorly indurated. Most types of shale are too weak to be suited to the ordinary uses of crushed and broken stone.

Slate. This stone is a thinly foliated metamorphic rock composed essentially of muscovite (sericite), quartz, and graphite, all in grains of microscopic or submicroscopic size. Slate is formed by compaction and partial recrystallization of shale, and it is commonly dark-colored and moderately hard. Slate is desired mainly for use as dimension stone. Because slate has been subjected to intense pressure during formation, it has a low porosity and, consequently, a high strength. Its modulus of rupture is relatively high, and it is also resistant to weathering and to mechanical abrasion. The average unit weight ranges from approximately 26.7 to 28.3 kg/m^3 (170 to 180 lb/ft^3).

Tuff. The term "tuff" includes pyroclastic volcanic types, most of which would be classed as rhyolite or dacite tuffs or tufaceous sediments. Most tufaceous rocks are only moderately hard, although on exposure to air they commonly harden appreciably. Because of its softness, tufaceous rock is unsuited to most uses of crushed and broken stone.

Coral. In southeastern United States and certain tropical island areas it is often necessary to make use of coralline limestones for coastal construction, since more durable stone (such as granite or basalt) is unavailable. These materials are partially recrystallized coral formations which have sufficient resistance to breakdown and erosion to be acceptable for breakwater or revetment construction (Fig. 2.1). They are less resistant to mechanical breakage than denser, harder stone and therefore require special care in blasting and handling. However, it has been possible to produce large stones 89 to 267 kilograms (kg) (10 to 30 tons) from coralline limestones.

2.2   Categories of Stone Size and Gradation

The crushed and broken stone category includes all quarried stone that is not cut or shaped to specified dimensions. It ranges in size from granite blocks weighing 222 to 267 kilograms (kg) (25 to 30 tons) used as quarrystone armor units to shale ground to smaller than 200 mesh (0.075 mm) (3 mils) used as insecticide carrier. The larger categories of crushed or broken stone (greater than 75 mm or 3 in) are generally used in coastal construction.

2.2.1   Fill

Most fill material is natural earth obtained as surplus from excavations or from borrow pits, and it would not be classed as stone. Crushed stone is used for special types of fill, such as the cores of dikes or jetties. Only the least-expensive grades of crushed stone, commonly crusher run fines or unclassified waste from production of riprap, are

Figure 2.1   Coral riprap in Kosrae, Micronesia. *(Courtesy of Woodward-Clyde Consultants)*

used for fill. Miscellaneous soil, rock, and rubble fills may be used as random or temporary shore protection (Fig. 2.2).

2.2.2 Rockfill

Sound rock is ideal for producing rockfill materials. Some weathered or weak rocks, including sandstones and cemented shales (but not clay shales), may also be suitable (U.S. Army, Corps of Engineers, 1971a). Rocks or stones that break down to fine sizes during blasting, excavation placement, or compaction are unsuitable as rockfill; such materials should be treated as soils. Processing by passing rockfill materials over a grizzly may be required to remove excess fine sizes or oversize material. Quarry run and quarry waste materials are commonly used as core materials for breakwaters and jetties. This material should be reasonably well graded, and no skip grading or scalping of certain sizes should be allowed. The material should also generally contain no more than 5 to 10 percent fines.

2.2.3 Riprap

Riprap is generally heavy irregular fragments of broken stone or other resistant substances, well graded within wide size limits, and

Figure 2.2 Random miscellaneous dumped soil, rock, and rubble fill used as shore protection, Sunset Cliffs, San Diego, California. *(Courtesy of the city of San Diego)*

randomly placed without mortar to provide protection for an embankment or bluff toe from the physical erosive action of water. The stability of the stone layer depends on the density and mass of the stones
and on the evenness of their gradation. The stability increases as the
stones become more well graded. This means that, for riprap of small
stones, finer material is included in the gradation, making the voids
smaller, the face smoother, and wave reflectance higher. As long as
the physical requirements for stone are fulfilled, any type of rock may
be used for riprap. Chemical and mineral composition of the rock is
generally of minor importance.

2.2.4   Armor stone

Armor stone, chosen to be of nearly uniform size and of compact, sometimes blocky shape, depends on density and mass to resist the force of
waves or currents. The voids between the armor rock, to a certain extent, absorb energy through creation of turbulence. To a lesser extent,
wave energy is absorbed by wave run-up on the sloping outer face.

2.2.5   Underlayers

The underlying rock layers are usually randomly placed and serve to
support the armor rock. By size gradation, and sometimes in several
layers or zones, they may also absorb wave or current energy through
turbulence in the voids. Rocks used for these underlayers are considerably less costly per unit volume than armor rock.

2.2.6   Bedding layers

In breakwaters, jetties, and groins constructed on relatively flat sand
or mud bottoms, a bedding layer consisting of smaller well-graded
stones is required to prevent the fine bottom material from piping upward through the structures. Piping and subsequent erosion of the
foundation soils could result in settlement of the upper layers of rock,
particularly the large, heavy armor stones. Settlement could in turn
cause ultimate collapse of that part of the structure.

2.2.7   Filter layers

In revetments, seawalls, and bulkheads a layer of filter material is required. Although this layer of fine well-graded stones may in part act
as a supporting bedding layer for sloped revetment or seawall structures, its primary function is to relieve hydrostatic pressure due to
changes in water elevation on the seaward side or changes of ground
water elevation on the landward side. It is designed as a true filter to

permit the passage of water in either direction but prevent the disturbance on the sand or clay foundation. The gradation of the filter material depends on the characteristics of the backfill core or beach materials and on the voids of the riprap or armor units. The filter
material should be uniformly graded from fine sands, coarse sands,
gravels, and stones such that it will not wash into the rubble. The material could be in two or more layers. It should also be noted that filter
cloth is frequently used in place of a graded granular filter.

2.2.8 Other categories

Filler stone, consisting of well-graded gravel size material, 6.3 to 100
mm (0.25 to 4 in), is commonly used to fill the voids in core stone. Toe
stone is used to protect the base of a coastal structure from erosion or
scour. This stone typically ranges from approximately 0.89 kg (200 lb)
to more than 8.9 kg (2000 lb) in weight and should be reasonably well
graded. Chink stone is used to fill voids in riprap or armor stone.
Coarse and fine aggregates are used for making concrete.

2.2.9 Fill for gabions and cribs

Gabions are wire baskets that can be connected and filled with stone.
The baskets must be solidly filled, or the wires will be abraded by
movement of loose stones. The stones must be large enough, generally
10 to 25 centimeters (cm) (4 to 10 in) in diameter, to prevent loss of
stone through the gabion mesh. Cribs may be filled with similar stone
materials to form a gravity-type structure.

2.3 Stone Specifications

The character of the stone to be used in coastal structures is of primary importance. Materials may be obtained from any approved local
source. Material from new sources should be tested by the government
for quality to determine acceptability. When the contractor desires
materials from an untested new source or if the government elects to
retest a previously approved source, suitable samples for quality evaluation should be taken by the contractor under the supervision of the
contracting officer. Samples are generally delivered by the contractor
to the nearest Corps of Engineers laboratory for testing.

No standard testing procedures have yet been developed for the determination of the quality of stone. The Waterways Experiment Station (WES) and some Corps Division laboratories have devised tests to
evaluate such materials, but the test procedures employed in these
different laboratories vary somewhat. In any case, judgment is neces-

sary in applying the test results. Any testing program for the determination of the quality of rock for use as stone in coastal structures should include petrographic examination.

- Determination of absorption and bulk and specific gravity [ASTM Standard C97-47 (77) or C127-77]
- A soundness test [American Association of State Highway and Transportation Officials (AASHTO) Test T-104-46 or ASTM Standard C88-76]
- An abrasion test [ASTM Standard 535-69 (75)]

Other tests that may also prove useful include a slaking or wetting-and-drying test, and a freeze-thaw test.

Properties contributing to the durability of stone may be both physical and chemical. Tests usually measure physical properties, and therefore the results provide only an indication of how chemical change has already affected the stone rather than the stone's susceptibility to future chemical change. Wetting and drying tests have been used to evaluate rock which disintegrates badly as a result of chemical change, but there is still a question as to what the test actually measures and what the results mean. Chemical changes can best be evaluated by experience.

The best data for the evaluation of stone to be used in coastal structures are service records. If a stone has not been previously used, the quarry should be visited and old surface outcroppings examined for signs of weathering.

2.3.1 Stone size

To make optimum use of local materials, designs should not only have a wide range of stone sizes to choose from but also have an adequate number of classes within the range. Each class available for a specific use should be limited in range. Physical limitations in the size of armor stone that is feasible to use must also be considered. These may be truck or highway capacity or the handling limits of the quarry equipment. The geological structure of the rock quarry may also limit the quantity and size of stone that can be obtained.

The total weight or size of the armor units, the side slopes, the density of armor material, and the degree of interlocking or wedging between units are interrelated and comprise the principal factors in the design of a stone structure. Armor stone may be rubble mound placed at random, individually placed, or it may be rectangular blocks of stone carefully fitted together. Several empirical formulas have been derived for determining the size of armor stone required for stability under wave action. These are contained in the appropriate design manuals. In the Hudson equation for design of armor stone [U.S. Army, Corps of Engineers, Coastal Engineering Research Center (CERC) 1977], the required size of individual armor stone is roughly inversely proportional to density. This flexibility of size vs. density of stone permits some latitude in choice between two quarry sites. Stones larger than about 22,727 to 27,273 kg (25 to 30.0 tons) are generally not easily handled. The greatest dimension of each individual large stone should be no greater than 3 times the least dimension.

The reverse of this size-density factor can be used for a more efficient consideration of choice of rock source for core and underlayers when density is not a critical design consideration. Frequently a savings in cost can be effected by bidding the armor stone in tons and the underlying stone in solid cubic yards. Underlayers placed beneath the armor units should be adequate in size to prevent withdrawal of the units through the interstices of the cover layer and to prevent excessive movement and subsequent breakage. The weight of underlayer stone may range from 3 to over 30 kg (6.6 to 66 lb).

The most frequently used core is a quarry run material, the gradation of which is governed by economics or by the desired degree of impermeability. A rubble structure may also need protection from settlement (resulting from leaching, piping, undermining, or scour) by use of a bedding layer or blanket. The gradation requirements of a bedding layer depend primarily on the littoral characteristics in the area and on the foundation conditions. However, quarry spoil ranging in weight from about 4 kg (10 lb) to about 220 kg (480 lb), will generally suffice.

Typical classifications from the state of California standard specifications for rock slope protection are shown in Table 2.1. The weights by which the classes are designated do not necessarily correspond to the weights called for by the various design formulas. For example, if a shore protection formula, such as Hudson's equation in the *Shore Protection Manual* (SPM) (U.S. Army, Corps of Engineers, CERC, 1977), should call for 4500-kg (5-ton) armor stone, it may be proper to use the 7200-kg (8-ton) class because approximately 80 percent of that class would be larger than 5-ton. For the same example, the SPM suggests that, for a cover layer with a two-stone thickness, approximately 75 percent of the stones may be greater than 5-ton with a range between approximately 333- and 556-kg (3.75- and 6.25-ton).

2.3.2 Stone shape

Stone block structures may be closely fitted seawalls, groins, jetties, or breakwaters such as are built along the coasts of Hawaii (Fig. 2.3), the Gulf of Mexico, and the Great Lakes. Rectangular stone blocks of suf-

TABLE 2.1   Classes of Rock Slope Protection

Rock sizes	Method A[a]					Method B[b]					Backing	
	8-ton	4-ton	2-ton	1-ton	½-ton	1-ton	½-ton	¼-ton	Light	Facing	No. 1	No. 2
8-ton	50	0										
4-ton	95	50	0									
2-ton	—	95	50	0	0							
1-ton	—	—	95	50	0	50	0					
½-ton	—	—	—	95	50	—	50	0				
¼-ton	—	—	—	—	95	90	—	50	0			
200-lb	—	—	—	—	—	—	90	—	50	0		
75-lb	—	—	—	—	—	—	—	90	—	50		
5-lb	—	—	—	—	—	—	—	—	90	90	0	
No. 4	—	—	—	—	—	—	—	—	—	—	50	0
No. 200	—	—	—	—	—	—	—	—	—	—	95	90

[a]Method A. A footing trench shall be excavated along the toe of slope. The larger rocks shall be placed in this toe trench. Rocks shall be placed with their longitudinal axis normal to the alignment of the embankment face, with attitude horizontal or dipped slightly inward, and arranged so that each rock above the foundation course has a 3-point bearing on the underlying rocks. Bearing on smaller rocks which may be used for chinking voids will not be acceptable. Placing of rocks by dumping will not be permitted. Local surface irregularities of the slope protection shall not vary from the planned slope by more than 1 ft measured at right angles to the slope.

[b]Method B. A footing trench shall be excavated along the toe of the slope. Rocks shall be so placed as to provide a minimum of voids, and the larger rocks shall be placed in the foundation course and on the outside surface of the slope protection. The rock may be placed by dumping and may be spread in layers by bulldozers or other similar equipment. Local surface irregularities of the slope protection shall not vary from the planned slopes by more than 1 ft measured at right angles to the slope.

Source: California Department of Public Works, 1960.

ficient uniformity in size and shape closely fitted together should be used, since such structures prevent loss of the underlying rubble stone through the interstices of the fitted armor stones. However, there must be sufficient space or openings between the armor stones to relieve the hydrostatic uplift pressures that occur during storm waves.

Rounded stones.   Particularly for the armor layers, rounded stones are to be avoided whenever possible. They are difficult to place and are not stable because of either wave forces or their inherent instability on steep slopes. For those reasons, field or streambed stone are generally not acceptable.

Angular stones.   Angular stones (Fig. 2.4), particularly for the armor layer, have two advantages: (1) the voids are maximized, which increases energy losses through turbulence and (2) the stones tend to interlock with their neighbors, which increases their resistance to movement by wave or current forces.

Figure 2.3 Fitted stone blocks along a coast in Hawaii.

Although angularity is most desirable, the stones should be approximately rectangular in shape. The greatest dimension of large stone should be no greater than 3 times the least dimension. Sharp points should be avoided because they may cause a stone to wobble or may break off under stress and thereby disturb the stability of the section.

Figure 2.4 Angular block stones.

Even if dislodged or partially dislodged, an angular stone will tend to find a stable position, whereas a rounded stone will tend to roll to the toe of the structure.

2.3.3 Specific gravity

Required stone size is a function of the specific gravity, and specific gravity, unless unreasonably low, should not be a limitation on stone use. In fact, excellent results have been reported for lava with a specific gravity of 1.5. Once the stone size has been selected on the basis of a certain specific gravity, specifications should then prohibit the use of stones with specific gravity appreciably lower. Low-density stone may be used in the core (and underlayers), and high-density stone can be used, provided an adjustment is made in the thickness, for the armor layer. Armor is generally priced by the ton. Core and underlayer materials may be priced by volume to control the use of overweight stone in these zones.

2.3.4 Absorption

There is a general correlation between absorption and weathering. The absorption test is more significant when the rock is to be used in areas subject to freezing and thawing. A limit of 2 percent absorption is reasonable.

2.3.5 Soundness

Rocks that are laminated, fractured, porous, or otherwise physically weak are subjected to a soundness test (use of sodium sulfate). Stones that show a loss of less than 5 percent should be satisfactory.

2.3.6 Abrasion

The Los Angeles rattler test and Wetshot rattler test measure resistance to abrasion. These tests are more significant when the rock is to be used in shore protection where it is subjected to a pounding surf carrying sand, gravel, and smaller stones. Stones having relatively high losses in these tests have performed satisfactorily in shore protection, and therefore a rather lenient value is permissible. A 40 percent maximum loss for the Wetshot rattler and 45 percent for the Los Angeles rattler are considered reasonable.

2.4 Quarrying Methods

Quarries produce most of the rock required for construction in this country. In most quarries, all the material mined is usually consumed as an end product. The size and quality of material obtained from a quarry depend largely on the geology of the site and the method of blasting (du Pont de Nemours and Lo, 1977). Depending on the area's topography, a quarry will generally be developed either as a sidehill or as a pit-type operation. When the area is hilly and the rock outcrops, the quarry will be developed by opening a face into the side of the hill. The point of entry is usually at the bottom of the rock seam or in a very thick seam. A convenient point should be chosen to provide an almost level floor with just enough slope for natural drainage. When the terrain is almost flat, it is necessary to ramp down into the rock to create a pit that is entirely below the surface of the surrounding terrain.

The height of the quarry face may be determined by the thickness of the formation. However, since most formations being mined exceed the practical limits of bench heights, the determining factor in the choice of bench heights is usually that of safety. Bench heights must be selected to be compatible with the loading equipment so that the broken rock can be safely removed from the muck pile. If a single piece of loading equipment, such as a shovel or front-end loader, is incapable of reaching high enough to remove all unstable rock from the broken bench face, then it is customary to use a dozer to work the rock down to a safe height for the loader. The use of the proper blast design will result in the utilization of the maximum safe bench heights and the optimal use of loading equipment.

A somewhat special method used in quarry development is called coyote tunneling. This is a method whereby tunnels are excavated into a bluff or hillside and partially filled with explosives. In a crude sense, they might be considered large-diameter horizontal boreholes, with the additional option of turning corners or excavating T-sections. With this method, it is extremely important that the natural jointing characteristics of the rock have the capability of producing the desired material, since there is relatively limited additional fragmentation by this blasting method itself. The method is economically attractive if there is a vertical or steep rock face and the rock has the desired jointing characteristics. For example, it has proved to be very successful for producing riprap in columnar basalts and for producing crusher feed in diced basalts.

Dimension stone quarries may be developed in some formations such as sandstone, marble, or granite. Drill holes are generally spaced close together along the desired break line, and small charges of blasting powder are used in these quarries. The purpose is to move a block of stone a short distance in one piece without any damaging cracks.

2.4.1 Blast design

In general, the blasting method is determined by the geology of the material to be broken, the fragmentation required, the hole diameter, and the type of explosive. The type of equipment available for handling and loading should also be taken into consideration (e.g., in determining bench height).

Geology. The geological and physical characteristics of the material to be broken are the most important factors in determining the end product stone size and the overall blast design. A half dozen or more factors are reported by researchers as being related to the manner in which rock is fragmented under the action of explosives. They include such factors as the maximum sonic velocity, the minimum sonic velocity, the ratio of the two (the sonic anisotrophy), the lowest tensile strength, the specific gravity, and the number of joints intersecting a blasting round. Factors such as hardness and brittleness may also be included, but they are related to those previously mentioned. At least a part of these properties are measured or estimated in a typical site investigation for a quarry. Even the most cursory of quarry investigations will provide an estimate of rock quality and jointing. In turn, rock quality and jointing are probably the two most important factors used in the preliminary design of blasting tests. However, it is generally true that both the design of blasting rounds and the chances of their success largely depend on the previous experience and skill of the blaster.

In many cases, a very hard, brittle rock will break with less difficulty than a soft, spongy rock. That is dramatically true if the hard rock is closely jointed and the soft rock is massive. The orientation of the primary joint system in a formation is a very important factor in the blast design. When the primary jointing is dipping at a steep angle, it is usually advantageous to develop the quarry face at no less than 45° (and preferably 90°) to the jointing angle, but that is not universally true. In determining the angle of the quarry face, consideration should also be given to the desired end product (e.g., crusher feed vs. riprap), slope stability questions, ease of development, traffic flow, and equipment. Frequently, the direction of development is not a controllable factor. In such cases, it is important to make certain that blast designs and excavating procedures take into account the geometric relations between the jointing, explosives action, excavating sequences, and final surfaces.

Fragmentation. The degree of fragmentation desired depends on the use of the product mined. In quarrying, when the stone will be sized for construction use, it is usually undesirable to produce a large percentage of stone less than 5 cm (2 in) in size. Even under the best conditions, it should be anticipated that up to 5 percent fines (fine sand, silt, and clay) will be generated by the blasting operation.

Because many of the various parameters involved in blast design are strongly interrelated, it is difficult to isolate each factor and discuss its relation to the final product. However, it is possible to make certain statements which are generally true about some parameters. For example, a high powder factor (quantity of explosive per unit volume of rock) will produce a greater degree of fragmentation than a lower powder factor if other factors are kept the same. Similarly, greater fragmentation will be achieved in a massive rock by using a larger number of smaller-diameter holes at closer spacing rather than larger-diameter holes at greater spacings. Also, it appears that the best fragmentation is achieved when holes are detonated individually rather than simultaneously.

Hole diameter. The proper hole diameter depends largely on the physical properties of the formation, the fragmentation required, and the height of the quarry face. The hole diameter should be selected to be compatible with the geological and physical characteristics of the formation, which are the only factors in the overall blast design that cannot be altered. Unfortunately, the selection is sometimes made on the basis of the total volume of rock to be mined, the duration of the project, production rates, capital costs, and depreciation rates, with no consideration given to geology.

Type of explosive. On large-scale projects, it can be reasonably assumed that the first choice of explosive will be made for economical rather than technical reasons; for dry quarry work, explosives will usually be ammonium nitrate fuel oil (ANFO). If water is encountered, the choice would normally be a specially packaged ANFO or some form of slurry or water gel. Despite its relatively low price, ANFO is a good, general-purpose explosive. Its low brisance is compensated for by its high gas volume; hence, it is capable of producing as much work as some of the more expensive brisant explosives.

The majority of blast-induced fractures produced in the rock are radial from the charge location, and are associated primarily with the propagating stress waves (du Pont de Nemours and Lo, 1977). Spalling at the bench face from reflected stress waves produces very little fragmentation with the burdens normally used under typical field conditions. Thus it is apparent that the inherent fracture planes in the rock are important and should be considered in determining the fragmentation; hence, they should also be considered in determining the blast design. If the inherent fracture planes are closely spaced, the material can be broken more easily and with larger-diameter holes at greater spacing.

Production of large stones is usually a more difficult problem than production of crusher feed. If the rock is overblasted and the particles are too small, there is no way, of course, to make them larger. However, if rock sizes are too large, it is at least physically possible to break them to smaller sizes, although at greater cost. For the production of large stone, it is customary to detonate simultaneously a row of holes behind an open face, using relatively light charges.

2.4.2   Loading equipment

The degree of fragmentation required is also related to the type and size of the loading equipment and the size and type of crusher available. Obviously, larger equipment can tolerate larger rock fragments. However, the main advantage of larger equipment is the ability to handle a larger volume of material, not simply larger particles. It is poor economy to use larger blasthole spacings and burdens to produce larger rock sizes because large loading equipment has been acquired. Under those circumstances, the loading and crushing equipment is not being utilized to its maximum capabilities. It is much less expensive to do more work with explosives. Of course, if the needed product is large-size stone, such as riprap, it is essential to have equipment capable of handling it.

2.4.3   Processing of materials

The properties and quantities of the various types of materials are dependent not only on the onsite geological conditions but also on the construction procedures; therefore, it is imperative that suitable processing methods be used in the quarry operation. Further, it should be anticipated that variations and changes from the anticipated geologic conditions in the quarry may occur and methods of construction may vary.

Gradations and sizes.   The large number of material types generally obtained from a quarry require separating materials into various sizes and gradations. Very large material, up to 26,700 kg (30 tons), may require special equipment. It is anticipated that large stone material, e.g., 1- to 8-ton stone, will be separated by the excavating equipment. The material larger than 0.6 meter (m) (24 in) and smaller than 38 mm (1.5 in) is generally removed and separated by some type of screening process. Removal of fines (material passing a No. 200 mesh sieve) may require some type of washing process. The various operations could have a significant effect on the material gradation and generation of fines, which would ultimately affect the available quantity of some of the material types. It is therefore important that the sequence of operations, type of equipment, rate of production, and number of times the material is handled be evaluated and controlled.

The best time to control the gradation of quarried stone is during the quarrying. Control of gradation to meet specifications is usually carried out by visual inspection. In order to calibrate the judgment of the inspector, it is very helpful to establish, at a convenient location in the quarry, a pile of stone with the desired gradation. This standard pile should contain 4500 to 9090 kg (5 to 10 tons) of material and should be formed by measuring and selecting the individual stones to be combined in the correct proportion. In the case of riprap or other large stone, there may need to be a larger volume in the pile used for visual calibration.

Some projects may permit the use of "quarry run": whatever quality and gradation of stone that may come from the quarry without any special control or processing. However, when specifications are being prepared, it is very important that any limitations be carefully identified to the project owner. The owner may expect quarry run to be the product of very cautious blasting by using controlled methods and may encounter either a design problem or contractual problem if entirely different methods are used by the contractor.

2.4.4   Minimizing degradation

It should be anticipated that even fresh rock will contain up to 5 percent fines due to blasting and careful handling and that the quantity of fines can be increased substantially by the equipment and methods used in handling. Commonly, another 5 percent fines is generated in crushing and processing. Of course, the greater the number of times the material is handled and processed, the greater the percentage of fines generated or material lost (as in stockpiling).

To minimize further degradation of the materials, the contractor should use large-size loading equipment in the borrow area and keep rehandling to a minimum. With large-size loading equipment, it should not be necessary to use dozers on the muck pile. Such use can be expected to cause a significant degradation of the stone particles. Further, the use of large loading equipment minimizes the need to use excessive quantities of explosives to provide excessive rock fragmentation.

2.4.5   Equipment and transportation

Excavation of materials is usually carried out with power shovels, draglines, scrapers, and front-end or side-delivery loaders (Sherard et al., 1963). In quarries, the most common items of equipment are shovels and front-end loaders. Power shovels provide the greatest reaching capacity and greatest digging capacity in poorly fragmented or poorly loosened materials (Fig. 2.5).

Digging capacity is important when production of large stone for riprap is desired. The reaching capacity often contributes greatly to overall efficiency by permitting higher benches. Large front-end and side-delivery loaders are probably the most economical types to use in a medium-size operation producing crusher feed, where blasting is designed to produce fine fragmentation, and on projects in which high mobility is desirable. (It is a slow process to move a large shovel any distance.) Small front-end loaders are probably the least efficient, since they can handle only small-size materials in a loosened state. To do so requires heavy blasting or extensive reworking with a dozer. In all cases, the type of equipment, the capacity, the general handling procedures, and the rates of excavations should be documented and the effect of the overall operation on the material gradation should be evaluated. This is generally done by observing the operation and then sampling and determining the gradation of the material from the muck pile before excavation and after handling by the equipment.

The basic methods of transporting the stone materials are by scrapers, trucks, and belt conveyors. The type of truck generally has little influence on the properties or processing of the embankment material. In recent years, there has been a dramatic increase in the use of con-

Figure 2.5   Power shovel loading trucks at a quarry in Catalina.

veyor belts, partly because of improvements in belting materials and designs which permit the conveyors to be moved more rapidly. Clearly, lack of mobility is a serious limitation. It is also important to consider carefully the size and gradation of the stone products and the elevation changes required for transport. Most conveyors are easily damaged by large stones, and special designs are required in steep topography.

It is not common on construction projects to require the cutting and splitting of dimension stone. When such products are required (usually for facing of completed structures), it is customary to purchase them as would be done for any other manufactured product. Only a very limited number of sites are suitable for producing dimension stone, and the production requires equipment and skills that ordinarily are not readily available.

2.5 Placement Methods

Stone should be placed by equipment and methods suitable for handling materials of the size specified. Placement of the stone should begin at the bottom of the section and should continue in such a manner as to produce a graded mass of material with maximum interlocking and minimum voids. In general, the larger stones should be so placed that vertical joints are broken with the long axis of the stone set approximately normal to the structure slope and pointing inward toward the center of the structure section. Stone should be placed to the lines and grades shown on the contract drawings.

Typical extreme tolerances for finished surfaces, as currently contained in various Corps of Engineers specifications, are ± 30 cm (12 in) when placing under water and ± 15 cm (6 in) when placing in the dry. Tolerances of ± 7.5 cm (3 in) may be required for smaller underlayer and bedding layer stone. Up to 45 cm (18 in) may be allowed for large armor stone. The extremes of tolerances for underlayer stones and large armor stone should generally not be continuous over areas of structure surface greater than approximately 18.5 and 93 square meters (m^2) [200 and 1000 square feet (ft^2)], respectively (U.S. Army, Corps of Engineers, North Central Division, 1978).

Rubble mound structures exposed to wave action during construction should be completed in sections (including placement of armor stone) to minimize damage. Particular care must be taken when building structures such as groins and jetties that pass through the surf zone. Placement of extra stone on and around the end of the structure as the structure progresses seaward may be required to prevent damaging scour in the surf zone during construction. Damage to unprotected dikes is generally the responsibility of the contractor.

To place stone firmly, particularly in the armor layer, it must be placed or seated on the underlying stones so that it does not tend to slip, tilt, or wobble either under wave attack or from the weight of stone placed on top of it. This is commonly known as *seating the stone*. Small armor stone and armor stone placed in areas of very rough seas may have to be randomly placed. If so, design allowance must be made for a less stable structure. Controlled placement methods are preferred for the best use of armor stone; this depends to a great extent on the skill of the contractor personnel.

2.5.1   Filter, bedding, and core material

The method used in placement of filter, bedding, or core material should be such that the soft and organic materials on the bottom are displaced outward toward the extreme outside toes of the required sections of the structure and in the direction of the construction. The stone should be handled and placed in such a manner as to minimize segregation and provide a well-graded mass. If the materials are placed by clamshell, dragline, or similar equipment, the stone should not be dropped from a height exceeding about 0.6 m (24 in) above the existing bottom or previously placed material. The use of bottom-dump scows and self-unloading vessels may be permitted when the vessel is in motion along the centerline of the structure and the material is dropped as near to and directly over its final location as possible. The finish surface of the material should be free of mounds or windrows.

In areas where the stone is to be placed on geotextile filter cloth, care should be taken not to rupture the cloth and the stone should not be dropped from a height greater than about 0.3 m (1 ft). Maximum heights from which stones may be dropped on geotextile filter cloth are specified for various sizes of stone in Table 8.1.

2.5.2   Underlayer stone

Underlayer stone should be placed to a full-zone thickness in one operation in a manner to avoid displacing the underlying material or placing undue impact force on the underlying material and supporting subsoils. The underlayer stone should be placed in a manner to produce a resultant graded mass of stone with minimum voids. Rearranging of individual stones may be required to achieve this result. Placement by any method which is likely to cause segregation of the various sizes is not permitted. Unsegregated stone can be lowered in a bucket or container and placed in a systematic manner directly on the underlying material. Casting or dropping stone more than 0.6 m (24 in) or moving by drifting and manipulating down the slope is generally not permitted (U.S. Army, Corps of Engineers, North Central Division, 1978).

2.5.3 Cover (armor) stone

The SPM (U.S. Army, Corps of Engineers, CERC, 1977) and White (1948) show the placement of cover or armor stone as either uniform, special, or random.

Uniform placement. Uniform placement is applicable only to concrete armor units and cut or dressed quarrystones in that they are of a uniform size and shape and thus lend themselves to an orderly placement pattern. Since quarrystones (as opposed to cut stone) are of random size and shape, uniform placement of quarrystone is impracticable.

Special placement. Special placement is applicable only to parallel pipe-shaped stone, and it involves the longest axis being placed perpendicular to the slope of the structure face. For special placement, the longest axial dimension of the stone should be at least twice as great as either of the other two dimensions. The special placement method and the associated stability coefficient should not be used unless quarrystone meets these dimensional specifications and prospective contractors for the project can assure the developer that they can obtain the quarrystone and place it with the long axis normal to the face of the structure slope.

In general, because of the turbidity of the water at a construction site, the special placement method can be used only above the water surface; it is not possible to observe or place stones accurately below the water surface by using this method. Even then, special care must be taken to ensure proper orientation and seating at the interface of the change in placement and at the slope-crown interface.

The special placement method will require close inspection and clear instruction to the contractor to ensure proper placement procedures. The method requires more time than random placement and should, therefore, increase the selection, handling, and placement costs of the quarrystone.

Random placement. "Random (formerly pell-mell) placement" is a term used to describe a variety of placement techniques ranging from dumping the armor stone under water from a scow to careful, individual placing of angular quarrystone in the above-water section. Quarrystone placement by a contractor can vary not only above and below the water level and along the axis of the rubble structure, but also vary from one job to the next. Placement also can vary from one contractor to another.

The variables and difficulties in placing armor units one at a time or dumping by skiff above and below water present the engineer with a difficult design problem. The extent of interlocking achieved is unpredictable when the quarrystone used is of random size but still within specified limits. Generally, in specifying quarrystone armor units, the dimensions of the maximum axis is no greater than 3 times the minimum axis. This applies only to armor stone; the ratio was specified to forestall the use of flat or platelike stone that, if laid flat on the structure slope, would be less stable than a more cubic stone. Because of the unpredictable variables, all methods of placement, except for uniform and special placements, have been lumped together as "random placement" to encompass the range of placement methods.

Cover or armor stone should generally be placed individually and in a manner to avoid displacing underlying materials, avoid placing undue impact force on underlying material, and minimize chipping the stones (Fig. 2.6) (U.S. Army, Corps of Engineers, North Central Division, 1978). The stones should be placed with minimum voids and with maximum interlocking of stones. All stone when placed should be stable, keyed, and interlocked, with no overhanging or "floaters."

"Keying" is the wedging and interlocking of the individual piece of

Figure 2.6 Placing armor stone on jetty, Marina del Rey, California. *(Courtesy of William J. Herron)*

armor stone so that the piece is not only firmly seated but is wedged in by the adjacent armor stones. "Keying" should not be confused with "chinking" by small nondesign stones. These will be removed by the first severe storm and provide little or no stability. The various sizes of cover stone should be so distributed as to produce a uniform well-graded mass. Adjacent stones should be selected as to size and shape and carefully keyed in to provide a compact and integrated surface course. Smaller stones should be used to fill the space between larger ones so as to leave a minimum of voids. Equipment used for placing large stone should be capable of placing the stone near its final position before release and capable of moving the stone to its final position if necessary. Dragline buckets and skips should generally not be used for placement of armor stone. Casting or dropping of stone over 30 cm (12 in) or moving by drifting or manipulating down the slope should not be permitted (U.S. Army, Corps of Engineers, North Central Division, 1978). Final shaping of the slope should be performed during placement of stones.

2.5.4 Riprap

Stone for dumped riprap should be placed on the filter bedding layer or filter cloth in such a manner as to produce a reasonably well-graded mass of rock with the minimum practicable percentage of voids. Riprap should be placed to its full course thickness in one operation and in such a manner as to avoid displacing the bedding material. The larger stone should be well distributed, and the entire mass of stones in the final positions should be roughly graded to conform to the specified gradation. The finished riprap should be free from objectionable pockets of small stones and clusters of larger stones. Placing riprap in layers should not be permitted.

Placing riprap by dumping into chutes or by similar methods that are likely to cause segregation of the various sizes should not be permitted. The desired distribution of the various sizes throughout the mass may be obtained by selective loading of the material at the quarry or other source, by controlled dumping of successive loads during final placing, or by other methods of placement which will produce the specified results. Rearranging individual stones by mechanical equipment or by hand may be required to obtain a reasonably well-graded distribution of stone sizes. Pushing material up the slope or dumping down the slope should not be permitted.

2.5.5 Other stones

Stones to be used as filler material or toe protection should be distributed evenly over the required area. Filler stone can be placed on and dumped with core material. Toe protection stone should be so placed as to produce a reasonably well graded mass of stone with the minimum practicable percentage of voids. Larger stone should be well distributed.

Chink stone should be spread uniformly to form a fairly flat surface even with the top of the riprap. Placing of materials which tend to segregate particle size should not be permitted. Placement of chink stone may be by hand. In that case the stone should be forced into the riprap voids by rodding, spading, or other satisfactory methods.

2.6 Repair of Structures

One of the advantages of stone structures is that they are relatively flexible and are not easily impaired or weakened by slight movement resulting from settlement or other minor adjustments. Damage to stone structures generally consists of wearing, erosion, dislodging, or removal of the stone. The repair consists primarily of rebuilding the stone structure or replacing the stone with new material. In some cases repair can be achieved with concrete or asphalt grout.

2.7 Environmental Considerations

In a natural environment, most of the varieties of stone normally used in coastal structures are very durable and take centuries to erode and become part of the sediments of the earth. In coastal structures, this property of durability, plus density, makes stone a valuable material. Except for the environmental considerations specifically discussed, stone is not significantly affected by the coastal environment.

2.7.1 Periodic wetting and drying

The gain or loss of moisture when stone is alternately exposed to a damp or wet situation and then dried will be most rapid if the pores are large or straight and least rapid if they are small or tortuous. The leaching action of water which is not chemically combined may remove cementing materials from the stone and weaken it.

2.7.2 Freezing and thawing

Freezing and thawing, whether in fresh or salt water, can affect the durability of stone. For example, if water is absorbed in the pores of stone and is subsequently frozen, it will expand. The forces developed in filled pores will cause the stone to crack or spall, and the porosity will be increased. As the pores grow, repetition of each cycle will damage the rock. Stone can be disintegrated by freezing and thawing only if the pores are virtually filled with water.

2.7.3   Chemical attack

Calcareous stones are subject to decomposition by acids which may be formed by the combination of moisture and gases, such as sulfur dioxide, which may be present in the air. A sandstone in which the cementing material is calcium carbonate may also disintegrate under such action, whereas a silicate would be more resistant.

2.7.4   Rock borers

Rock borers found in salt water may penetrate and destroy soft rock such as unconsolidated sandstone and hard clays.

2.7.5   Exposure to waves

Coastal structures are often so located that parts of the structure may be subjected to nonbreaking, breaking, and broken wave forces. Pressure due to nonbreaking waves will be essentially hydrostatic. Breaking and broken waves exert additional pressures that are due to the dynamic effects of the turbulent water in motion and the compression of entrapped air pockets. These pressures may be much greater than those that are due entirely to hydrostatic forces. Therefore, structures or parts of structures located in areas in which storm waves may break should be designed to withstand much greater forces and moments than structures which would be subjected only to nonbreaking waves. Coastal structures may be damaged more by the dislodging of stones from the structure by wave action than by disintegration or breaking of individual stones. The deterioration of stone by abrasion may be most significant when the stone is subject to a pounding surf carrying sand, gravel, and smaller stones.

2.7.6   Exposure to ice floes

No analytical method of determining ice forces on coastal structures is known (U.S. Army, Corps of Engineers, 1971), but some data on ice pressures are available. Floating ice fields may exert a major pressure on maritime structures when driven by strong winds or currents that pile up huge ice masses. Also, stones frozen to ice sheets may be carried away by drifting ice. These factors should be considered when designing structures on the Great Lakes and in other northern locations.

2.7.7   Effects of temperature and fire

Stones, like most other materials, expand when heated and contract when cooled. Unlike most materials, however, they do not quite return to their original volume when cooled but show a permanent swelling. This swelling is determined by using a coefficient of temperature expansion, which is the change in length per unit length per degree temperature change, for various stones (Mills, Hayward, and Radar, 1955). These coefficients per degree Celsius and Fahrenheit range as shown in Table 2.2.

Practically all stones are injured if they are exposed to high temperatures such as may be encountered in fires, and particularly if they are exposed to the combined action of fire and water. The disintegration is usually attributed to internal stresses resulting from unequal expansion of unequally heated parts of the material.

Granites. As experience has shown, granites have particularly poor resistance to fire and are susceptible to cracking and spalling. That is probably due to the irregularity of the stone structure and the complexity of the mineral composition. Coarse-grained granites are most susceptible to the action of fire and water, and gneisses often suffer even more severely because of their banded structure.

Limestones. Limestones suffer little from heat until a temperature somewhat above 100°C (212°F) is reached; at that point the decomposition of the stone begins because of the driving out of carbon dioxide. The stone then tends to crumble because of the flaking of the quicklime formed.

Marble. Marble, because of the coarseness of its texture and its purity, suffers from heat more than limestone does. The cracking is irregular, and the surface spalls off in a way similar to that of granites.

Sandstones. Especially if they are of a dense, nonporous structure, sandstones suffer from high temperature and sudden cooling less than most other stones. The cracking of sandstones that does occur appears mostly in the planes of the laminations. Sandstones in which the ce-

TABLE 2.2   Coefficients of Temperature Expansion

Material	Coefficients of expansion per °C (°F)
Granites	0.00000560 to 0.00000734
	(0.00000311 to 0.00000408)
Limestones	0.00000675 to 0.00000760
	(0.00000375 to 0.00000420)
Marbles	0.00000650 to 0.00001012
	(0.00000361 to 0.00000562)
Sandstones	0.00000902 to 0.00001196
	(0.00000501 to 0.00000622)

menting component is silica or lime carbonate are more fire-resistant than those in which the grains are bound by iron oxide or clay.

2.8 Uses in Coastal Construction

A few uses of stone in coastal construction are cited below to illustrate the typical ways in which stone is used in the coastal zone. Generally, stone is used as a mass material to break up wave energy and protect slopes against erosion.

2.8.1 Breakwaters

Stone is one of the principal materials used in breakwater building. It is used from the core to the armor in various lift and layers each of which has a different gradation. Not all cores are made of stone; but when a stone core is used, it usually is made of impermeable quarry run stone. The core is covered by a blanket of filter material graded to protect it from being eroded away by the action of waves and currents and to allow changes of hydrostatic pressure in the core without loss of core material. The next layer is usually the underlayer graded to be stable against the anticipated surge and current action. The final layer of armor stone is placed in the area where waves impinge on the breakwater. Armor stone is graded and sized to remain stable under the impact of unbroken, breaking, and broken waves. When storm waves may overtop the breakwater, armor stone must be placed on the backslope as well as the seaward face. The elevations and width of crest will depend on the desired use as well as the degree of porosity. Porosity or void ratio is important in dispersing the wave energy and reducing the impact load of the waves striking the breakwaters.

The design size of armor rock for a breakwater is a function of slope, density, and wave height (U.S. Army, Corps of Engineers, 1971). Hence, the primary concerns in the selection of armor stone are density, durability, and available size. Armor stone may be required in pieces varying from about 900 to 27,000 kg (1 to 30 tons). It is usually difficult to quarry, transport, and place stones larger than 27,000 kg in size. Therefore, as design wave heights increase, it becomes more economical to use the more efficient concrete shaped structures such as tetrapods, tribars, and dolosse (see Chap. 4).

2.8.2 Fill material for caissons

Because of its density and generally low cost, stone fill material is frequently used to ballast caissons and sheet pile cells. Rockfill should be well graded and free of loam and organic material in order to have the highest density and minimize settlement or, in the case of a perforated caisson, minimize the loss of material due to currents or wave action.

2.8.3   Toe protection

One of the major causes of failure, or structural damage, of breakwaters has been the undercutting of the toes of the structures. When waves impinge on these structures, they not only exert large forces on the armor stone, or face of a vertical structure but may also impose strong uplift forces on the lower armor stone and toe stone. Thus the armor stone must be carried to sufficient depths to resist wave forces. An additional problem is the turbulence created in such depths, particularly in the case of waves breaking directly on the structure. This can create scour of the sandy bottom and result in undermining the toe stone, which will cause general collapse of the armor layer and exposure of the smaller stone of the underlayers. Scour can be controlled either by carrying the armor and bedding layers to sufficient depth or overbuilding the toe section in anticipation of the quarrystone settling into scour holes. The design of such toe protection is dependent upon wave height and the relative depths of the toe protection as compared to the depth of the natural bottom (U.S. Army, Corps of Engineers, CERC, 1977). The same care must be taken as in the rubble structure.

2.8.4   Jetties and groins

The primary difference between a breakwater and a jetty or a groin is that the jetty or groin must have a sand-tight core to prevent the passage of littoral materials or currents through the structure, whereas breakwaters may be designed to be either permeable or impermeable. All these structures may be subjected to very large breaking or nonbreaking waves. Overtopping of breakwaters and groins in the breaker or uprush zone may be acceptable, but overtopping of jetties must be restricted to prevent passage of sand into navigation channels. Breakwaters connected to shore would otherwise be designed to use stone in the same manner as offshore breakwaters.

Jetties. Jetties are usually constructed from the shoreline through the breaker zone seaward to 12- to 18-m (40- to 60-ft) depths. They are generally perpendicular to the shoreline. However, because of perhaps as much as 30° skew or because of variable wave directions, their alignment may vary from 0° to 90° from the direction of wave travel. Because of variable depths, different parts of the structure may be exposed to unbroken, breaking or broken waves. Thus with careful design, the elevation total cross section and size of armor rock can be varied to produce an economical structure. Since jetties are used to define harbor or river access to the sea, they may be subjected to major tidal or river currents or a combination of tidal and river flow. This factor must be carefully considered, particularly in the design of the inner toe of the structure. Since a jetty's primary purpose is to prevent the passage of littoral material through the littoral drift zone, the uprush area must be impermeable or sand-tight. Other design considerations are much the same as for a breakwater.

Groins. The structural design and the selection of stone sizes and gradation for a groin are much the same as for a jetty. The primary difference is that, whereas in a jetty no sand should pass through or over the structure, most groins are part of a beach stabilization program and it is usually desirable to permit some littoral sand to pass around, over, or through them. This is not to imply that the permeability of armor stone can be designed to allow a selected amount of sand to bypass, since current procedures are not capable of designing successfully functioning permeable groins.

Many attempts have been made to design groins with a varying degree of permeability. For rubble stone groins, it is usually adequate to design the elevation of the impermeable core through the nearshore and foreshore area to the desired beach profile. The voids in the armor rock will generally be adequate to pass the surplus sand through to the downdrift beach. A groin usually terminates just seaward of the breaker zone in from 1.8 to 3.6 m (6 to 12 ft) of water (MLW). The seaward end is designed against the largest breaking wave possible at that depth, taking tidal elevations into account. The breaker zone is an area of constant turbulence, and care must be taken to properly place, as well as design, the bedding layer or the structure will fail. Considerable success has been experienced in recent years in replacing the bedding layer of stone or by combining it with filter cloth.

Seawall or revetment. A stone rubble seawall or revetment is used to protect the shore, or a shore structure, against erosion by wave action or currents (Fig. 2.7). It may be a trapezoidal gravity seawall-type structure backfilled by shore material, or it may be a form of sloped revetment against a shore bank of earth, wood, steel, or concrete.

Current protection. A revetment designed to protect against tidal currents or river currents is designed much the same except that, in the case of tidal currents, the flow may be reversible whereas river currents are unidirectional. When river and tidal currents combine, tidal elevations must be considered to determine the stage of maximum or critical velocities. Also, in bays or large river mouths, consideration must be given to local wind waves, residual swell, or switching from

Figure 2.7   Stone rubble revetment at Jacksonville Beach, Florida.

the open sea. In the case of river mouth entrances, the revetment may simply represent a transition section from the steady-flow river revetment to the wave-exposed jetty. In other cases, where a jettied entrance connects the open sea to a wide mouth or bay, the channel must be revetted to the point where sea waves are attenuated and currents are diminished below the scouring velocity. In the same manner as for a breakwater, there must be a layer of armor rock, an underlayer, and a filter layer. Special consideration should be given to ensure stability of the toe of the structure because of the unidirectional flow of most currents.

Wave protection.   The design of a rock rubble face of a seawall or revetment against wave forces is similar to that of the seaward face of the breakwater. The armor stone must be designed against the force of breaking waves, unbroken waves, or broken waves. The design size of armor stone will be a function of density, slope, and wave height. In contrast to some breakwaters, almost all seawalls or revetments must be designed to an elevation to prevent wave overtopping. Care must also be taken to construct an adequate toe structure to prevent undermining of the entire structure during severe wave action. This may not be a serious problem in lakes or bays, where advantage can be taken of prolonged periods of small or no wave action to construct the toe trench. Conversely, along the open seacoast, where the action of the surf is continuous, it is generally not possible to excavate to a sufficient depth to reduce scouring velocities. The usual alternative is to overbuild the toe structure in the anticipation that, as sand is scoured from under the toe, the excess rock will drop into place and maintain toe support of the structure.

Fixed structures generally have smooth and vertical, or near vertical, faces on the seaward side. The effects of turbulence due to wave action or scouring velocities due to currents can be stronger and more serious on such structures than on rubble structures. In these cases the toe structure can serve two functions:

1. Designed as a submerged rubble structure, it may rise an appreciable height above the natural bottom and serve to reduce wave or current stresses on the fixed or solid structure (U.S. Army, Corps of Engineers, CERC, 1977).

2. Designed as a filter blanket or bedding layer, it may be used to prevent scouring of the natural bottom material, which might result in undercutting of the wall.

Piers and wharves.   Revetments at pier abutments must be protected the same as any open revetment and in addition must be designed to protect the abutment from loss of foundation. Piers in sandy areas are subject to scour around pilings where strong currents also exist. They can be protected by laying down a quarrystone blanket under the pier in the scour area.

2.8.5 Anchors

A deadweight anchor can be any object that is dense, heavy, and resistant to deterioration in water. The type of ocean operation and the availability of materials usually dictate the shape, form, size, and weight of a deadweight anchor. Common examples include stones, concrete blocks, individual chain links, sections of chain links, and railroad wheels. In most instances, a deadweight anchor functions simply as a deadweight on the sea floor that resists uplift by its own weight in water and resists lateral displacement by its drag coefficient with the sea floor. The use of stone as a deadweight anchor becomes increasingly impractical as the holding capacity requirement exceeds 680 kg (1500 lb).

3.1 Component Types and Classes of Soils

The words "earth" and "soil" cover a large assortment of materials of various origins; for engineering purposes the materials are generally classified as gravel, sand, silt, clay, and organic material. Most soils are composed of mixtures containing two or more of those materials. Different geological processes (such as alluvial, residual, glacial, and loessial) and parent materials (sedimentary, igneous, and metamorphic) will affect the type and nature of the soils formed. A soil can be described by its grain-size classification, appearance and structure, and compactness or hardness.

There are several soil classification systems, but the most widely used in engineering is the Unified Soil Classification System (USCS). It is defined in ASTM Standard D2487 and MIL-STD-619A. A summary of the classification system is presented in Table 3.1, and the general soil characteristics are discussed in the following paragraphs. Table 2-3 in TM S-818-1 is another useful version of USCS. A more detailed presentation of the classification systems and soil properties can be found in the report entitled "Geotechnical Engineering in the Coastal Zone," Callender and Eckert.

3.1.1 Coarse-grained materials

Gravels and sands are known as coarse-grained soils. Coarse-grained materials are such that 50 percent or more of the materials by weight are retained on the No. 200 sieve. They are recognized either visually and manually or more formally by:

- Effective grain size (D_{10}): grain size such that 10 percent by weight of the materials are finer

---

This chapter was written by Ross A. Morrison and Louis J. Lee.

TABLE 3.1 United Soil Classification System (ASTM D-2487)

Major Divisions			Group Symbols	Typical Names	Laboratory Classification Criteria	
Coarse-grained soils (More than half of material is larger than No. 200 sieve size)	Gravels (More than half of coarse fraction is larger than No. 4 sieve size)	Clean gravels (Little or no fines)	GW	Well-graded gravels, gravel-sand mixtures, little or no fines	$C_u = \dfrac{D_{60}}{D_{10}}$ greater than 4. $C_c = \dfrac{(D_{30})^2}{D_{10} \times D_{60}}$ between 1 and 3	Determine percentages of sand and gravel from grain-size curve. Depending on percentage of fines (fraction smaller than No. 200 sieve size), coarse-grained soils are classified as follows: Less than 5 percent GW, GP, SW, SP More than 12 percent GM, GC, SM, SC 5 to 12 percent Borderline cases requiring dual symbols [b]
			GP	Poorly graded gravels, gravel-san mixtures, little or no fines	Not meeting all gradation requirements for GW	
		Gravels with fines (Appreciable amount of fines)	GM [d] u	Silty gravels, gravel-silt mixtures	Atterberg limits below A line or P.I. less than 4 / Atterberg limits below A line with P.I. greater than 7	Above A line with P.I between 4 and 7 are borderline cases requiring use of dual symbols
			GC	Clayey gravels, gravel-sand-clay mixtures		
	Sands (More than half of coarse fraction is smaller than No. 4 sieve size)	Clean sands (Little or no fines)	SW	Well-graded sands, gravelly sands, little or no fines	$C_u = \dfrac{D_{60}}{D_{10}}$ greater than 6: $C_c = \dfrac{(D_{30})^2}{D_{10} \times D_{60}}$ between 1 and 3	
			SP	Poorly graded sands, gravelly sands, little or no fines	Not meeting all gradation requirements for SW	
		Sands with fines (Appreciable amount of fines)	SM [d] u	Silty sands, sand-silt mixtures	Atterberg limits above A line or P.I. less than 4 / Atterberg limits above A line with P.I. greater than 7	Limits plotting in hatched zone with P.I. between 4 and 7 are borderline cases requiring use of dual symbols
			SC	Clayey sands, sand-clay mixtures		

TABLE 3.1 (Continued)

Plasticity Chart

					Description
Fine-grained soils (More than half material is smaller than No. 200 sieve)	Silts and clays (Liquid limit less than 50)	ML	Inorganic silts and very fine sands, rock flour, silty or clayey fine sands, or clayey silts with slight plasticity
		CL	Inorganic clays of low to medium plasticity, gravelly clays, sandy clays, silty clays, lean clays
		OL	Organic silts and organic silty clays of low plasticity
	Silts and clays (Liquid limit greater than 50)	MH	Inorganic silts, micaceous of diatomaceous fine sandy or silty soils, elastic silts
		CH	Inorganic clays of high plasticity, fat clays
		OH	Organic clays of medium to high plasticity, organic silts
	Highly organic soils	Pt	Peat and other highly organic soils

a Division of GM and SM groups into subdivisions of d and u are for roads and airfields only. Subdivision is based on Atterberg limits; suffix d used when L.L. is 28 or less and the P.I. is 6 or less; the suffix u used when L.L. is greater than 28.

b Borderline classifications, used for soils possessing characteristics of two groups, are designated by combinations of group symbols. For example: GW-GC, well-graded gravel-sand mixture with clay binder.

- Uniformity coefficient $(C_u) = \dfrac{D_{60}}{D_{10}}$

- Coefficient of curvature $(C_c) = \dfrac{(D_{30})^2}{D_{10} \cdot D_{60}}$

Because most soils are composed of more than one constituent, the USCS makes the following distinctions for sands and gravels:

- *Well-graded gravel (GW) or sand (SW).* All particle sizes are represented within the constituent limits; C_u is greater than 4 or 6, respectively; C_c is between 1 and 3; and the fraction smaller than the No. 200 sieve size does not exceed 5 percent.
- *Poorly graded gravel (GP) or sand (SP).* Some particle sizes are missing or are in excess within the constituent limits; gradation requirements for (GW) or (SW) are not met; and the fraction smaller than the No. 200 sieve size does not exceed 5 percent.
- *Silty gravel (GM) or sand (SM).* More than 12 percent by weight is finer than No. 200 sieve, and the fines have little or no plasticity.
- *Clayey gravel (GC) or sand (SC).* More than 12 percent by weight is finer than the No. 200 sieve, and the fines are plastic.

When the fraction smaller than the No. 200 sieve size is greater than 5 percent and less than 12 percent, a dual symbol should be used.

Well and poorly graded gravels and sands are further referred to as clean gravels or sands. Silty or clayey gravels and sands may be referred to as dirty gravels or sands. It should also be noted that the particle shape has an influence on the density and the stability of the coarse-grained soils.

Gravel (G). The USCS defines a material as gravel when its size ranges between 76.2 millimeters (mm) (3 in) and the No. 4 sieve. Materials larger than 76.2 mm are designated as cobbles. Gravels may be manmade (crushed stone) or may come from natural deposits (bank run). Gravels are cohesionless materials.

Sand (S). A material is defined as sand when its grain size is between 4.76 and 0.075 mm (188 and 3 mils) (No. 4 and No. 200 sieves, respectively). The USCS developed a further classification: The sand is coarse when its grain size varies between 4.76 and 2.00 mm (188 and 79 mils) (No. 4 and No. 10 sieves, respectively); medium when between 2.00 and 0.42 mm (79 and 17 mils) (No. 10 and No. 40 sieves, respectively); and fine when between 0.42 and 0.075 mm (17 and 3 mils) (No. 40 and No. 200 sieves, respectively). Sands are cohesionless materials, but they present an apparent cohesion when damp or moist because of the surface tension effects of pore fluids. The effects disappear when the sand is dry or completely saturated.

3.1.2 Fine-grained materials

Silts and clays are known as fine-grained soils. Fine-grained soils are such that 50 percent or more of the material by weight passes the No. 200 sieve. They are distinguished either visually and manually or by means of the Atterberg limits. The USCS, contrary to most other classification systems, does not make any size distinction between silt and clay. That is because the engineering properties of fine-grained soil are more closely related to plasticity characteristics than to grain size.

The USCS distinguishes the following:

- Silt, clay, and organic silt and clay having liquid limits less than 50 percent: ML, CL, and OL, respectively.

- Silt, clay, and organic silt and clay having liquid limits greater than 50 percent: MH, CH, and OH, respectively.

Fine-grained soils usually have a low permeability [10^{-7} to 10^{-9} centimeters per second (cm/s)] with silty soils being somewhat more permeable than clayey ones. Organic materials tend to lower the strength characteristics of the soil, lower the maximum density, increase the time for consolidation, and increase the optimum water content.

Silt (M). Silt is a fine-grained soil of low plasticity which may exhibit an apparent cohesion that is due to capillary forces. Silts have relatively poor strength characteristics, except when they are dry or in the form of siltstones. They make poor foundation materials in cold climates because of frost heave. Confined, relatively dense silts may perform satisfactorily as foundation soil, but they must be evaluated on a case-by-case basis. Most coastal silts are found in combination with some clay, which will increase cohesion and improve foundation characteristics.

Clay (C). Clay is distinguished by its fine particle size and cohesive strength, which is inversely related to its water content. For that reason, clay's performance as a foundation material is strongly influenced by its stress history. In situ overconsolidated clays, clays which have been loaded to higher stresses than the present load, may perform quite well in foundations. Normally consolidated or underconsolidated clays typical of estuaries will generally experience large settlements when loaded. The minerals included in the clay composition influence the properties of the soil; e.g., montmorillonite is a highly active mineral, and soils containing such a mineral will present high swelling and shrinkage characteristics. Two other commonly occurring minerals are illite (less active than montmorillonite and commonly found in marine clays) and kaolinite (the least active mineral).

Organic materials (O). Peat, organic mulch, and muskeg are highly organic soils which usually have a spongy nature and a fibrous texture. Organic materials come from the decay of vegetable matter. They are recognized by their odor, which is intensified by heating, and by their dark color (although some dark soils may be inorganic). Usually, organic soils have high moisture and gas contents and relatively low specific gravities.

3.2   Properties and Characteristics of Soils

The major significant engineering properties of soil are shear strength, compressibility, and permeability. Geotechnical problems encountered in the design of coastal structures which involve these characteristics are slope stability, bearing capacity, settlement, and erosion. A detailed discussion of the properties and characteristics of soils and the tests required to determine them can be found in *Geotechnical Engineering in the Coastal Zone* (Callender and Eckert). The presence of pollutants derived from industrial wastes, such as toxic heavy metals (mercury, cadmium, lead, and arsenic), chlorinated organic chemicals (DDT and PCBs), and pathogens (bacteria, viruses, and parasites) should also be considered in the evaluation of the use of any soil in coastal structures. Polluted soils should not generally be used.

3.2.1   Shear Strength

The three types of tests commonly performed to determine soil strength are designated as:

- Unconsolidated-undrained triaxial test, commonly known as a UU-test or Q-test
- Consolidated-undrained triaxial test, commonly known as a CU-test or R-test
- Consolidated-drained triaxial test, commonly known as a CD-test or S-test

The Q, R, and S designations are standard use in Corps literature. The descriptions are indicative of the conditions under which the tests are run. From the results of the tests, the stress-strain characteristics under various loading conditions and the conditions of failure for the soil are established. The strength of a soil is usually defined in terms of the stress developed at the peak of the stress-strain curve and is presented in the form of Mohr circles and a Mohr failure envelope. The strength is then expressed in terms of cohesion and the angle of internal friction.

3.2.2 Compressibility

The simplest compressibility or consolidation test is the one-dimensional, laterally confined compression test (often referred to as the odometer test). In this test the soil sample is placed within a restraining ring and loaded with special types of plates on either top or bottom or both. The change in sample height is measured by a deflection gage and is used to calculate the change in void ratio e at different normal pressures P. If the soil is saturated, the sample is placed between two porous disks that permit the water to drain away during compression. That in turn leads to information which permits plotting of the so-called $e \log P$ relation. From $e \log P$ plots for sands, silts, clays, or mixtures of them, factors are quantified for consolidation and settlement estimates. It should be emphasized that in such tests the lateral expansion is restrained. In real situations constraint is only approximated by the loading of relatively thin layers of compressible soil through load distribution over a large area.

Sand compressibility. The most important property of sand, which governs the stiffness of the sand, is relative density. The relative density of a sand is usually determined in the field by means of standard penetration tests or Dutch cone penetration tests.

Clay and silt compressibility. Predictions of static settlement on silts and clays are usually made on the basis of consolidation or odometer tests. The rate of settlement and the time for essential completion of primary consolidation can be predicted on the basis of the test. Typically, silts are less compressible than clays.

3.2.3 Permeability

Permeability is the soil property that indicates the relative ease with which a fluid will flow through the soil. The coefficient of permeability k of a soil is defined as the average percolation velocity v divided by the hydraulic gradient i in the soil at that particular point. It is seen then that the coefficient of permeability has units of velocity; they are commonly stated as centimeters per second or feet per minute. Permeability depends on the characteristics of both the pore fluid and the soil. Viscosity, unit weight, and polarity are the major pore fluid characteristics. Particle size, void ratio, composition, fabric, and degree of saturation are the major soil characteristics.

In general, a qualitative approximation of the permeability of the materials can be made on the basis of grain size. For example, clean gravels will have permeabilities ranging from 10^1 to 10^2 centimeters per second (cm/s). Clean medium to coarse sands will have permeabilities ranging from 10^{-2} to 1 cm/s. Very fine sand will have permeabilities ranging from 10^{-5} to 10^{-4} cm/s. Organic and inorganic silts, mixtures of sand, silt, and clay, glacial till, and some stratified clay deposits will have permeabilities ranging from 10^{-6} to 10^{-5} cm/s. Clays, which are practically impervious and commonly used for core materials in water-retaining embankments, will have permeabilities ranging from 10^{-9} to 10^{-7} cm/s.

3.2.4   Other properties and characteristics

Other soil properties and characteristics needed for the proper design of coastal structures include dry density, water content, specific gravity, resistivity and corrosion potential, grain-size distribution, plasticity characteristics, chemical properties, and durability.

3.3   Methods of Soil Improvement

Methods of soil improvement generally include densification, drainage, changing soil properties at depth by grouting or injection, surface stabilization by admixtures, and reinforcement with metal or fabric strips or mesh. All these methods have been utilized in one manner or another to improve soils used in coastal structures. The most widely used, and generally the most practical, are densification and drainage. A somewhat newer method developed in Europe and now being more widely used in the United States is reinforced earth (Figs. 3.1 and 3.2). Some of the methods available for improvement of soils are classified according to the basis of soil improvement as shown in Table 3.2 (Mitchell, 1976).

Figure 3.1   Reinforced-earth seawall construction at Petersburg, Alaska. *(Courtesy of The Reinforced Earth Company)*

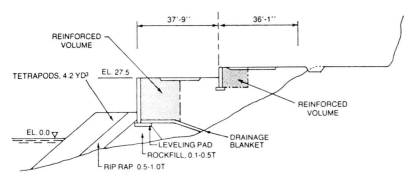

Figure 3.2 Cross section of reinforced-earth seawall at La Reunion Island. *(Courtesy of The Reinforced Earth Company)*

3.4 Placement of Soil for Coastal Structures

Earthfills for coastal developments can be placed from the land by dumping or shoving into place, from water by deck and bottom-dump barges, or hydraulically from dredging operations. The method and equipment used depend on the type and source of material, the area of placement (above or below the water level), the depth of water (for operating bottom-dump barges), the purpose and design of the structure or fill, the availability of equipment, the impact on the environment, and the economics of the operation. The material may also be placed with or without compaction to alter the soil density.

3.4.1 Dumped loose fills

Soils obtained from land sources are typically transported by and dumped from trucks, scrapers, conveyor belts, or possibly shovels. The materials are then shoved into place and leveled. Soils obtained from marine sources are typically excavated by shovels, draglines, or bucket dredge, and they are either dumped directly into place or placed in trucks or barges and transported to the site. The soils placed in trucks are dumped the same as in a landfill. It should be noted that leakage from the trucks is unacceptable in most urban and environmentally sensitive areas. Soils placed in barges are transferred to shore equipment or dumped through the water (Fig. 3.3). Soils placed in this latter manner will typically have low to medium relative densities.

The placement of fill landward of a bulkhead driven into a mud bottom may result in the formation of mud waves. They progress ahead of the advancing fill and can overload the bulkhead and cause its failure before the landfill reaches it. Either the mud should be removed before fill placement or the bulkhead should be designed for the mudwave pressures. Care should also be taken to place select fill behind the bulkhead before placing the general fill to ensure that the active pressure zone has the shear strength planned for in the design.

TABLE 3.2 Improvement of Soils for Coastal Construction

Method	Principle	Most suitable soil conditions/types	Special materials required	Special equipment required	Properties of treated material	Special advantages and limitations
Blasting	Shock waves and vibrations cause liquefaction, displacement, remolding	Saturated, clean sands, partly saturated sands and silts after flooding	Explosives; backfill to plug drill holes	Jetting or drilling machine	Can obtain relative densities from 70 to 80%; may get variable density	Rapid and inexpensive; can treat small areas; variable properties; no improvement near surface; dangerous
Terraprobe	Densification by vibration; liquefaction induced settlement under overburden	Saturated or dry clean sand	None	Vibratory pile-driver and 750-mm-diam open steel pipe	Can obtain relative densities of 80% or more	Rapid and simple; good under-water; soft under-layers may damp vibrations; difficult to penetrate; stiff over-layers; not good in partly saturated soils
Vibratory rollers	Densification by vibration; liquefaction induced settlement under roller weight	Cohesionless soils	None	Vibratory roller	Can obtain very high relative densities; upper few decimeters not densified	Best method for thin layers or lifts

TABLE 3.2 (Continued)

Method	Principle	Most suitable soil conditions/ types	Special materials required	Special equipment required	Properties of treated material	Special advantages and limitations
Compaction piles	Densification by displacement of pile volume and by vibration during driving	Loose sandy soils; partly saturated clayey soils; loess	Pile material (usually sand or soil plus cement mixture)	Pile driver	Can obtain high densities, good uniformity	Useful in soils with fines; uniform compaction; easy to check results; slow; limited improvement in upper 1 to 2 m (39 to 78 in)
Heavy tamping (dynamic consolidation)	Repeated application of high-intensityimpacts at surface	Cohesionless soils best; other types can also be improved	None	Tamper of 44 to 176 t (10 to 40 tons) high-capacity crane	Can obtain high relative densities, reasonable uniformity	Simple and rapid; suitable for soils with fines; usable above and below water; requires control; must be away from existing structures
Vibrofloflation	Densification by vibration and compaction of backfill material	Cohesionless soils with less than 20% fines	Granular backfill	Vibroflot, crane	Can obtain high relative densities, good uniformity	Useful in saturated and partly saturated soils; uniformity

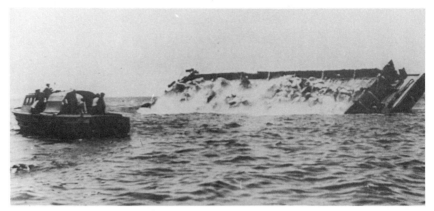

Figure 3.3 Fill being placed by tilt barge at Redondo Beach, California. *(Courtesy of Woodward-Clyde Consultants)*

In dumping fill material through water, avoid excessive turbulence in order to prevent segregating the materials or spreading them over extremely flat slopes at the edge of the fill. Uncontrolled bottom dumping from barges through great depths of water will encourage segregation and spread fill over a wide area. Berms or dikes of coarse-grained material or stone can be used to confine the material.

3.4.2 Compacted fills

When compaction of fills above the water level is desired, either the method of compaction or the desired compaction density can be specified. A test section is usually required to determine the effectiveness of the methods before specifying a particular method. Usually, the required density, moisture limits, and lift thickness are specified and the contractor is allowed some selection in compaction methods.

Coarse-grained, cohesionless soils with less than 4 percent passing a No. 200 sieve when well graded or less than 8 percent when uniformly graded are generally insensitive to compaction moisture. These soils should be placed when they are at the highest practical moisture content and compacted by vibratory methods. When materials with sizes up to 150 mm (6 in) maximum are used, the large sizes will interfere with the compaction of soil smaller than a No. 4 sieve or 19-mm (0.75 in) size. Where large parts (more than 30 percent by weight) of gravel and cobbles are present, a slight reduction (several percent) in the required density of the minus No. 4 sieve size may be tolerated.

3.4.3 Hydraulic fills

Hydraulic fills are placed on land or under water by pumping material through a pipeline or by water sluicing through a conveyor. Borrow materials used for these fills are generally obtained by dredging (Fig. 3.4). The characteristics of such hydraulic fills may be generally classified according to the nature of the borrow material (Whiteman, 1970). This classification is shown in Table 3.3.

Generally, material with more than 15 percent nonplastic fines or 10 percent plastic fines passing a No. 200 sieve should not be placed under water.

The wash water of fills hydraulically placed on land should run off in such a manner that fines are not concentrated in pockets. This may require starting the pipeline discharge point at one side or corner of the area and advancing it along previously deposited material in an attempt to force any soft fines out ahead of the fill as it is placed. Hydraulic fills placed behind walls or bulkheads should be placed in lifts thin enough to permit wash water to run off without building up a full height of hydrostatic pressure.

Dredging and placing operations will produce significant textural differences between original bottom sediments and sediments deposited at the fill site. In general, there will be an increase in the mean grain size as fines are lost and a decrease in the uniformity coefficient (Table 3.2) of delivered vs. bottom sands. The volumetric losses resulting from winnowing associated with this process appear to be of the

Figure 3.4 Dredge *Chequinquira* at dock. *(Courtesy of Woodward-Clyde Consultants)*

TABLE 3.3   Characteristics of Fill Based on Nature of Borrow Material

Nature of borrow material	Characteristics of fill
Fairly clean sand	Reasonably uniform fill of moderate density (relative density of 40 to 60 percent)
Silty or clayey sand	Very heterogeneous fill of large void ratio (low relative density)
Stiff cohesive soil	Skeleton of clay balls, with matrix of sand and clay
Soft cohesive soil	Laminated normally consolidated or unconsolidated clay

order of 10 percent or more, depending on the original bottom material.

3.5   Artificial Beach Restoration

The placement of sandfill along a part of shore front for beach restoration may be done in one of three ways. It may be placed directly onto the shore to be protected or developed as a beach; it may be placed in an area adjacent to and updrift of the area to be protected; or it may be placed offshore. That is, it may be directly placed to form a feeder beach (designed to be eroded by waves and tidal currents and transported via longshore transport to the area to be protected), or placed as a bar (designed to be transported inshore by waves and currents). To be effective, offshore restoration must involve placing fill in the zone where shoreward movement of the sand occurs, piling it as high as possible, and placing it during the spring or early summer to take full advantage of the seasonal beach building before the annual period of high winter storm waves.

3.6   Repair of Earth Structures

Damage to earth structures or earth parts of structures generally consists of erosion or removal of the soil. The repair consists primarily of replacing the earth material or the protective layer. If permeability or stability of the structure is a problem, the voids can be filled with concrete or asphalt grout.

3.7   Environmental Effects on Soils

Soils are sensitive to a variety of environmental conditions, and the environmental effects cannot be ignored when the soils are used in the coastal environment.

3.7.1 Physical effects

Both the nature of the soil pore fluid and the temperature can influence the behavior of clay soils. A loss in shear strength of marine clays can be realized by removal of the salt through leaching by fresh water. Changes in moisture content can cause swelling or shrinking of clay soils. Decreasing the temperature of a cohesive soil can cause an expansion of the soil. Fine-grained soils are also susceptible to frost heave.

The most significant environmental effect on the physical properties of soils is liquefaction, which causes loss of strength. The liquefaction can be due to a seismic event, tsunami (tidal wave), or the continual impact of high breaking waves, and it can result in failure of the structure. The soil properties generally related to these phenomena are saturation, grain size, relative density, and permeability. Problems are generally associated with loose fine sands and silts below the water level at sites in highly seismic areas or areas subject to high breaking waves. Liquefaction of foundation soils under gravity ocean structures offshore has been attributed to water wave forces on the structure. Liquefaction can generally be minimized or mitigated by densification or treatment of the soils or by providing drainage (rock drains). A discussion of soil response to both seismic- and water-wave-induced dynamic loads is presented in Callender and Eckert.

3.7.2 Erosion effects

The erosion and subsequent deterioration of both natural landforms and manmade coastal structures is of concern. Shore erosion is a major problem along coastlines of the oceans and the Great Lakes. Erosion is caused principally by storm-induced wave action and associated longshore currents. The processes are further complicated by erosive forces that may come from ice, wind, rain, burrowing animals, or human activity. Shore erosion problems become more critical when beaches become eroded or submerged and subject adjoining highly erodible upland areas to direct wave attack (Fig. 3.5). Unconsolidated sands and silts are generally the most easily eroded; clays and gravels are slightly more resistant to erosion; and cemented soils and rock are the least erodible earth materials.

The soils may be protected from erosion by various devices such as revetments, seawalls, and bulkheads. Groins may be used to maintain beaches. A detailed discussion of beach erosion is presented in chapter 4 of the SPM (U.S. Army, Corps of Engineers, CERC, 1977).

3.8 Uses of Soils in Coastal Construction

Earth can be used for almost any kind of coastal structure. Coastal structures are generally associated with three types of projects: port

Figure 3.5 Wave erosion of Sunset Cliffs in San Diego. *(Courtesy of the city of San Diego)*

and harbor development (including marinas), land reclamation, coastal protection, and beach replenishment. Design considerations and criteria are discussed in detail in Callender and Eckert. Soils are used in coastal structures for backfill materials, core materials, slopes, and beach restoration. The use of the material is generally associated with the type of development and the type of structure. The choice of material depends on the economy and availability of the material, depth of water, expected water or wave forces, and the purpose of the structure. Some of the various uses are presented in the following paragraphs.

3.8.1 Breakwaters

In rubble mound breakwaters, submerged reefs, and other coastal rock structures, sand can be used as a core material provided other materials are used to protect the sand against wave damage and piping. Clay can be mixed with the sand to reduce permeability, but it is generally not adequate core material by itself. The structures can be designed as permeable or impermeable. The soils are usually dredged materials and are placed hydraulically. The core can be covered with filter material and quarrystone riprap or armor units; the number of covering layers depends on the water depth, the design storm waves, and the desired degree of permeability.

3.8.2 Caissons

Concrete caissons and sheet pile cells can be filled with sand and clay. If the surface is to be paved and used for load bearing, sand is preferable to clay because of its higher bearing capacity. A filter layer and armor rock must be provided to cover the fine soils where waves or water currents are expected to impinge on earthfill.

3.8.3 Jetties and groins

Rubble mound jetties and groins, like caissons and sheet pile cells, can be filled with soil as described for breakwater construction.

3.8.4 Low-cost shore protection, fabric bags

Low-cost groins and breakwaters have also been constructed by means of fabric bags filled with medium sand or sand cement. The bags are generally made of nylon, and they may be coated with polyvinyl chloride or acrylic to delay fiber degradation by ultraviolet rays. The bags can be filled with available beach sand and used as a low-cost shore protection device. The bag filling can be done by using a 19-mm (¾-in) diaphragm pump or, for more efficiency, a small front-end loader, a hopper, and a jet pump. Typical filled bags measure approximately 3 by 1.5 by 0.5 m [10 by 5 by 1.5 feet (ft)], hold about 1.9 cubic meters (m^3) [2.5 cubic yards (yd^3)] of sand, and weigh about 3180 kg (7000 lb). The sand used should be saturated to eliminate air pockets, which would cause a buoyant force on the bags. Small bags filled with sand (Fig. 3.6) or concrete (Fig. 3.7) can be used for protection against erosion.

Figure 3.6 Sea cliff upper-slope erosion protection by cloth bags filled with sand and supported by wood bulkheads. *(Courtesy of Woodward-Clyde Consultants)*

Figure 3.7 Sea cliff shore protection by cloth bags filled with concrete and placed on soil bank, Solano Beach, California. *(Courtesy of Woodward-Clyde Consultants)*

3.8.5 Bulkheads, quaywalls, and seawalls

Usually, soil materials are used as backfill or foundation materials for bulkheads and walls. Their primary purpose is to provide a level surface or to fill a void behind the structure. The materials can be placed from land, end-dumped from trucks or conveyor belts, or be placed hydraulically from dredges, but the relative density achieved by each method may be widely different. The backfill can be composed of a mixture of all or part of the four component types of soil (gravel, sand, silt, clay), but not all soil mixtures are equally effective in a given situation. Organic materials are usually considered detrimental and are not used, since they tend to be more compressible and have lower shear strengths. Highly expansive clay also should generally not be used for backfill of coastal structures.

A filter layer of gravel or crushed stone is often used (with or without a geotextile filter) under and behind walls to provide for relief of hydrostatic pressure and to prevent piping. The material should meet filter design criteria.

3.8.6 Wharves and piers

Earth is generally used for wharves and piers as fill behind or for slopes underneath the facility. The natural in-place soils generally provide support for the structures, and foundations are designed in accordance with conventional geotechnical procedures. The purpose of the fill is to help provide stability and rigidity to the structure and to provide useful working areas behind the structure. Most soils can be used for this purpose, but they generally need some type of protection, such as rock riprap and a filter, to mitigate erosion. Coarse-grained granular soils are preferable for use as backfill materials, since they are typically stronger and less compressible. The properties of the backfill soils of use in the design include dry density, water content, shear strength, and compressibility. Other important properties may be compaction characteristics, permeability, and corrosive characteristics.

3.8.7 Land reclamation

Land reclamation may include dredging for marinas, construction of fills for water-oriented land developments, enlargement of streams, and related activities in waters and wetlands. These activities generally involve discharge of fill material onto the adjacent shoreline or into waters or wetlands for construction of structures; site development fills for recreational, industrial, commercial, residential, and other uses; causeways or road fills; dams and dikes; artificial islands; property protection; groins and beach restoration; levees and artificial reefs. These fills may be obtained from land sources and dumped by land methods or from water sources and placed hydraulically by dredging. The properties of soils most useful to know in land reclamation projects are the strength characteristics, consolidation characteristics, and chemical properties after placement of the fill.

3.8.8 Dikes

Dikes can be constructed of sand, clay, or a combination of both. Earth dikes are usually utilized as containment structures for dredged materials, but they may also be used as protective devices such as hurricane barriers. They generally require some type of protection when subjected to wave action. Clay dikes and dikes with sand cores and clay covers have also been built with seaward slopes of 1:6 to 1:10 with a grass cover.

3.8.9 Protective beach and dune restoration

The placement of sandfill along a part of shore front is a nonstructural erosion control technique, referred to as beach nourishment, that is utilized for the protection of beach areas or for the creation of protective beaches in areas where none exist. Artificial restoration projects should generally define the source of material, the method of placement, and the grain-size distribution and amount of sand. The source of material may be either on land or offshore. The methods of placement include placing sand directly on the beach along the entire length of the project, placing it in stockpiles at a feeder beach at one end of the site, and placing it as an offshore bar. The grain size of the materials used should be larger than, or at least the same size as, that of the original beach material. If coarser sands are used in the beach restoration, the equilibrium slope will be steeper than the existing one, and vice versa for finer particles. A more detailed presentation on beach nourishment can be found in chapters 5 and 6 of the SPM (U.S. Army, Corps of Engineers, CERC, 1977).

A subsand filter system (gravel filter bedding layer placed in the foreshore or offshore zones) may be useful in the stabilization of offshore profiles. Preliminary studies have indicated that such a filter may have a stabilizing effect on the bed material in the offshore zone and that it may be effective in speeding accretion in the foreshore zone. The latter use may be employed for berm building or berm replacement (Machemehl, French, and Huang, 1975).

The construction of dunes is another approach to nonstructural erosion mitigation. Dunes are constructed or enhanced by the placement of sandfill and by the planting of stabilizing vegetation. Snow fences may also be used to physically retain initial sand. Dunes are generally constructed parallel to and behind the beach proper, and they serve to trap and absorb sand which is transported by onshore winds, storm overwash, or offshore winds blowing over overwash plains. The construction and stabilization of sand dunes is discussed in chapters 5 and 6 of the SPM (U.S. Army Corps of Engineers, CERC, 1977).

Portland Cement Concrete

4.1 Introduction

Concrete, a diversified construction material, exists in two physical states: the first as a semifluid or plastic state while being mixed, transported, and placed in final forms, the second as a solid after having set and cured. These features give concrete wide application in coastal and waterfront structures under many special conditions. Ingredients for making concrete exist in virtually all areas of the world, and the use of concrete in coastal structures depends only on understanding and knowledge of the materials, design, and processes required for the end use. Concrete has proved to be an excellent construction material, and it is adaptable to many coastal structures. With good planning it will probably find many additional uses that take advantage of its physical qualities.

Concrete can be considered to be made of two components: aggregates and paste. Aggregates are generally classified into two groups: fine and coarse. Fine aggregates consist of sand with particle sizes smaller than 6 mm (0.25 in); coarse aggregates are those with particle sizes greater than 6 mm. Aggregates make up about 60 to 80 percent of the concrete. The paste is composed of cement, water, and sometimes admixtures and entrained air. Cement paste ordinarily constitutes 25 to 40 percent of total volume of concrete, cement being 7 to 15 percent and water 14 to 21 percent. The content of air and admixtures may range up to 8 percent.

The durability of portland cement concrete, defined as ability to resist weathering action, chemical attack, abrasion, or any other process of deterioration, is a major factor in the excellence of concrete as a

---

This chapter was written by Carrol M. Wakeman and Lawrence L. Whiteneck.

coastal construction material. Durable concrete will retain its original form, quality, and serviceability when exposed to its environment.

This chapter discusses the materials used in making concrete, including additives that enhance its properties, the mixing of concrete, and the more important causes of concrete deterioration in coastal structures. It gives suggestions on how to prevent such damage, with particular attention to damage caused by freezing and thawing, aggressive chemical exposure, abrasion, reactive aggregates, and corrosion of embedded materials. Methods for repairing concrete that has not withstood the forces of deterioration and the use of protective coatings to enhance durability also are discussed.

4.2   Types of Portland Cement

The types and characteristics of portland cement used for coastal structures are specified in ASTM Standard C150-78:

- Type I cement is used in ordinary structural concrete for foundations, roads, curbs, and ongrade foundations not subject to marine exposure or freezing and thawing conditions.

- Type IA is a normal air-entraining portland cement for use in the same concrete structures as type I cement but which are subject to freezing climates.

- Type II cement is a mild sulfate-resisting cement. It is used for concrete in a marine environment not subject to freezing and thawing. It contains a maximum allowable tricalcium aluminate (C_3A) of 8 percent, and no excess over that limit should be permitted.

- Type IIA cement should be used for concrete in freezing atmospheres.

- Type III cement is used when a high early strength is needed. Concrete made with this cement will attain a strength in 7 days equivalent to that made with type I cement in 28 days. It should not be used for marine concrete.

- Type IIIA cement contains an air-entraining agent but otherwise is the same as type III.

- Type IV cement provides a low heat of hydration in concrete where a heat buildup may occur. It is used in structures such as dams or in other mass concrete where such heat may be undesirable.

- Type V cement has a greater resistance to sulfates than all the others and should be used in marine environments when available. It has a maximum C_3A content of 5 percent. Seawater contains approximately 26,000 parts per million (ppm) of chlorides and 2600 ppm of SO_3. Addition of an air-entraining agent is mandatory in freezing climates.

4.2.1 Special Qualities

Low-alkali cement. Federal specifications state that the sum of the percents of Na_2O plus 0.658 of the percent of K_2O shall not exceed 0.6 percent of the total cement content. Low-alkali cements are usually well below the 0.6 percent limit.

Concrete for piles driven in soil containing from 0.10 to 0.20 percent water-soluble sulfate (as SO_4) or driven in water containing from 150 to 2000 ppm of SO_4 should be made with cement containing not more than 8 percent C_3A, such as type II, or a moderate sulfate-resistant (MS) cement. In environments where the water-soluble sulfate exceeds 0.20 percent or the sulfate solution contains more than 1000 ppm, portland cement with the C_3A content limited to 5 percent (e.g., type V) should be used. For very severe sulfate exposure (more than 10,000 ppm), type V cement with a fly ash admixture should be used.

Reactive elements. When silica in the aggregates is reactive with alkali of the cement, a cement containing less than 0.60 percent alkali should be used. The foregoing alkalis, if present in sufficient amounts, may cause swelling of certain aggregates such as opal and chalcedony. Such swelling can often be minimized by using low-alkali cement.

Special-purpose cement. Many cements used for special purposes, such as expansive cements, cause the concrete to expand after setting (not used for marine concrete), waterproof portland cement, plastic cement (used for making stucco and plaster), oil well cement, and special-blended custom and polymer cements.

4.3 Properties

The basic properties of concrete are placability, consistency, strength, durability, and density. These properties will vary with the specific components and ratios of components added during mixing. Well-established relations governing these properties are discussed below.

4.3.1 Placability

Placability (including satisfactory finishing properties) encompasses traits loosely subsumed by the terms "workability" and "consistency." Workability is considered to be that property of concrete which determines the capacity to be placed and consolidated properly and to be finished without harmful segregation. It is affected by the grading, particle shape, and proportions of aggregate, the amount of cement, the presence of entrained air, admixtures, and the consistency of the mixture. These factors are to be taken into account to achieve satisfactory placability economically.

4.3.2 Consistency

Consistency, loosely defined, is the wetness of the concrete mixture. It is measured in terms of slump—e.g., the higher the slump the wetter the mixture—and it affects the ease with which the concrete will flow during placement. In properly proportioned concrete, the unit water content required to produce a given slump will depend on several factors. Water requirement increases as aggregates become more angular and rough textured, but this disadvantage may be offset by improvements in other characteristics such as bond to cement paste. The amount of mixing water required decreases as the maximum size of the well-graded aggregate is increased. It also decreases with the entrainment of air. Mixing water requirement may often be significantly reduced by certain admixtures.

4.3.3 Strength

Strength is an important characteristic of concrete, but other characteristics such as durability, permeability, and wear resistance are often equally or more important. For a given set of materials and conditions, concrete strength is determined by the net quantity of water used per unit quantity of cement. The net water content excludes water absorbed by the aggregates. Differences in strength for a given water-cement ratio may result from changes in maximum size of aggregate; grading, surface texture, shape, strength, and stiffness of aggregate particles; differences in cement types and sources; air content; and the use of admixtures which affect the cement hydration process or develop cementitious properties themselves. However, in view of the number and complexity, accurate predictions of strength must be based on trial batches or experience with the materials to be used.

4.3.4 Durability

The ability of concrete to withstand environmental exposure is called durability. Concrete must be able to endure the exposures which may deprive it of its serviceability—e.g., freezing and thawing, wetting and drying, heating and cooling, chemicals, and deicing agents. Use of a low water-cement ratio will prolong the life of concrete by reducing the penetration of aggressive liquids. Resistance to severe weathering, particularly freezing and thawing, and to salts used for ice removal is greatly improved by incorporation of a proper distribution of entrained air.

Entrained air should be used in all exposed concrete in climates where freezing occurs. By using a suitable cement and a properly proportioned mix, concrete will resist sulfates in soil, ground water, or seawater, provided that concentrations are not in excess of 0.05 M [7 grams per liter (g/L)] Na_2SO_4. High-quality concrete will resist mild acid attack, but no concrete has good resistance to strong acids; special protection is necessary in that case.

Sometimes concrete surfaces will wear away as the result of abrasive action. In hydraulic structures, particles of sand or gravel in flowing water can erode surfaces. The use of high-quality concrete and, in extreme cases, a very hard aggregate may provide longer durability under such exposures. More detailed discussion of the exposures that impact the durability of concrete and the techniques to resist those impacts is given in Sec. 4.6.

4.3.5 Density

For certain applications concrete can be used primarily for its weight characteristic. To the extent possible, selection of concrete proportions should be based on test data or experience with the materials actually to be used. When such a background is limited or not available, estimates given herein can be employed.

4.4 Components

Concrete is composed principally of cement, aggregates, and water. It will contain some amount of entrapped air, and it may also contain purposely entrained air obtained by use of an admixture or air-entrained cement. Admixtures are also frequently used for other purposes such as to accelerate, retard, or improve workability, reduce mixing water requirement, and increase strength, durability, density, and appearance. The required characteristics are governed by the use to which the concrete will be put and by conditions expected to be encountered at the time of placement. These are often, but not always, reflected in specifications for the job.

4.4.1 Mixing water

Almost any natural water can be used as mixing water for making concrete. Potable fresh water is usually acceptable as satisfactory mixing water but should meet ASTM Standard C94. Water suitable for making concrete, on the other hand, may not be fit for drinking. Water high in chlorides should not be used in concrete containing steel reinforcement.

Water of questionable suitability can be used for making concrete if mortar cubes made with it have 7- and 28-day strengths equal to at least 90 percent of the strengths of comparison specimens made with tap water. Mortar cubes should be made and tested according to ASTM Standard C109. In addition, Vicat needle tests (ASTM Standard C191), should be made to ensure that impurities in the mixing water do not adversely shorten or extend the setting time of the concrete. Excessive impurities in mixing water also may cause efflorescence, staining, or corrosion of reinforcement. Therefore, certain optional limits may be set on chlorides, sulfates, alkalies, and solids in the mixing water. A water source comparable in analysis to any of the waters in Table 4.1 is probably satisfactory for use in concrete.

Water containing less than 2000 ppm of total dissolved solids can generally be used satisfactorily for making concrete. Water containing more than 2000 ppm of dissolved solids should be tested for its effect on strength and time of set. Water containing 2000 to 3000 ppm, not including Na_2SO_4, of dissolved solids is acceptable if it is free of organic matter. American Concrete Institute (ACI) Committee 201

TABLE 4.1   Chemical Limits for Mixing Water

Chemicals	Maximum concentration,[a] ppm	Test method[b]
Chloride, as Cl		ASTM Std. D512
Prestressed concrete or concrete in bridge decks	500[c]	
Other reinforced concrete in moist environments or containing aluminum embedments or dissimilar metals or with stay-in-place galvanized metal forms	1,000[3]	
Sulfate, as SO$_4$	3,000	ASTM Std. D516
Alkalies, as ($Na_2O + 0.658K_2O$)	600	
Total solids	50,000	AASHTO T26 (Sec. 3.1)

[a]Wash water reused as mixing water in concrete can exceed the listed concentrations of chloride and sulfate if it can be shown that the concentration calculated in the total mixing water, including mixing water on the aggregates and from other sources, does not exceed the stated limits.
[b]Other test methods that have been demonstrated to yield comparable results can be used.
[c]For conditions allowing use of $CaCl_2$ accelerator as an admixture, the chloride limitation can be waived by the purchaser.

(1977), limits chloride ions to percentages of weight according to types of concrete (Table 4.2).

Prestressed work.   Water for use in prestressed work should be more definitely restricted in salt, silt, and organic contents.

1. No impurities that will cause a change in time of set greater than 2.5 percent or a reduction in 14-day strength greater than 5 percent as compared with distilled water
2. Less than 650 ppm of chloride ion (some authorities permit up to 1000 ppm)
3. Less than 1300 ppm of sulfate ion (some authorities limit this to 1000 ppm)
4. Free from oil

Seawater may be used if no other water is available and no steel reinforcement is present. The early strength of seawater concrete will be somewhat greater than that of concrete made with fresh water but after about a month the strength of the fresh water concrete will be stronger. At the Port of Los Angeles, thousands of specimens for long-time testing were made by using seawater for gaging and tap water for control specimens. Storage environments were as follows: fog room for controls, air, fresh water, and seawater. Compression tests were made at increments of 1 day, 7 days, 28 days, 6 months, 1 year, and thereafter each 5 years through 35 years.

The results showed that, in the early phases of the program (within the first year), strength gains for the seawater-gaged concrete (compressive strength, modulus of rupture, and modulus of elasticity) slightly exceeded those of the tap water controls. However, beginning at about 1 year, the tap water control increased above that of the seawater specimens. At the end of the 35-year period the tap water series

TABLE 4.2   Permissible Chloride Ions

Type of concrete	Maximum chloride ion, %
Prestressed concrete	0.66
Conventionally reinforced concrete in moist environment and exposed to chloride	0.10
Aboveground building construction where concrete will stay dry (does not include locations where concrete will be occasionally wetted such as waterfront structures)	No limit for corrosion

was roughly 15 percent stronger than the seawater series. All concrete mixes were of excellent quality, structural grade concrete, and, of course, no form of reinforcement was used. Table 4.3, compares tap water to seawater for total dissolved solids.

Seawater containing up to 35,000 ppm of dissolved salts is generally suitable as mixing water for unreinforced concrete. This strength reduction can be compensated for by reducing the water-cement ratio. Quality concrete can be made with seawater if the mix is properly adjusted.

When suitable fresh water is not available, seawater can also be used for making reinforced concrete. Its use may increase the risk of corrosion, but the risk is reduced if the reinforcement has sufficient cover and if the concrete is watertight and contains an adequate amount of entrained air. Reinforced-concrete structures made with seawater and exposed to marine environment should have a water-cement ratio of less than 0.45, and the reinforcement cover should be at least 75 mm (3 in) thick. Seawater should not be used to make prestressed concrete in which the prestressing steel is in contact with the concrete. Sodium or potassium salts present in seawater used for mix water can produce substances that combine with alkali-reactive aggregates in the same manner as when combined with cement alkalies. Therefore, seawater should not be used as mixing water for concrete with known potentially alkali-reactive aggregates even when the alkali content of the cement is low.

Impurities.   The following is a résumé of the effects of certain impurities in mixing water on the quality of plain concrete.

TABLE 4.3   Typical Analyses of City Water Supplies and Seawater

Dissolved solids	Dissolved solids, ppm, in tapwater, analysis number						In seawater[a]
	1	2	3	4	5	6	
Silica (SiO_2)	2.4	0.0	6.5	9.4	22.0	3.0	—
Iron (Fe)	0.1	0.0	0.0	0.2	0.0	0.0	—
Calcium (Ca)	5.8	15.3	29.5	96.0	3.0	1.3	50–480
Magnesium (Mg)	1.4	5.5	7.6	27.0	2.4	0.3	260–1410
Sodium (Na)	1.7	16.1	2.3	183.0	215.0	1.4	2190–12,200
Potassium (K)	0.7	0.0	1.6	18.0	9.8	0.2	70–550
Bicarbonate (HCO_3)	14.07	35.8	122.0	334.0	549.0	4.1	—
Sulfate (SO_4)	9.7	59.9	5.3	121.0	11.0	2.6	580–2810
Chloride (Cl)	2.2	3.0	1.4	280.0	22.0	1.0	3960–20,000
Nitrate (NO_3)	0.5	0.0	1.6	0.2	0.5	0.0	—
Total dissolved solids	31.0	250.0	125.0	983.0	564.0	19.0	35,000

[a]Different seas contain different amounts of dissolved salts.

Alkali carbonate and bicarbonate. Sodium carbonate can cause very rapid setting; bicarbonate can either accelerate or retard set. In large concentrations, the salts can materially reduce concrete strengths. When the sum of these dissolved salts exceeds 1000 ppm, tests for setting time and 28-day strength should be made.

Chloride and sulfate. Concern over a high chloride content in the water is chiefly due to the possible adverse effect of chloride ions on the corrosion of reinforcing steel or prestressing strands. The chloride level at which corrosion begins is about 7.6 kilograms per cubic meter (kg/m^3) [1.3 pounds per cubic yard (lb/yd^3)]. Placing an acceptable limit on chloride content for any one ingredient, such as mixing water, is difficult in view of the several sources of chloride ions in concrete. An acceptable limit in the mixing water depends upon how significantly mixing water contributes to the total chloride content. Suggested limits are given in Table 4.1. Water containing less than 500 ppm of chloride ion is generally considered acceptable. However, the contribution of chlorides from other ingredients also should be considered.

Iron salts. Iron salts in concentrations up to 40,000 ppm do not usually affect mortar strengths adversely.

Miscellaneous inorganic salts. Salts of manganese, tin, zinc, copper, and lead in mixing water can cause a significant reduction in strength and large variations in setting time. Of these, salts of zinc, copper, and lead are the most active. Other salts that are especially active as retarders include sodium iodate, sodium phosphate, sodium arsenate, and sodium borate. All can greatly retard both set and strength development when present in concentrations of a few tenths percent by weight of the cement. Generally, concentrations of these salts up to 500 ppm can be tolerated in mixing water. Another salt that may be detrimental to concrete is sodium sulfide; even the presence of 100 ppm warrants testing.

Acid waters. Generally, mixing waters with hydrochloric, sulfuric, and other common inorganic acids in concentrations as high as 10,000 ppm have no adverse effect on concrete strength. Acid waters with pH less than 3.0 may create handling problems and should be avoided.

Algae. Water containing algae is unsuited for making concrete because the algae can cause excessive reduction in strength either by influencing cement hydration or by causing a large amount of air to be entrained in the concrete. Algae may also be present on aggregates, in which case the bond between the aggregate and cement paste is reduced.

4.4.2 Polymers in concrete

The following three types of concrete materials utilize polymers to form composites: (1) polymer-impregnated concrete (PIC), which is a hydrated portland cement concrete that has been impregnated with a monomer and subsequently polymerized in situ, (2) polymer–portland cement concrete (PPCC), which is produced by adding either a monomer or polymer to a fresh concrete mixture and subsequently curing and polymerizing the material in place, and (3) polymer concrete (PC), which is a composite material formed by polymerizing a monomer and aggregate mixture.

A *monomer* is an organic molecular species which is capable of combining chemically with molecules of the same or other species to form a high-molecular-weight material known as a polymer. A polymer consists of repeating units derived from the monomers which are linked together in a chainlike structure. The chemical process through which these linkages occur is known as polymerization.

Polymer-impregnated concrete.   The selection of suitable monomers for polymer-impregnated concrete (PIC) is based on the impregnation and polymerization characteristics, availability and cost, and the properties of the resultant polymer and PIC. Liquid vinyl monomer systems have generally been used. As supplied, monomers normally contain an inhibitor to prevent premature polymerization of the monomer. Since polymerization begins immediately on adding a promoter, use of a promoter in PIC would be restricted to shallow impregnations.

The basic method of producing PIC consists of the fabrication of precast concrete specimens, ovendrying, saturation with monomer, and in situ polymerization.

For full impregnation, good-quality concrete having a cross section of up to 305 mm (12 in) will require soaking in monomer for about 60 min under a pressure of 69 kilopascals (kPa) [10 pounds per square inch (lb/in^2)]. Concrete may be only partially impregnated when improved strength is not needed but greater durability is desired.

Polymer–portland cement concrete.   Polymer–portland cement concrete (PPCC) has been prepared with both premixed and postmixed polymerized materials. The premixed polymerized materials include latexes and polymer solutions or dispersions. The postmix polymerized PPCC has been made with a number of resins and monomers.

At present, latex-modified concretes represent the large majority of commercial applications of polymer-modified concretes in the United States. Suitable latex formulations greatly improve the shear bond, tensile, and flexural strength of cements and mortars.

Thermosetting water-soluble polymers which have been added to fresh concrete include epoxies, amino resins, polyesters, and formaldehyde derivatives. Thermoplastic materials include polyvinyl alcohol and polyacrylamides.

PPCC process technology is based upon overcoming the incompatibilities of most organic polymers and their monomers with mixtures of portland cement, water, and aggregate. The mix proportioning of latex PPCC will vary in much the same way as do normal concretes and mortars.

Polymer concrete. Most of the work on polymer concrete has been with polyester-styrene resin systems and to a lesser extent with furan, epoxy, and vinylester resin systems. The polyester resins are attractive because of moderate cost, availability of a great variety of formulations, and moderately good PC properties.

Most of the monomer and resin systems for PC are polymerized at ambient temperatures. Vinyl monomer systems can be polymerized with catalysts such as benzoyl peroxide with an amine promoter. The polyester-styrene systems are polymerized with promoter-catalyst systems such as methylethyl ketone peroxide with cobalt naphthanate promoter. Other systems include amine curing agents for epoxy resins.

Mixing and placing techniques for PC are based on adaptation of existing equipment and methods for producing portland cement concrete. A knowledge of polymer chemistry is helpful but not essential; directions for curing mixes are available from the resin manufacturers. Curing of PC may be performed by thermal catalytic, promoter-catalyst, or radiation techniques. Promoter-catalyst systems are frequently best suited for PC, with curing times varied, as needed, between a few minutes and several hours. Full strength is attained when polymerization is completed. PC has been made with epoxy, polyester, and furan resins, and more recently with styrene monomer systems.

4.4.3 Aggregates

Normal aggregates. Normal aggregates consist of clean sand, river-washed gravel, and crushed rock. In certain locations volcanic rock, such as basalt, may be used. The aggregates should have clean, hard and uncoated particles and comply with ASTM Standard C33. Other ASTM tests for concrete aggregates are given in Table 4.4.

Harmful substances may be present in aggregates. They include organic matter, rubbish of all kinds, silt, clay, coal, lignite, dolomitic

TABLE 4.4   Characteristics and Tests of Aggregates

Characteristic	Significance	ASTM Standard	Requirement of item reported
Resistance to abrasion	Index of aggregate quality; wear resistance of floor pavements	C131 C295 C535	Maximum percent of weight loss
Resistance to freezing and thawing	Surface scaling, roughness, loss of section, and unsightliness	C295 C666 C682	Maximum number cycles or frost immunity; durability factor
Resistance to disintegration by sulfates	Resistance to weathering action	C88	Weight loss, particles exhibiting distress
Particle shape and surface texture	Workability of fresh concrete	C295 D3398	Maximum percent of flat and elongated piece
Grading	Workability of fresh concrete; economy	C117 C136	Minimum and maximum percent passing standard sieves
Bulk unit weight or density	Mix design calculations; classifications	C29	Compact weight and loose weight
Specific gravity	Mix design calculations	C127, fine aggregate C128, coarse aggregate C29, slag	—
Absorption and surface moisture	Control of concrete quality	C70 C127 C128 C566	—
Compressive and flexural strength	Acceptability of fine aggregate failing other tests	C39 C78	Strength to exceed 95% of strength achieved with purified sand
Definitions of constituents	Clear understanding and communications	C125 C294	—

limestones, chalcedonic cherts, opal, cristobalite, some types of volcanic glass, and pyrites. An aggregate containing these substances may be considered as reactive. Materials finer than the No. 200 sieve may form coatings on the aggregate which weaken the bond between the aggregate and the cement paste. Soft particles of aggregate affect the wear resistance and durability of the concrete.

Tests to qualify aggregates to be used to make durable concrete are:

- Abrasion resistance tests
- Sulfate soundness tests (used for many years as an index of quality although experience has shown that it does not correlate well with the actual performance of aggregates in concrete)
- Tests for organic impurities (on aggregates from new sources)
- Laboratory freezing and thawing tests (of limited value, but they do furnish useful information for new source material)
- Tests to determine presence of opal and chalcedony (on aggregates from new sources) by making mortar bars and testing them according to ASTM Standard C342

It is an interesting fact that the water requirement for concrete of a given consistency decreases in inverse proportion to the maximum size of the coarse aggregate. For example, a 19-mm (0.75-in) aggregate would require about 152 kg (335 lb) of water per 0.76 m^3 (1 yd^3) of concrete; a 50.8-mm (2-in) aggregate would need only about 125 kg (275 lb) of water per 0.76 m^3 (1 yd^3). The latter would lower the water-cement ratio and, of course, produce concrete of a greater strength than that containing the 19-mm aggregate. Table 4.5 gives maximum aggregate sizes for various uses.

Gap-graded aggregates can often be used effectively in areas where ASTM C33 standards cannot be met. A typical gap-graded aggregate may contain only one size of coarse aggregate together with sand. In this respect, the resultant concrete would resemble prepacked concrete.

When natural aggregates are found to be unacceptable through service records or tests, they may sometimes be improved by removing lightweight, soft, or otherwise inferior particles by processing.

Lightweight aggregate. A number of materials are used to produce lightweight concrete. Among the natural aggregates are tuff, pumice, volcanic cinders, scoria, and diatomite. Pumice is frequently used in structural concrete. For example, parts of the large concrete counterweights on a bascule bridge (Fig. 4.1), were composed of a combination

TABLE 4.5 Maximum Size of Aggregate Recommended for Various Types of Construction

Section	Use or features	Aggregate size, mm (in)
190.5 mm (7.5 in) wide	Heavily reinforced floor and roof slabs; parapets, cobels and where space is limited	19 (0.75)
190.5 (7.5 in) mm wide with clear distance between reinforcement bars 57 mm (2.25 in)		31.75 (1.15)
305-mm- (12-in-) wide unreinforced sections and 407-mm- (16-in-) wide reinforced sections with clear distance between reinforcement bars 114.3 mm (4.5 in) to 229 mm (9 in)	Piers, walls, baffles, and basin floor slabs in which satisfactory placement of 152.4-mm or cobble concrete cannot be accomplished even though reinforcement spacing would permit the use of larger aggregates	76.2 (3)
Massive sections with clear distance between reinforcing bars 229 mm (9 in)	Retaining walls, piers and baffles in which suitable provisions are made for placing concrete containing the larger-size aggregate without producing rock pockets or other undesirable results	152.4 (6)

of normal structural concrete about 3409 kg/m^3 [150 pounds per cubic foot (lb/ft^3)], lightweight pumice concrete 1570 kg/m^3 (98 lb/ft^3) and heavyweight concrete 3613 kg/m^3 (225 lb/ft^3). Concrete made with pumice weighs from 1445 to 1606 kg/m^3 (90 to 100 lb/ft^3).

Perlite is a member of the artificial lightweight aggregate family. It produces a poor grade of concrete that weighs between 803 and 1285 kg/m^3 (50 and 80 lb/ft^3), and is often used as an underlayer for built-up roof decks. It will not produce structural grade concrete. Expanded clay aggregates produce a lightweight, structural quality concrete with densities ranging from 1438 to 1745 kg/m^3 (90 to 109 lb/ft^3). There are other manufactured lightweight aggregates for making concrete, but they are not considered here. Concrete made with vermiculite and used extensively as an insulating material weighs from 562 to 1205 kg/m^3 (35 to 75 lb/ft^3).

Heavy aggregate. Heavyweight concrete is made with normal coarse aggregates (ASTM Standard C33) and heavy natural or manufactured aggregates such as magnetite (specific gravity 4.2 to 4.4), limonite (specific gravity 5.0 to 5.5), and barite (specific gravity 2.5 to 3.5). Some of these minerals could contain pyrite, which can decompose on

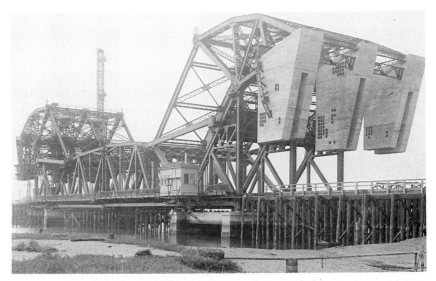

Figure 4.1   Bascule bridge with lightweight concrete counterbalances.

weathering and should not be used in concrete. Magnetite and limonite should be tested for the presence of pyrite before being used in concrete.

Manufactured heavy aggregates are usually iron and steel products. Concrete made from them can weigh more than 4805 kg/m^3 (300 lb/ft^3). More information on heavy aggregates is provided in *Design and Control of Concrete Mixtures* (Portland Cement Association, 1979).

Regional aggregates.   For the sake of economy, it may be desirable to use aggregates from the nearest source even if they do not fully meet the requirements of ASTM Standard C33, unless they contain harmful minerals such as pyrite or chalcedony. If it is essential to use aggregates containing reactive minerals, pozzolan admixtures are used to reduce or eliminate potential expansion from alkali-reactive aggregates.

Coral.   Coral deposits are found in many oceans of the world. When mined and prepared for use as aggregates for making concrete, the physical and chemical properties of coral may vary widely. When coral aggregates are used to produce structural concrete, a strength factor is established by using trial mixes until the proper strength has been attained. Cores were taken in 1972 from coral aggregate concrete made by the Japanese many years prior to 1941 on the island of Kwajalein.

The average compressive strength of the cores was 13.8 kPa (2000 lb/in^2).

4.4.4 Chemical reactions of aggregates

Chemical reactions of aggregates in concrete can affect the performance of the concrete. Some reactions may be beneficial, but others result in serious damage to the concrete by causing abnormal expansion, cracking, and loss of strength (Woods, 1968). The reaction that has received greatest attention and was the first to be recognized is one between alkalies (Na_2O and K_2O), from the cement or from other sources with hydroxyl and certain siliceous constituents that may be present in the aggregate. This phenomenon was originally, and is still sometimes, referred to as "alkali-aggregate reaction," but in recent years it has been more properly designated as "alkali-silica reaction."

Deterioration of concrete has occurred in certain sand-gravel aggregates. The deterioration has been regarded as a chemical phenomenon and is a reaction between the alkalies in cement and some siliceous constituents of the aggregates, complicated by environmental conditions that produce high concrete shrinkage and concentration by drying (Hadley, 1968). It has also been clearly demonstrated that certain carbonate rocks participate in reactions with alkalies that, in some instances, produce detrimental expansion and cracking. Detrimental reactions are usually associated with argillaceous dolomitic limestones which have somewhat unusual textural characteristics (Hadley, 1964). This reaction is designated as "expansive alkali-carbonate reaction."

Other damaging chemical reactions involving aggregates include the oxidation or hydration of certain unstable mineral oxides, sulfates, or sulfides that occur after the aggregate is incorporated in the concrete, e.g., the hydration of anhydrous magnesium oxide, calcium oxide, or calcium sulfate, or the oxidation of pyrite (Mielenz, 1964). Still other reactions may result from organic impurities (such as humus and sugar). Engineers should be aware of these possibilities and supply corrective measures when necessary. Careful testing and examination of the aggregates will usually indicate the presence of reactive impurities, and then their use in concrete can be avoided. The alkali-silica, cement-aggregate, and expansive alkali-carbonate reactions are most important.

The alkali-silica reaction. The alkali-silica reaction can cause expansion and severe cracking of concrete structures and pavements. The phenomenon is complex, and various theories have been advanced to explain the field and laboratory evidence (Diamond, 1976). Unanswered questions remain. Apparently, reactive material in the presence of potassium, sodium, and calcium hydroxide derived from the cement reacts to form either a solid nonexpansive calcium-alkali-silica complex or an alkali-silica complex (also solid) which can expand by imbibition of water.

Laboratory tests should be made on aggregates from new sources and when service records indicate that reactivity may be possible. The most useful are:

- Petrographic examination (ASTM Standard C295)

- Mortar bar test for potential reactivity (ASTM Standard C227)

- Chemical test for potential reactivity (ASTM Standard 289)

Petrographic examination. Petrographic examination provides a recommended practice for the evaluation of aggregates. Recommendations that are available show the amounts of reactive minerals that can be tolerated. The reactive rocks and minerals that have been more frequently encountered since 1960 appear to have larger potassium proportions and are harder to recognize in petrographic examination. Highly deformed quartz with an angle of undulatory extinction of 35° to 50° or more with deformation lamellae appears to be characteristic of the reactive quartz-bearing rocks. Relatively coarse-grained micas have also been regarded as reactive constituents; fine-grained micas are reactive in argillites (Dolar-Mantuani, 1969).

Mortar bar test for potential reactivity. The mortar bar test is the method most generally relied on to measure potential alkali reactivity. Acceptance criteria for evaluating the test results are given by ASTM Standard C33. The procedure is useful for the evaluation of not only aggregates but also specific aggregate-cement combinations. However, criteria have not been developed for the evaluation of metamorphic siliceous and silicate rocks.

Chemical test for potential reactivity. The chemical test is the method used primarily for a quick evaluation of natural aggregates. The results are obtainable in a few days as compared with 3 to 6 months or more with the mortar bar test. Acceptance criteria for this test are given by ASTM Standard C33. Care must be exercised in interpreting the results of the test. The test method has given questionable results for the evaluation of lightweight aggregates; therefore, it is not recommended for that purpose (Ledbetter, 1973).

General criteria for judging reactivity. When available, the field performance record of a particular aggregate, if the aggregate has been used with cement of high alkali content, is the best means for judging the reactivity of the aggregate. If such records are not available, the most reliable criteria are petrographic examination with corroborating evidence from the mortar bar test (U.S. Army, Corps of Engineers, 1971b), sometimes supplemented by tests on concrete even though they have not been standardized. The chemical test results should also be used in conjunction with results of the petrographic examination and mortar bar test. It is preferable not to rely on the results of only one kind of test in any evaluation (U.S. Army, Corps of Engineers, 1971b).

Recommended procedures with alkali-reactive aggregates. If aggregates are shown by service records or laboratory examination to be potentially reactive, they should not be used when the concrete is to be exposed to seawater or alkali environments if nonreactive aggregates are available (Highway Research Board, 1958). When reactive aggregates must be used, it should be only after thorough testing, and preferably after service records have established that, with appropriate limits on the alkali content of the cement, or with the use of appropriate amounts of an effective pozzolan, or both, satisfactory service can be anticipated. When seawater or alkaline soil environments are not involved and no sound materials are available economically, reactive materials may be used provided certain limits are set:

- Specify a low-alkali cement having a maximum of 0.6 percent equivalent Na_2O.
- Prohibit the use of seawater or alkali soil water as mixing water.
- Avoid addition of sodium or potassium chloride.
- When low-alkali cements are not economically available, use a suitable pozzolanic material as prescribed by ASTM Standard C618 and tested in accordance with ASTM Standard C441 to determine its effectiveness in preventing excessive expansion due to the alkali-silica reaction.

Whenever the use of pozzolanic materials is considered, it should be remembered that if the materials increase water demand, they may cause increased shrinkage in concrete exposed to drying. Increased water demand results from high fineness and poor particle shape. The rate of strength development in correctly proportioned pozzolanic concrete can equal that of portland cement concretes, which is 28 days.

Cement-aggregate reaction. Recent research indicates that the cement-aggregate reaction is mainly a reaction between the alkalies in the cement that produce high pH and abundant hydroxyl and siliceous constituents of the aggregates. However, the field performance of concretes made with reactive sands and gravels does not correlate well with cement alkali content. The concrete deterioration results from moderate interior expansion caused by alkali-silica reactivity and surface shrinkage caused by severe drying conditions. Evaporation at the surface of the concrete causes an increase in alkali concentration in the pore fluids near the drying surface, and a net migration of alkali toward that surface. Under those conditions even a low-alkali cement may cause objectionable deterioration, particularly near the surface. This alkali distribution is altered by the leaching of alkalies near the surface during periods of heavy rain (Hadley, 1968). Although special tests, such as ASTM Standard C342, have been devised to indicate potential damage from this phenomenon, their reliability is doubtful.

The use of potentially deleterious cement-aggregate combinations should be avoided when possible. If they must be used, however, a suitable pozzolan that does not increase drying shrinkage and 30 percent or more (by weight) of coarse limestone should be used with potentially deleterious cement-aggregate combinations. Concrete tests should be used to determine whether the resulting combination is satisfactory.

Expansive alkali-carbonate reactivity. Certain limestone aggregates, usually dolomitic, have been reported as reactive in concrete structures. There are many unanswered questions, and more than one mechanism has been proposed to explain expansive carbonate reactivity. The affected concrete is characterized by a network of pattern or map cracks that are usually most strongly developed in areas of the structure where the concrete has a constantly renewable supply of moisture, such as close to the waterline in piers, from the ground behind retaining walls, beneath road or sidewalk slabs, or by wick action in posts or columns. A feature distinguishing the alkali-carbonate reaction from the alkali-silica reaction is the general absence of silica gel exudations at cracks. Additional signs of the severity of the reaction are closed expansion joints with possible crushing of the adjacent concrete (Hadley, 1964).

4.4.5 Laboratory testing

The most useful laboratory tests are discussed below.

Petrographic examination. The petrographic examination of aggregates can be used to identify the features of the rock. The presence of all or any dolomite in a fine-grained carbonate rock makes it desirable to perform the rock cylinder test (ASTM Standard C586). The test is recommended whether or not the texture is believed to be typical and whether or not insoluble residue including clay amounts to a substantial part of the aggregate. As expansive rocks are recognized from more areas, the more variable the textures and compositions appear to be.

Expansion of concrete prisms. The expansion test is performed with prisms made of job materials and stored at 100 percent relative humidity at 23°C (73°F). To accelerate the reaction the prisms may be made with additional alkali and stored at elevated temperature. The comparison is usually made with the expansion of prisms containing a nonreactive control aggregate.

Petrographic analysis. A petrographic analysis of the concrete can confirm the type and characteristics of the aggregate present. Distress that has occurred in the aggregate and surrounding matrix, such as micro- and macrocracking, may be observed. Reaction rims, which do not necessarily signify harmful results, may be observed in certain aggregate particles and can be identified as negative or positive by acid etching. Secondary deposits of calcium carbonate, calcium hydroxide, and ettringite (calcium sulfoaluminate) may be found in voids within the concrete, but no deposits of silica, hardened or in gel form, are associated with the suspect aggregate pieces.

Additional tests of the alkali-carbonate reaction include identifying sawed or ground surfaces by visual observation. X-Ray examination of reaction products is also sometimes useful.

Criteria for judging reactivity. Several criteria are available for judging the reactivity of aggregates. They include definitive correlations between expansions occurring in the laboratory in rock cylinders or concrete prisms and deleterious field performance which has not yet been established. The factors involved are complex and include the heterogeneity of the rock, coarse aggregate size, permeability of the concrete, and seasonal changes in environmental conditions in service, principally availability of moisture, level of temperature, and possibly the use of sodium chloride as a deicing chemical.

It is not certain that rapid determination of potential reactivity can always be made by using the rock cylinder test, because some rocks showing an initial contraction may develop considerable expansion later on. Expansions greater than 0.10 percent in the rock cylinders are usually taken as a warning that further tests should be undertaken to determine expansion of the aggregate in concrete. Fortunately, many carbonate rocks that expand in rock cylinders do not expand in concrete.

Recommended procedures to minimize alkali-carbonate reactivity. Procedures that can be employed to mitigate the effects of the reaction include:

- Avoiding reactive rocks by selective quarrying.
- Dilution with nonreactive aggregates or use of a smaller maximum size.
- Use of low-alkali cement (probably 0.4 percent combined alkali or lower), which will prevent harmful expansions in most cases. In pavements where sodium chloride is used as a deicing chemical, however, this cannot be taken as certain.

Of these measures, the first is the safest and usually the most economical.

Preservation of concrete containing reactive aggregate. There are no known methods of adequately preserving existing concrete that contains the elements that contribute to the previously described chemical reactions. Water or moisture is partly involved in at least two of the reactions. The destructive effects of freezing and thawing are more pronounced after the initial stages of destruction by the chemical reactions. Therefore, any practicable means of decreasing the exposure of such concrete to water may extend its useful life.

4.4.6   Admixtures

An admixture is defined as a material other than water, aggregates, and hydraulic cement that is used as an ingredient of concrete or mortar and is added to the batch immediately before or during its mixing. ASTM Standard C494 classifies certain chemical admixtures in terms of function as follows:

- Type A, water-reducing admixtures
- Type B, retarding admixtures
- Type C, accelerating admixtures
- Type D, water-reducing and set-retarding admixtures
- Type E, water-reducing and set-accelerating admixtures

These admixtures are discussed according to the types of materials constituting the admixture or the characteristic effects of their use.

Water-reducing and set-retarding admixtures. Water-reducing admixtures are used to improve the quality of concrete, to obtain specified strength at lower cement content, or to increase the slump of a given mixture without increase in water content. They may also improve the properties of concrete containing aggregates that are harsh, poorly graded, or both, or may be used in concrete that must be placed under difficult conditions. Set-retarding admixtures delay the onset of hardening, prolonging the period when the concrete is workable. Both types of admixtures are useful when placing concrete by means of a pump or when using a tremie process. The materials that are generally available for use as water-reducing admixtures and set-retarding admixtures fall into four general classes:

1. Lignosulfonic acids and their salts
2. Modifications and derivatives of lignosulfonic acids and their salts
3. Hydroxylated carboxylic acids and their salts
4. Modifications and derivatives of hydroxylated carboxylic acids and their salts

Hydroxylated carboxylic acid salts act as water-reducing, non-air-entraining retarders. Lignosulfonates are available as the calcium, sodium, or ammonium salts. Admixtures of classes 1 and 3 can be used either alone or combined with other organic or inorganic, active or essentially inert substances. They are water-reducing, set-retarding admixtures.

Admixtures of classes 2 and 4 are water-reducing admixtures offered as combinations of substances designed either to have no substantial effect on rate of hardening or to achieve varying degrees of acceleration or retardation in rate of hardening of concrete. The admixtures may include an air-entraining agent.

The composition of the portland cement affects the air-entraining properties of lignosulfonate admixtures in concrete. Concrete containing a lignosulfonate retarder generally requires 5 to 10 percent less water than comparable concrete without the admixture. Compressive strengths at 2 or 3 days are usually equal to or higher than those of corresponding concrete without the admixture, and the strength at 28 days or later may be 10 to 20 percent higher.

Lignosulfonic acid salts, carboxylic acid salts, or modifications or derivatives thereof can be mixed or reacted with other chemicals that entrain air, modify setting time, or affect the strength development of concrete. Calcium chloride, neutralized wood resins, alkyl aryl sulfonates, and triethanolamine are examples of additives that have been used. The use of compounded or modified water reducers usually causes a water reduction of 5 to 10 percent at equal air content. Compressive strengths at ages greater than 2 days are usually from 10 to 20 percent higher than those of similar concretes without admixture.

Accelerating admixtures. Accelerating admixtures are added to concrete either (1) to increase the rate of early strength development, (2) to shorten the time of setting, or (3) for both purposes. Chemicals which accelerate the hardening of mixtures of portland cement and water include some of the soluble chlorides, carbonates, silicates, fluorosilicates, and hydroxides (Steinour, 1960) and also some organic compounds such as triethanolamine (Newman et al., 1943). Calcium aluminate cements and finely ground hydrated portland cement also have been advocated.

Some of the soluble chlorides, particularly calcium chloride (Highway Research Board, 1952) and to a much lesser extent triethanolamine, have general applicability as admixtures in concrete. Some of the other materials are suitable only for use in the preparation of quick-set cements.

By far the best known and most widely used accelerator is calcium chloride. Many other materials have been found to accelerate the strength gain of concrete but, in general, they are seldom used and only limited information concerning their effect on the properties of concrete is available. Most of the information given on accelerators applies mainly to the use of calcium chloride. The effects of accelerators on some of the properties of concrete are as follows:

- The setting time, initial and final, is reduced. The amount of reduction varies with the amount of accelerator used, the temperature of the concrete, and the ambient temperature. Excessive amounts of the accelerator may cause rapid setting.

- Less air-entraining admixture is required to produce the required air content. However, in some cases larger bubble sizes and higher spacing factors are obtained.

- Earlier heat release is obtained, but there is no appreciable effect on the total heat of hydration.

- Compressive strength is increased substantially at early ages. The ultimate strength may be reduced slightly. The increase in flexural strength is usually less than the increase in compressive strength.

- Volume change is increased for both moist-curing and drying conditions. There is a question of the degree of the effect caused by the accelerators as opposed to other factors influencing volume change.

- Resistance to freezing and thawing and to scaling caused by the use of deicing salts is increased at early ages, but it may be decreased at later ages.

- Resistance to sulfate attack is decreased.

- Expansion produced by the alkali-silica reaction is greater. This can easily be controlled by the use of low-alkali cement or pozzolans.

Corrosion of metals may occur, especially in the use of calcium chloride when steam curing is employed. The use of calcium chloride in recommended amounts does not cause progressive corrosion of conventional steel reinforcement in typical reinforced concrete under normal conditions when the bars have sufficient concrete cover. Stannous chloride when properly used acts as an accelerator and does not cause corrosion of the steel even when steam curing is used.

Calcium chloride.   Calcium chloride is available in two forms. Regular flake calcium chloride, ASTM Standard D98 (type 1), contains a minimum of 77 percent $CaCl_2$. Concentrated flake, pellet, or granular calcium chloride ASTM Standard D98 (type 2), contains a minimum of 94 percent of $CaCl_2$. Calcium chloride can generally be used safely in amounts up to 2 percent by weight of the cement (McCall and Claus, 1953). Larger amounts may be detrimental, and, except in rare instances, they provide little additional advantage. The benefits of the use of calcium chloride are usually more pronounced when the salt is employed in co ιcrete with a mixing and curing temperature below 21°C (70°F). Aι high mixing and curing temperatures, long-term strength, especially flexural strength, may decrease and shrinkage and cracking may increase.

Laboratory tests have indicated that most increases of compressive strengths of concrete resulting from the use of 2 percent calcium chloride by weight of cement are in the range of 2760 to 6890 kPa (400 to 1000 lb/in^2) at 1 to 7 days for 21°C curing. At 4.4°C (40°F) curing the increases in strengths obtained at 1 and 7 days with calcium chloride are in the same range as for 21°C curing. The increase in strength usually reaches its maximum in 1 to 3 days and thereafter generally decreases. At 1 year, some increase is still evident in concrete made with most cements. The specific effect of the use of calcium chloride varies, however, for different cements as is indicated by the range of strength increases cited above for the early ages.

The relative increase in flexural strength of concrete resulting from the use of 1 or 2 percent of calcium chloride is not as great as the increase in compressive strength. Calcium chloride increases the flexural strength at 1 and 3 days but decreases the flexural strength at 28 days or at later ages (McCall and Claus, 1953).

Flexural strengths of concretes containing 1 to 2 percent calcium chloride are usually increased over the strengths of similar concrete without the admixture by 40 to 90 percent at 1 day and by 5 to 35 percent at 3 days, respectively, when moist-cured at 21°C. At 28 days, decreases of up to 12 percent have been reported from laboratory tests of moist-cured concrete.

The use of 1 percent calcium chloride by weight of the cement is sufficient in most cases to accelerate setting and increase strength sufficiently for cold weather concreting, with the understanding that cold weather protection is provided. The selection of the optimum amount should be based on the type of cement, the temperature of the concrete, and the ambient air temperature.

Calcium chloride may promote corrosion of the usual reinforcement in concrete even though adequate concrete cover is provided for the steel. However, it should not be used where stray electric currents are expected and should not be used in prestressed concrete because of possible stress corrosion of the prestressing steel (Arber and Vivian, 1961). Calcium chloride in concrete may be expected to aggravate corrosion of embedded galvanized metal and of galvanized forms that are left in place. Combinations of metals, such as aluminum alloy electrical conduit and steel reinforcing, should not be used in concrete exposed to water.

Calcium chloride may be especially beneficial for concrete exposed to low or freezing temperatures at early ages if used as recommended in the ACI Standard ACI 604-56. Calcium chloride increases the rate of early heat development and accelerates the set, but it lowers the freezing point of the water in concrete only to an insignificant extent.

Air-entraining admixture. Many materials, including natural wood resins, fats, and oils, may be used in preparing air-entraining admixtures. These materials are usually insoluble in water and generally must be chemically processed before they can be used as admixtures. Since not all such materials produce a desirable air-void system, air-entraining admixtures should meet the requirements of the ASTM Standard C260.

Air-entrained concrete containing a large number of very small air bubbles is severalfold more resistant to frost action than non-air-entrained concrete made of the same materials. Air-entrained concrete should be a dense, impermeable mixture that is well-placed, protected, finished, and cured if maximum durability is to be obtained.

Air entrainment, while improving both workability and durability, may reduce strength. Within the range of air content normally used, the decrease in strength usually is about proportional to the amount of air entrained. For most types of exposed concrete a slight reduction in strength is far less significant than the improved resistance to frost action. The reduction in strength will rarely exceed 15 percent in the case of comprehensive strength and 10 percent in the case of flexural strength.

In some installations of precast concrete units such as cribbing and curbing, there is considerable exposure to freezing and thawing action. The use of adequately prepared and controlled air-entrained concrete is the best way to improve resistance to freezing and thawing.

Air-detraining admixtures. There have been cases in which aggregates have released gas into, or caused excessive air entrainment in, plastic concrete which made it necessary to use an admixture able to dissipate the excess air or other gas (MacNaughton and Herbich, 1954). Also, it is sometimes desirable to remove part of the entrained air from a concrete mixture. Compounds such as tributyl phosphate, dibutyl phthalate, water-insoluble alcohols, and water-insoluble esters of carbonic and boric acids, as well as silicones, have been proposed for this purpose, but tributyl phosphate is the most widely used material.

Admixture to reduce alkali-silica expansion. Test data indicate that small additions of certain chemical substances may be effective in decreasing expansion resulting from the alkali-silica reaction (McCoy and Caldwell, 1951). Outstanding reductions in expansion of laboratory mortar specimens have been reported for additions of 1 percent by weight of the cement of lithium salts and for additions of about 2 to 7 percent of certain barium salts. Moderately reduced expansions were also obtained with certain protein air-entraining admixtures and with some water-reducing, set-retarding admixtures. It was found that some of these substances were more effective in reducing expansion than others. The results reported are limited, and further work is needed. There is some evidence that expansions due to the alkali-silica reaction are slightly lowered by air entrainment and the use of low-alkali cement.

Expansion admixtures. Admixtures which during the hydration period of concrete or grout themselves expand or react with other constituents of the grout to cause expansion are used to minimize the effects of dry shrinkage. They are used in both restrained and unrestrained placement. The most common admixtures for this purpose are finely divided iron and chemicals to promote oxidation of the iron. This use is generally limited to relatively small projects. Expansive cements are most often used on large projects.

Shrinkage preventing admixtures. Three different shrinkage-compensating cements are described in ASTM Standard C845 and are designated as types K, S, and M. The expansion of each of these cements when mixed with sufficient water is due principally to the formation of ettringite. Most shrinkage-compensating cements consist of constituents of conventional portland cement with added sources of aluminate and calcium sulfate. The three types of expansive cements differ from each other in the form of the aluminate compounds from which the expansive ettringite is developed. The principal constituents of these cements are:

- Type K portland cement, calcium sulfate, and portland-like cement containing anhydrous tetracalcium trialuminate sulfate
- Type M portland cement, calcium sulfate, and calcium aluminate cement
- Type S portland cement high in tricalcium aluminate and calcium sulfate

An important requirement is the selection of material proportions such that the Ca, S_3, and especially the Al_2O_3 become available for ettringite formation during the appropriate period after the mix water is added. Determination of these proportions should be based on test results in accordance with ASTM Standard C806.

Bond-improvement admixtures. Bonding admixtures are water emulsions of several organic materials that are mixed with portland cement or mortar grout for application to an old concrete surface just prior to placing topping or patching mortar or concrete or are mixed with the topping or patching material. Common bonding admixtures are made from polymers that include polyvinyl chloride, polyvinyl acetate, acrylics, and butadiene-styrene copolymer. Bonding agents usually cause entrainment of air and a sticky consistency in grout mixtures.

Penetration and plasticity admixtures. Admixtures which improve the ability of freshly mixed concrete and grout to penetrate into voids and cracks also increase the plasticity of the mix. The degree of plasticity of fresh concrete, the amount of surface area of the solids per unit of water volume, will determine the bleeding characteristics and workability of concrete and grout. A low ratio of surface area of solids to volume of water results in a thin and watery paste; consequently, the aggregate particles are only slightly separated and the mixture lacks plasticity and tends to segregate. The ratio of surface area of solids to volume of water may be increased by increasing the amount of cement or by adding a suitable mineral admixture to the mix. Admixtures that are relatively chemically inert, such as ground quartz and limestone, cementitious materials such as natural cements, hydraulic limes, and slag cements, and pozzolans are commonly used.

Impermeability admixtures. Concrete and grout are not impermeable to the penetration of water, but the terms "waterproofing" and "dampproofing" have come to mean a reduction of rate of penetration of water into dry concrete and grout. Admixtures comprised of fatty acids, usually calcium or ammonium stearate or oleate, cause air entrainment during mixing. Also used are mineral oils, asphalt emulsions, and certain cutback asphalts.

Corrosion-inhibiting admixtures. In the manufacture of certain concrete products containing steel, it might be desirable to accelerate the rate of strength development by use of both a chemical accelerator and heat. The latter is usually in the form of steam at atmospheric pressure. When calcium chloride is used as the accelerator in this type of curing, laboratory studies have found the rate of corrosion of the steel to be accelerated. However, Arber and Vivian (1961) found that certain compounds containing an oxidizable ion such as stannous chloride, ferrous chloride, and sodium thiosulfate act as accelerators as does calcium chloride but also appear to cause less corrosion than the latter. Stannous chloride appeared to be the best of the products tried, and 2 percent of the salt by weight of cement was more effective than 1 percent and was as effective as greater amounts from the standpoint of both acceleration and resistance to corrosion. For effective use, the salt must be added to the concrete in the stannous form and a dense concrete must be used.

Color admixtures. Pigments are often added to produce color in the finished concrete. The requirements of suitable coloring admixtures include:

- Color fastness when exposed to sunlight
- Chemical stability in the presence of alkalinity produced in the set cement
- No adverse effect on setting time or strength development of the concrete
- Stability of color in autoclaved concrete products during exposures

to the conditions in the autoclave

Pigments frequently used to color concrete are:

- Grays to black—black iron oxide, mineral black, carbon black
- Blue—ultramarine blue, phthalocyanine blue
- Red—red iron oxide
- Brown—brown iron oxide, raw and burnt umber
- Cream or buff—yellow iron oxide
- Green—chromium oxide, phthalocyanine green
- White—titanium dioxide

4.5   Concrete Mixes

There are standard methods for selecting proportions for concrete made with aggregate of normal density and workability suitable for usual cast-in-place construction. The methods provide a first approximation of proportions and are intended to be checked by trial batches in the laboratory or field and adjusted, as necessary, to produce the desired characteristics of the concrete.

4.5.1   Mix proportions

The procedure for selection of mix proportions given in this section is applicable to normal-weight concrete. Estimating the required batch weights for the concrete involves a sequence of logical, straightforward steps which, in effect, fit the characteristics of the available materials into a mixture suitable for the work. Regardless of whether the concrete characteristics are prescribed by the specifications or are left to the individual selecting the proportions, estimation of a total batch weight per cubic unit of concrete can best be accomplished by referring to Table 4.11.

Choice of slump.   If slump is not specified, a value appropriate for the work can be selected from Table 4.6. The slump ranges that are shown apply when vibration is used to consolidate the concrete. Mixes of the stiffest consistency that can be placed efficiently should be used.

Maximum size of aggregate.   Large maximum sizes of well-graded aggregates have fewer voids than smaller sizes. Hence, concretes with the larger-sized aggregates require less mortar per unit volume of concrete. Generally, the maximum size of aggregate should be the largest

TABLE 4.6   Recommended Slumps for Various Types of Construction

	Slump, mm	
Types of construction	Maximum[a]	Minimum
Reinforced foundation walls and footings	76	25
Plain footings, caissons, and substructure walls	76	25
Beams and reinforced walls	102	25
Building columns	102	25
Pavements and slabs	76	25
Mass concrete	51	25

[a]May be increased 25 mm (1 in) for methods of consolidation other than vibration.

that is economically available and is consistent with the dimensions of the structure. In no event should the maximum size exceed one-fifth of the narrowest dimension between sides of forms, one-third the depth of slabs, or three-fourths of the minimum clear spacing between individual reinforcing bars, bundles of bars, or pretensioning strands.

Mixing water and air content. The quantity of water per unit volume of concrete required to produce a given slump depends on the maximum size, particle shape, and grading of the aggregates and the amount of entrained air. It is not greatly affected by the quantity of cement. Table 4.7 provides estimates of required mixing water for concretes made with various maximum sizes of aggregate with and without air

TABLE 4.7   Approximate Mixing Water Requirements for Different Slumps and Maximum Sizes of Aggregates[a]

Maximum size of aggregate, in	Recommended average total air content, %[b]	Air-entrained concrete			Non-air-entrained concrete			
		Slump, in			Approximate amount of entrapped air, %	Slump, in		
		1 to 2	3 to 4	5 to 6		1 to 2	3 to 4	5 to 6
		Water, lb/yd^3 of concrete[c]				Water, lb/yd^3 of concrete[c]		
3/8	7.5	310	340	360	3.0	350	385	410
1/2	7.5	300	325	340	2.5	335	365	385
3/4	6.0	275	300	315	2.0	310	340	360
1	6.0	260	285	300	1.5	300	325	340
1½	5.0	240	265	285	1.0	275	300	315
2	5.0	225	250	265	0.5	260	285	300
3	4.0	210	235	—	0.3	240	265	—
6	3.0	185	200	—	0.2	210	235	—

[a]Adapted from *Recommended Practice for Selecting Proportions for Concrete* (ACI 613-54).
[b]Plus or minus 1 percent.
[c]These quantities of mixing water are for use in computing cement factors for trial batches. They are maximums for reasonably well-shaped angular coarse aggregates graded within limits of accepted specifications.
SOURCE: Portland Cement Association, 1979.

entrainment. Depending on aggregate texture and shape, mixing water requirements may be somewhat above or below the tabulated values, but they are sufficiently accurate for the first estimate. Such differences in water demand are not necessarily reflected in strength.

Table 4.7 indicates the approximate amount of entrapped air to be expected in non-air-entrained concrete on the right side of the table and shows the recommended average air content for air-entrained concrete on the left side of the table. The use of normal amounts of air entrainment in concrete with a specified strength near or about 34 MPa (5000 lb/in^2) may not be possible because each added percent of air lowers the maximum strength obtainable with a given combination of materials (Tuthill, 1960).

When trial batches are used to establish strength relations or verify the strength-producing capability of a mixture, the least favorable combination of mixing water and air content should be used. This is, the air content should be the maximum permitted or likely to occur, and the concrete should be gaged to the highest permissible slump. This will avoid developing an overly optimistic estimate of strength on the assumption that average rather than extreme conditions will prevail in the field.

Water-cement ratio. The required water-cement ratio is determined not only by strength requirements but also by factors such as durability and finishing properties. The average strength selected must, of course, exceed the specified strength by a sufficient margin to keep the number of low tests within specified limits (Table 4.8). For severe conditions of exposure, the water-cement ratio should be kept low even though strength requirements may be met with a higher value. Table 4.9 gives limiting values.

TABLE 4.8   Relations between Water-Cement Ratio and Compressive Strength of Concrete[a]

Compressive strength at 28 days, psi[a]	Water-cement ratio, by weight	
	Non-air-entrained concrete	Air-entrained concrete
6000	0.41	—
5000	0.48	0.40
4000	0.57	0.48
3000	0.68	0.59
2000	0.82	0.74

[a]Values are estimated average strengths of concrete containing not more than the percent of air shown in Table 4.7. For a constant water-cement ratio, the strength of concrete is reduced as the air content is increased. Strength is in accordance with Section 9(b) of ASTM Standard C31.

Source: American Concrete Institute (ACI), 1979.

TABLE 4.9   Maximum Permissible Water-Cement Ratios for Air-Entrained Concrete in Severe Exposures

| | Exposure | |
| | Continuously or frequently wet and exposed to freezing and thawing | Exposed to seawater or sulfates[a] |
Type of structure		
Thin sections (railings, curbs, sills, ledges, ornamental work) and sections with less than 1 in cover over steel	0.45	0.40
All other structures	0.50	0.45

[a]If sulfate-resisting cement (type II or type V of ASTM Standard C150) is used, permissible water-cement ratio may be increased by 0.05.
SOURCE: American Concrete Institute, 1979.

Cement content.   The amount of cement per unit volume of concrete is fixed by the determinations made above. The required cement is equal to the estimated mixing water content divided by the water-cement ratio. If, however, the specification includes a separate minimum limit on cement in addition to requirements for strength and durability, the mixture must be based on whichever criterion leads to the larger amount of cement. The use of pozzolanic or chemical admixtures will affect properties of both the fresh and hardened concrete.

Coarse aggregate content.   Varying the volume of coarse aggregate used per unit volume of concrete changes the workability of the mix. Given coarse and fine aggregates of available maximum size and gradation, respectively, the correct volume of coarse aggregate per unit volume of concrete must be chosen to produce satisfactorily workable concrete. The fine aggregate is characterized by its fineness modulus, a measure of the part of the coarse and medium sand described in ASTM Standard C125. The fineness modulus increases with coarseness and is usually restricted to values between 2.3 and 3.1. The weight and volume characteristics of coarse aggregate are determined by tests on dry aggregate placed in thin layers and compacted by rodding.

   The volume of coarse aggregate, in cubic feet on a dry-rodded basis, for a cubic yard of concrete is equal to the value from Table 4.10 multiplied by 27. This volume is converted to the dry weight of coarse aggregate required in a cubic yard of concrete by multiplying it by the dry-rodded weight per cubic foot of the coarse aggregate.

TABLE 4.10 Volume of Dry-Rodded Coarse Aggregate[a] per Unit of Volume of Concrete

Maximum size of aggregate, in	Fineness modulus of sand			
	2.40	2.60	2.80	3.00
3/8	0.50	0.48	0.46	0.44
1/2	0.59	0.57	0.55	0.53
3/4	0.66	0.64	0.62	0.60
1	0.71	0.69	0.67	0.65
1½	0.75	0.73	0.71	0.69
2	0.78	0.76	0.74	0.72
3	0.82	0.80	0.78	0.76
6	0.87	0.85	0.83	0.81

[a]As described in ASTM Standard C29.
SOURCE: American Concrete Institute, 1979.

Fine aggregate content. After the coarse aggregate size and quantity have been chosen, all ingredients of the concrete have been estimated except the fine aggregate. Its quantity is determined by the method of difference. Either of two procedures may be employed—i.e., the weight method or the absolute-volume method. If the weight of the concrete per unit volume is assumed or can be estimated from experience, the required weight of fine aggregate is simply the difference between the design weight of fresh concrete and the total weight of the other ingredients. Often the unit weight of concrete is known with reasonable accuracy from previous experience with the materials. In the absence of such information, Table 4.11 can be used to make a first estimate.

TABLE 4.11 First Estimate of Weight of Fresh Concrete

Maximum size of aggregate, in	Weight, lb/yd^{3a}	
	Non-air-entrained concrete	Air-entrained concrete
3/8	3840	3690
1/2	3890	3760
3/4	3960	3840
1	4010	3900
1½	4070	3960
2	4120	4000
3	4160	4040
6	4230	4120

[a]Values are calculated for concrete of medium richness (550 lb of cement per cubic yard) and medium slump with an aggregate having a specific gravity of 2.7. Water requirements are based on values for 76 to 102 mm (3- to 4-in) slump in Tables 4.6 and 4.7.
SOURCE: American Concrete Institute, 1979.

Even if the estimate of concrete weight per cubic yard is rough, mixture proportions will be sufficiently accurate to permit easy adjustment on the basis of trial batches.

Adjustments for aggregate moisture. The aggregate quantities actually to be weighed out for the concrete must allow for moisture in the aggregates. Generally, the aggregates will be moist and their dry weights should be increased by the percent of water, both absorbed and surface, they contain. The mixing water added to the batch must be reduced by an amount equal to the free moisture contributed by the aggregate, i.e., total moisture minus absorption.

Trial batches. The calculated mixture proportions should be checked by means of trial batches prepared and tested in accordance with ASTM Standard C192, "Making and Curing Concrete Compression and Flexure Test Specimens in the Laboratory" or a full-size field batch. Only sufficient water should be used to produce the required slump regardless of the amount assumed in selecting the trial proportions. The concrete should be checked for unit weight and yield (ASTM Standard C138) and for air content (ASTM Standard C138, C173, or C231). It should also be carefully observed for proper workability, freedom from segregation, and finishing properties. Appropriate adjustments should be made in the proportions for subsequent batches.

4.6 Curing

Curing is essential in the production of quality concrete. Potential strength and durability will be fully developed only if the concrete is properly cured for an adequate period of time before being placed in service. Proper curing prevents loss of moisture for the time necessary to obtain necessary hydration of the cement. Excess mixing water is allowed to escape, but the appearance of plastic shrinkage cracks in the surface of the concrete about the time the concrete is ready for finishing indicates that the paste is losing water too rapidly.

Concrete should be cured by keeping the concrete damp for not less than 7 days if made of normal portland cement and for not less than 3 days if made of high- (early-) strength cement. For each decrease in the average curing temperature of 2.7° below 21°C (5° below 70°F), the curing period should be increased by 4 days for units made of normal portland cement and by 2 days for units made of high- (early-) strength cement or until the concrete has attained its design strength. When units are cured by high-pressure steam, steam vapor, or other approved processes used to accelerate the hardening of the cement, the curing time may be reduced provided the compressive strength of the concrete is equal to the 28-day strength obtained by damp curing. Concrete units should not be moved from the casting bed until the curing period is complete.

4.6.1 Methods of curing

There are two general methods of retaining the required water for hydration furnished by the mixing water in concrete.

Moist environment. A moist environment can be maintained through water ponding, water sprays, steam, or saturated cover materials such as burlap or cotton mats, carpets (some carpets may contain certain dyes which inhibit the settling of concrete), earth, sand, sawdust, or straw, all of which must be kept continuously wet.

Sealing materials. Curing can also be accomplished by preventing the loss of mixing water by means of sealing materials or curing compounds. Sealing is accomplished by the use of impervious sheets of paper or plastics or by the application of an impervious membrane-forming curing compound applied to the freshly placed concrete. Compounds consisting essentially of waxes, resins, chlorinated rubber, and solvents of high volatility at atmospheric temperatures are used extensively for curing concrete. The formulation must be such as to provide a moisture seal shortly after being applied and must not be injurious to portland cement. All compounds should comply with the requirements of ASTM Standard C309.

Before applying curing compound, tops of joints that are to receive sealant should be tightly closed with temporary material to prevent entry of the compound and to prevent moisture loss during the curing period. The compound should be applied on damp surfaces as soon as the moisture film has disappeared. The curing compound should be applied by power-spraying equipment using a spray nozzle equipped with a wind guard. The compound should be applied in a two-coat continuous operation at a coverage of not more than 10 square meters per liter (m^2/L) [400 square feet per gallon (ft^2/gal)] for each coat. When applied by hand sprayers, the second coat should be in a direction approximately at right angles to the direction of the first coat. The compound should form a uniform, continuous, adherent film that will not check, crack, or peel and should be free from pinholes or other imperfections. Surfaces subjected to rainfall within 3 hours (h) after compound has been applied or surfaces damaged by subsequent construction operations within the curing period should be immediately resprayed at the rate specified above.

Membrane curing compound should not be used on surfaces that are to receive any subsequent treatment that depends on adhesion or bonding to the concrete or on surfaces that are maintained at curing temperatures with free steam. When membrane-forming curing compounds are permitted, permanently exposed surfaces should be cured by use of a nonpigmented membrane-forming curing compound containing a fugitive dye. When nonpigmented-type curing compounds are used, the concrete surface should be shaded from the direct rays of the sun for the curing period. Surfaces coated with curing compound should be kept free of foot and vehicular traffic and from other sources of abrasion and contamination during the curing period.

4.6.2 Special conditions

There are some conditions of curing concrete for coastal structures that frequently occur and require special consideration.

Hot weather concreting. High temperatures impact on concrete by more rapid hydration of cement, greater mixing water demand, increased evaporation of mixing water, reduced strength, and a tendency to crack either before or after hardening. Special precautions are necessary; they include cooling the aggregate, adding ice to the concrete mix, and covering the curing concrete to keep it moist. Certain water-reducing retarders may counteract the accelerating hardening of concrete at high temperatures and reduce the need for additional mixing water. Curing concrete above 32.2°C (90°F) is undesirable.

Cold weather concreting. Fresh concrete should be maintained at a minimum temperature of 10°C (50°F) until initial strength is attained. This requirement may require heating the aggregate and mixing water, not adding admixtures until the mixing water temperature is 32.2°C (90°F) or below, and protecting the concrete surface from freezing temperatures until safe strength has developed in the concrete. Most of the hydration is developed in the first 3 days of hardening, but it may be necessary to provide housing or additional heat to ensure adequate temperature and moisture for curing to obtain the strength and durability intended of the concrete.

Underwater concrete curing. No special precautions are usually feasible for curing concrete placed under water except for temperature control. Concrete will cure best in a temperature range of 10 to 24°C.

Higher temperatures will accelerate curing; lower temperatures will delay curing.

4.6.3 Preferred curing method

When physical conditions permit, the following curing methods, in order of performance, are used to obtain a highly durable concrete: (1) continuous drenching with water, (2) burlap, blankets, or carpets kept continuously wet, (3) membrane-forming curing compounds, (4) sand or straw randomly dampened, and (5) air curing.

4.7 Techniques to Enhance Durability

The designer and constructor-manufacturer share the responsibility to build concrete structures that remain essentially in their original state despite attack of the environment. The ability of the structure to withstand environmental attack is called durability.

Concrete is an extremely durable material and ranks high among all known structural materials for its resistance to the attack of natural environments. Freeze-thaw and salt water immersion tests have demonstrated the inherent durability of concrete. It is generally accepted that properly designed prestressed concrete piles are among the most durable piles for marine structures, even in a tropical salt spray environment.

Maintenance of durability is achieved only by proper design and construction. The consequences of disintegration and corrosion are potentially catastrophic. Corrosion and disintegration are not random or spot occurrences. Rather, when distintegration and disruption do take place, it is usually due to some fundamental error or neglect; and the damage often extends to the entire structure. Thus, except for some localized spot of impact or accident, a thorough investigation should be made of the entire structure if disintegration is found.

4.7.1 Impacts on durability

Disruption. Durability is affected by disruption of concrete structure, environmental attacks, and use of aggregates. Disruption may take several forms:

- Disintegration of the concrete
- Chemical replacement in the concrete with a consequent loss of strength
- Corrosion of reinforcing bars and ties or prestressing tendons, causing loss of strength, fracture, or lower resistance to fatigue

- Corrosion of the inserts, embedded fittings, and connections
- Corrosion of anchorages

In combination these forms may interact to intensify disruption. For example, corrosion of reinforcing bars increases bar volume and causes disintegration of the concrete cover.

Environmental attacks. Exposure to environmental elements may result in attacks that could severely impair the serviceability of a concrete structure if the structure were not made sufficiently durable. Among the more common environmental attacks are the following.

1. Those causing or accelerating disintegration of or change in the concrete:

- Reactive aggregates
- Unsound aggregates
- Cement containing a high percentage of alkalis or C_3A
- Freeze-thaw cycles
- CO_2 in air or surrounding water
- Erosion and abrasion from cavitation, ice, surf, or moving sand
- Acids, sulfates, nitrates, or organic substances in the mixing water or the surrounding water, as at discharge from chemical plants or in sewage-disposal structures

2. Those causing or accelerating corrosion of steel:

- Salt or alkalies on aggregates
- Chlorides in admixtures or water used for mixing and curing
- Chlorides in water surrounding concrete (salt water), salt spray, salt fog
- Oxygen
- Sulfides combined with moisture on stressed tendons before encasement or protection
- Stray electric currents
- Alkalies in surrounding soils
- High temperature
- Embedded metals other than steel, particularly copper and aluminum
- Inadequate thickness of concrete cover or permeability of cover

- Cracks
- Cement chemistry (e.g., too low C_3A)
- Deicing salts, acids, or other aggressive chemicals

4.7.2 Enhancement techniques

Fortunately, the steps to be adopted to overcome the many forms of environmental attack are complementary. Most have been adopted as standard good practice.

Aggregates. Although aggregate is commonly considered to be an inert filler in concrete, that is not always the case. Certain aggregates can react with portland cement and cause expansion and deterioration. Fortunately, care in the selection of aggregate sources, and use of low-alkali cement and pozzolans when they are appropriate, will reduce the problem significantly.

All aggregates should be sound, nonreactive, abrasion-resistant, and free from salt or alkalies. Particular care should be taken when working with aggregates from new sources, especially those with siliceous rocks and in desert areas. Sands from deposits several miles from the shores of the Persian Gulf are heavily contaminated with salt from salt fog; their use, unwashed, has led to serious corrosion in mild-steel reinforcing.

Aggregates should meet the requirements of ASTM Standard C33; in addition, they should be judged for their durability by an engineer on the basis of prior experience with them and also tests. Tests are especially necessary when working with new aggregates; they are listed in ASTM C33 and include tests for soundness (sodium sulfate soundness test), alkali-silica reactivity, cement-aggregate reactivity, and freeze-thaw durability. Washing aggregates with fresh water will remove salt and dust from sand and aggregates.

Reinforcing-steel protection. The spalling of concrete in bridge decks and marine structures, such as reinforced wharf decks, piles, groins, and concrete anchors, has been a serious problem for many years. The principal cause of spalling is corrosion of the reinforcing steel, which is largely due to the use of deicing salts, exposure to seawater, or inadequate embedding of reinforcing steel (Fig. 4.2). The corrosion products produce an expansive force which causes the concrete around the steel to spall. A minimum cover over the steel of 76 mm (3 in) and the use of a low-permeability, air-entrained concrete will ensure good durability in the great majority of cases, but more positive protection is needed in very severe exposures.

Figure 4.2   Steel reinforcement exposed by spalling of concrete cover.

Sufficient cover.   The concrete cover protects the steel by creating a passive condition of high pH at the surface of the steel. Too thin a cover allows carbonation, usually on the surface of the coarse aggregate particles. Carbonation lowers the pH. Oxygen is necessary to the corrosion mechanism; a thicker cover minimizes the movement of oxygen to the steel surface. In seawater, chloride ion movement also is inhibited by thicker covers. The cover should properly be related to the density and cement content. The exact relations have not been thoroughly established, so arbitrary values are usually used as guides or standards. Thicker covers make it possible to achieve better compaction, fewer voids, and less permeability.

Reinforcing-steel coating.   Concrete may not provide permanent protection of reinforcing bars under many conditions. Cracks in the concrete surface contribute to corrosion by providing access to moisture, air, and contaminants. Hydrated portland cement is subject to chemical reaction with carbon dioxide of the atmosphere. Carbonation reduces the alkalinity of concrete and thereby reduces the effectiveness of the concrete as a protecting medium. Concrete will also deteriorate from other causes such as freeze-thaw cycles, sulfate attack, and reactive aggregates; it will crack or weaken and thus become less able to protect embedded reinforcing. Corrosion of steel reinforcing can also result from stray electric current or corrosion cells that develop on the embedded steel. Electrical potential differences can occur at various spots on metals in concrete because of differences in moisture content, oxygen concentration, and electrolyte concentration and by contact of dissimilar metals. A corrosion cell results when regions of different electrical potential are interconnected by a conductive pathway. Loss of metal then occurs at the region of more positive potential, which acts as the anode.

One recently developed method of preventing the deterioration of reinforcing steel is to coat the bars with an epoxy material. The epoxy is applied in a mill or coating facility usually by a fusion-bonding process as a result of an irreversible heatcatalyzed chemical reaction. The careful application of a fusion-bonded or electrostatically applied epoxy coating has produced virtually pinhole-free coating protection of the steel bars from the moisture, chlorides, and other contaminates that may be in or enter into the concrete.

Fusion-bonded epoxy coatings have had a short but successful history of protecting reinforcing steel against corrosion in a highly alkaline and chloride-contaminated environment. The fusion-bonding epoxy coating is formed by combining an epoxy resin with appropriate curing agent, pigments, catalysts, flow control agents, etc., to achieve the desired application and performance characteristics.

"Fusion-bonded" means that the coating achieves adhesion as a result of a heat-catalyzed chemical reaction. When a fusion-bonded coating is exposed to heat, a chemical reaction occurs; and sufficient heat must be supplied for a given amount of time to allow the chemical reaction to reach completion. The reaction is irreversible. Unlike thermoplastic coatings, the fusion-bonded coating will not soften if it is heated after it is cured. The material is applied to rebars under controlled conditions at a mill away from the job site and is therefore not weather-dependent.

The coating system has four parts: surface preparation, material selection, application, and cure. Surface preparation requires sandblasting to white metal, since the surface must be completely clean and possess an anchor pattern. It is desirable that both physical and chemical adhesion be obtained. Materials selection can be made from commercially available materials that are selected in accordance with ASTM Standard D3415. Application is accomplished by heating the rebar with a noncontaminating heat source to approximately 232°C (450°F) but as recommended by the manufacturer. The resin application should be by electrostatic deposition to obtain an even coating that is 0.13 to 0.26 mm (5 to 10 mils) thick. Heat is continued until the gel time has been satisfied. The bars are then cooled, and an electrical holiday inspection is made. Following this coating application, the rebars can be transported to the job site and bent to necessary configurations with reasonable ease.

Prevention of cracks. Cracks in concrete allow carbon dioxide and oxygen to penetrate to the surface of the steel reinforcement. Cracks may also play a part in the formation of electrolytic cells in the concrete.

Elimination of voids at steel surface. Studies indicate that steel corrosion is associated with voids at the steel surface. It can be diminished by mix design and thorough consolidation. In posttensioning, grouting procedures which will prevent or minimize voids should be adopted.

Grouting of ducts in posttensioned concrete. Proper grouting is essential for corrosion protection and prevention of bursting during freezing.

Abrasion resistance. The abrasion resistance of concrete is defined as the "ability of a surface to resist being worn away by rubbing and friction." Research to develop meaningful laboratory tests of concrete abrasion has been underway for more than a century. The problem is complicated because there are several different types of abrasion, and no single test method which is adequate for all conditions has been found. Abrasion can be classified into four types:

- Wear on concrete floors due to foot traffic and light trucking, skidding, scraping, or sliding of objects on the surface (attrition)

- Wear on concrete road surfaces due to heavy trucks and automobiles with studded tires or chains (attrition, scraping, and percussion)

- Erosion in hydraulic structures such as dams, spillways, tunnels, bridge abutments, concrete breakwaters, and piling due to the action of abrasive materials carried by flowing water (attrition and scraping)

- Wear on concrete dams, spillways, tunnels, and other water-carrying systems in which high velocities and negative pressures are present (generally known as cavitation erosion, which is mainly the result of design and is not covered in this book)

To evaluate abrasion resistance properly, the type of concrete being tested must be considered. If the concrete is of the same mix throughout, the abrasion resistance can be expected to be a direct function of the concrete strength. If, however, metallic or other hardeners have been applied, the time required for the abrasion apparatus to penetrate the hard surface must be determined to evaluate the test results properly.

The abrasion resistance of concrete is affected primarily by compressive strength, aggregate properties, finishing methods, use of toppings, and curing. Tests and field experience have generally shown that compressive strength is the most important single factor controlling the abrasion resistance of concrete. Abrasion resistance increases with an increase in compressive strength. Compressive strength and abrasion resistance vary inversely with the ratio of voids (water plus air) to cement. Limiting the maximum size of the aggregate in rich mixes will result in improved compressive strengths and maximum abrasion resistance of concrete surfaces.

Proper finishing procedures and timing are essential if the quality of concrete near the surface of a slab is to be as good as that of the underlying section. Delaying the floating and troweling operations increases resistance to abrasion. Another highly important factor in producing wear-resistant, nondusting concrete surfaces is adequate curing (ACI 308-71). One study showed that a surface cured for 7 days is nearly twice as wear-resistant as one cured for only 3 days, and additional curing resulted in further improvement. The following measures will result in abrasion-resistant concrete surfaces.

4.7.3 Compressive strength

For a required concrete strength level, the strength selected should be appropriate to both the service exposure and the life of the structure. In no case should the compressive strength be less than 28 megapascals (MPa) (4000 lb/m^2). Suitable strength levels can be attained by:

- A low water-cement ratio.

- Proper grading of fine and coarse aggregate (meeting ASTM Standard C33) and limiting the maximum size to nominal 25 mm (1 in).

- Lowest consistency practicable for proper placing and consolidation with maximum slump of 75 mm (3 in) and 25 mm (1 in) for toppings.

- Minimum air content consistent with the exposure conditions. For indoor floors not subject to freezing and thawing, air contents of 3 percent or less are preferable. In addition to a detrimental effect on strength, high air contents can cause blistering, particularly when dry shakes are used.

- When wear conditions are severe, a high-strength [not less than 34

MPa (5000 lb/in^2)] topping layer, called a two-course floor. The maximum size of aggregate in the topping should be 12 mm (½ in).

4.7.4 Proper finishing

Floating and troweling should be delayed until the concrete has lost its surface water sheen or all free water on the surface has disappeared or been carefully removed. The delay period is usually 2 hr or more (depending on temperatures, mix proportions, and air content) after the concrete is placed. The recommendations of ACI Standards 302 and 304 with respect to finishing unformed surfaces should be followed.

Vacuum dewatering. Vacuum dewatering is a method of removing excess water from concrete immediately after placement. The process results in increased strength, hardness, and wear resistance of concrete surfaces, and it is primarily applicable to slab.

Special dry shakes and toppings. Where severe wear is anticipated, the use of special toppings or dry shakes (such as coats of cement and hard fine aggregate or of cement and iron aggregate) should be considered. If they are selected, the recommendations of ACI Committee 302 (1969), "Recommended Practice for Concrete Floor and Slab Construction," should be followed.

4.7.5 Proper curing procedures

Curing should start immediately after the concrete has been finished and be continued for at least 7 days if the cement is type I cement (5 days if it is type III). Curing with water by spray, damp burlap, or cotton mats is preferred, provided the concrete is kept continuously moist. Waterproof paper or plastic sheets are satisfactory, provided the concrete is first sprayed with water and then immediately covered with the paper or plastic with the edges overlapped and sealed with waterproof tape. Curing compounds meeting ASTM Standard C309 seal the moisture in the concrete and are economical and easy to apply; they can be used where other methods are impracticable. The curing compound should be covered with scuff-proof paper if the concrete is a floor area that must be used before curing is completed.

4.7.6 Freezing and thawing

Freezing and thawing damage to concrete is a serious problem in northern climates, but the mechanisms involved are now fairly well understood. Exposing damp concrete to freezing and thawing cycles is a severe test of the material, and poor concrete will certainly fail. In pavements the damage is greatly accelerated by the use of deicing salts, which often results in severe scaling at the surface. Fortunately, air-entrained concrete which is properly proportioned, manufactured, placed, finished, and cured will almost always resist cyclic freezing for many years. It should be recognized, however, that even good concrete may suffer damage from cyclic freezing in unusual conditions, particularly if the concrete is kept in a state of nearly complete saturation. Also, in cases when it is saturated on the back side and exposed to air on the front side, the concrete may exhibit extremely variable behavior ranging from complete freedom from damage to total failure.

There is general agreement that cement paste can be made completely immune to damage from freezing temperatures by means of entrained air unless special exposure conditions result in filling of the air voids. However, air entrainment alone does not preclude the possibility of damage of concrete due to freezing. Freezing phenomena in aggregate particles must also be taken into consideration (see test samples in Fig. 4.3).

Freezing in aggregate particles.   Most rocks have pore sizes much larger than those in cement paste, and they expel water during freezing. Size has been shown to be an important factor in the frost resistance of coarse aggregate. The critical size of rocks of good quality range up-

Figure 4.3   Test samples subjected to 150 cycles of freezing and thawing.

wards from perhaps 6 mm (¼ in). However, some aggregates (e.g., granite, basalt, diabase, quartzite, marble) have capacities for freezable water so low that they are not stressed when freezing occurs, regardless of the particle size. The role of entrained air in alleviating the effect of freezing in rock particles is minimal.

Without entrained air, the paste matrix surrounding the aggregate particles may fail when it becomes critically saturated and is frozen. However, if the matrix contains an appropriate distribution of entrained air voids characterized by a spacing factor less than about 200 μm (8 mi), freezing does not produce destructive stress. When absorptive aggregates (such as structural lightweight) are used and the concrete is in a continuously wet environment, concrete will probably fail if the coarse aggregate becomes saturated. The pressure developed when the particles expel water during freezing ruptures the particles and the matrix. If the particles are near the concrete surface, a popout can result.

Whatever the absorption characteristics of a given aggregate, the rate of absorption in concrete is limited by the rate at which water can pass through the envelope of hardened cement paste. Because the coefficient of permeability of hardened cement paste is lower as the cement content increases and the longer the paste has wet-cured, the rate of absorption of any kind of aggregate can be lowered by reducing the water-cement ratio of the paste and by requiring good curing.

4.7.7 Recommendations for durable structures

Concrete which will be exposed to a combination of moisture and cyclic freezing requires the following:

- Design of the structure to minimize exposure to moisture
- Low water-cement ratio
- Air entrainment
- Suitable materials
- Adequate curing
- Special attention to construction practices

Requirements for the water-cement ratio and air entrainment are described in detail below.

Water-cement ratio. Frost-resistant regular-weight concrete should have a water-cement ratio not to exceed the values given in Table 4.9.

Because the determination of the rate of absorption of lightweight aggregates is uncertain, it is impracticable to calculate the water-cement ratio of concretes containing lightweight aggregates. For those concretes, a specified 28-day compressive strength of 27.6 megapascals (MPa) (4000 lb/in^2) is recommended. For severe exposures, some have found it also desirable to specify a minimum cement content of 335 kg/m^3 (564 lb/yd^3) and only that amount of water necessary to achieve the desired consistency.

Entrained air. Too little entrained air will not protect cement paste against cyclic freezing; too much air will unduly penalize the strength. About 7 percent of air in the mortar for severe exposure and about 5 percent for moderate exposure are reasonable. Frequent determinations of the air content of the concrete should be made. For regular-weight concrete, the following test methods may be used: volumetric method (ASTM Standard C173), pressure method (ASTM Standard C231), and the unit weight test (ASTM Standard C138). An air meter may be used to provide an approximate indication of air content. For lightweight concrete, the volumetric method is recommended.

The air content and other characteristics of the air void system in hardened concrete can be determined microscopically (ASTM Standard C457). ASTM Standard C672 is often used to assess the resistance of concrete to deicer scaling.

Summary. Table 4.12 summarizes various techniques and practices that may be employed to obtain maximum durability.

4.8 Reinforcing and Prestressing Materials

In most modern concrete construction steel reinforcement is used to provide tensile strength, and so it is with structures in the coastal zone. The only difference from inland construction is that in the marine environment extra care must be taken to protect the steel from salt water. One of the best sources of information concerning the use of steel reinforcement is the American Concrete Institute, whose standard and specifications are the standard for the industry and are universally used. In addition, steel reinforcing should conform to the following ASTM requirements as applicable:

- Bar reinforcement, ASTM Standard A615 Grade 40 or 60
- Cold-drawn wire, ASTM Standard A185
- Welded-wire fabric, ASTM Standard A185, when zinc-coated (galva-

TABLE 4.12   Techniques to Obtain Maximum Durability

Aggregates	Sound nonreactive
	Abrasion-resistant
Cement	Low alkali content (less than 0.6% $Na_2O + K_2O$)
	Moderate C_3A (for marine environments)
	High cement factor
Water	Fresh water
	Free from chlorides and sulfates
Concrete mix	Low water-cement ratio
	Clean, small-size coarse aggregate
	Dense grading
	High-bond aggregate (where exposed to abrasion or cavitation)
	Limitation on chlorides from any source
Admixtures	Water-reducing
	Nonsegregating
	Air entrainment (for freeze-thaw environments)
Forms	Smooth-surfaced
	No reentrant angles, projections, etc.
	No sharp corners, edges
Placing	Thoroughly consolidated and compacted
	Minimum bleed holes
Cover	Adequate cover over mild steel
	Adequate cover over prestressing steel
Finish	Troweled
Construction joints	Well-prepared
	Well-bonded (by presoaking and grout or by epoxy)
Curing	Adequate water cure or steam cure
	Moisture available or sealed in during cooling period
	Drying after curing (for marine freeze-thaw environments)
	Water free from chlorides or sulfates
Mild-steel reinforcement	Free from pitting
	Well-distributed bars
	Use of very large bars avoided
Embedded metals	Avoid galvanic action, especially Cu and Al
	Ducts to be thoroughly cleaned and flushed with fresh water or fresh water containing inhibitor
Prestressing steel	Free from pitting and extensive surface rust
	Clean and dry
	No salt
	Galvanized or plastic- or epoxy-coated (for special cases only)
	Coating must not be abraded
	Kept free from corrosion by (a) VIP powder, (b) greases, (c) sealing, (d) limited time of exposure until finally grouted or protected
	High degree of prestress
Anchorages	Flush type with the anchors themselves in pockets
	Epoxy concrete is best material
Grouting	Follow best grouting practice
Coatings	Bitumastic
	Epoxy
	Metallic sheathing (for special cases only, e.g., chemical)
	Wood lagging

nized), not less than 28.54 grams per square meter (g/m^3) [0.8 ounce per square foot (oz/ft^2)] of grade 5 "Prime Western" conforming to ASTM Standard B6

4.8.1 Prestressing steel

High-tensile steel is almost the universal material for producing prestress and supplying the tensile force in prestressed concrete. Such steel can take any of three forms: wires, strands, or bars. At present, strands are most widely used; they are grouped, in parallel, into cables. Strands are fabricated in the factory by twisting wires together and thereby decreasing the number of units to be handled in the tensioning operations. Steel bars of high strength have also been developed and successfully applied to prestressed concrete with the result of considerable economy at times.

The prestressing wires now on the market are mostly high-tensile-strength wires obtained by cold-drawing high-tensile steel bars through a series of dies. The process of cold drawing tends to realign the crystals, and the strength of the wires is increased by each drawing: the smaller the diameter of the wires, the higher the ultimate unit strength. The ductility of wires is, however, somewhat decreased as a result of cold drawing. It must be recognized that the actual strength will vary with the composition and manufacture of the wire as well as with the diameter.

The as-drawn wire, although possessing a high ultimate strength, has a relatively low proportional limit, e.g., about 414 to 552 MPa $(60,000$ to $80,000$ lb/in$^2)$, above which the stress-strain curve flattens at an increasing rate. This is objectionable, since the deformation characteristics are relatively uncertain and the amount of elongation during prestress cannot be easily determined. Hence, various methods commercially known as the stress-relieving process have been used to increase the proportional limit of the as-drawn wire. Two common stress-relieving methods are discussed in the following subsections.

Time-stress treatment. The time-stress treatment consists of stretching the wire to a stress level higher than that to be used in the final application. The treatment increases the proportional limit to about 60 or 70 percent of the ultimate strength while the ultimate strength itself remains about the same. After this process of stretching, the wire will still have slight creep at an eventual stress of 50 percent of the ultimate, but, when the wire is stressed up to 70 percent, the creep will not be much more than 5 percent.

Time-temperature treatment. The time-temperature treatment consists of heating the wire to 399 to 427°C (750 to 800°F) for a period of 30 to 40 s. The heating is accomplished by drawing the wire through a molten lead bath or through a hot-air tunnel such as a ceramic tube with heat applied on the outside. This treatment will have an effect on the proportional limit and ultimate strength of the wire similar to that of the previous process. But this time-temperature-treated wire has practically no creep when subjected to 50 percent of the ultimate strength. At 60 percent of the ultimate strength it shows slightly more creep than the stretched wire, and at 70 percent and above the creep becomes excessive.

It is usually left to the engineer to specify the physical properties and sometimes the chemical composition desired. The chemical composition of prestressing wire may vary with the manufacturer. Some manufacturers may use a certain amount of silicon in the steel, although it is not included in the following sample analysis:

Carbon	0.60 to 0.85 percent
Manganese	0.70 to 1.00 percent
Phosphorus	0.050 percent maximum
Sulfur	0.055 percent maximum

In the United States, two kinds of high-tensile wire strands are available—one for pretensioning and another for posttensioning. Pretensioning strands are made of seven or more small uncoated wires as drawn. The strands are then drawn through a lead bath for stress-relieving and also to improve their bond characteristics. For posttensioning and unbonded work, strands consisting of 7 to 61 galvanized wires are produced. These strands are machine-fabricated and stress-relieved to increase their proportional limit and to minimize creep. When the strands are to be bonded to the concrete, the wires should preferably be ungalvanized.

High-strength bars up to 1034 MPa (150,000 lb/in^2) or more are made by cold-working special alloy steels. By alloying high-carbon steel with proper agents such as silicon and manganese, high strength is obtained. Then the proportional limit is raised by cold working. Again, the chemical contents of these bars may differ. A sample composition of high-strength steel bars is:

Carbon	0.6 percent
Silicon	2.0 to 2.5 percent
Manganese	0.7 to 1.0 percent

Phosphorus   0.2 percent

Sulfur   0.2 percent

To get a better bond between steel and concrete, various forms of surface indentation are used to afford direct mechanical keys with the surrounding concrete. It is assumed that the corrugations now commercially used will not alter the stress-strain properties of the wires, although some question of their fatigue strength in comparison with the straight ones has been raised. Some pretensioning factories pass their wires through a small machine that forms permanent waves which are believed to increase the bond resistance of the wire.

The ultimate strength of steel wires, strands, or bars varies with the manufacturer, so that it is frequently necessary to obtain sample tests for each lot of products. However, the general range of values is as listed in Table 4.13. While the ultimate strength of high-strength steel can be easily determined by testing, its elastic or proportional limit, or its yield point, cannot be so simply ascertained. First, there is no yield point for high-strength steel as there is for ordinary low-carbon steel. Second, the gradual curving of the stress-strain curve makes it difficult to fix a point for the proportional limit. Consequently, different methods for defining the yield point of high-tensile steel have been adopted.

Yield point and proportional limit must be obtained by testing the particular steel. But as a rough approximation Table 4.14 gives the usual values for high-tensile steels expressed in terms of the respective ultimate strengths. Approximate average values for the secant modulus at the proportional limit are shown in Table 4.15.

In order to avoid brittle failures in the prestressed concrete, a certain amount of ductility in the steel is desirable. It is measured by the amount of elongation in a certain gage length, generally 254 mm (10-in) gage in the United States. The average ultimate elongation is about 5 percent for wires and also for bars. For evident reasons, it cannot be easily measured for strands; only the wires making up the strands are measured for ductility.

TABLE 4.13   Strength of Steel Wires

Wires	Ultimate strength, MPa
Wires of varying diameters and makes	1379–2275
Strands of 7 uncoated small wires	1586–1862
Strands of 19 or more galvanized wires	1379–1517
Strands of 19 or more uncoated wires	1517–1655
Bars	965–1172

TABLE 4.14   Approximate Yield Point and Proportional Limit

	Yield point at 0.2% set, $f'_s{}^a$	Proportional limit, $f'_s{}^a$
Wires as drawn:	0.75	0.35
Prestretched wires	0.85	0.55
Time-temperature-treated wires	0.87	0.70
Galvanized wires	0.85	0.55
Strands, pretensioning as drawn	0.85	0.35
Stress-relieved strands	0.90	0.75
Strands, posttensioning and prestretched	0.85	0.55
Bars	0.90	0.60

^{a}Ultimate strength.

Steel, creep characteristics.   One of the important characteristics required of prestressing steel is minimum creep under maximum stress. Creep in steel is the loss of stress when the steel is prestressed and maintained at a constant strain for a period of time. It can also be measured by the amount of lengthening when the steel is maintained under a constant stress for a period of time. The two methods give about the same results when the creep is not excessive, but the constant-strain method is more often employed as a basis for measurement because of its similarity to the actual conditions in prestressed concrete. Creep varies with steel of different compositions and treatments; hence, exact values can be determined only by test in each individual case if previous data are not available.

Approximate creep characteristics are, however, known for most of the prestressing steels now on the market. Speaking in general, the percent of creep increases with increasing stress; and when a steel is under low stress, the creep is negligible. The following summarizes the creep characteristics of different steels. Compared to stress-relieved wires, the as-drawn wires have somewhat higher creep. Prestretched wires will have about 2 to 3 percent creep when subjected to $0.50f'_s$ (f'_s = ultimate strength); but when they are stressed to $0.70f'_s$, the creep will still be no more than 5 percent. Time-temperature-treated wires have practically no creep when subjected to $0.50f'_s$. At $0.60f'_s$, they have slightly more creep than "prestretched"

TABLE 4.15   Secant Modulus at Proportional Limit

	Secant modulus, MPa
Wires	186,159–206,843
Strands for pretensioning and stress-relieved	186,159–199,948
Strands for posttensioning and prestretched	165,474–179,264
Bars	172,369–193,053

wires; and at 0.70 to $0.80f'_s$, the creep becomes excessive. Galvanized wires have about the same creep characteristics as the time-temperature-treated wires, and preferably they should not be subjected to any stress above $0.60f'_s$ without carefully considering the effect of creep. Some limited tests seemed to show that, for stress up to about $0.55f'_s$, creep is not more than 5 percent in high-tensile bars.

Creep in steel is a function of time, but there is evidence that, under the ordinary working stress for high-tensile steel, creep takes place mostly during the first few days. Under constant strain, creep ceases entirely after about 2 weeks. If the steel is stressed to a few percent above its initial prestress and overstress is maintained for a few minutes, the eventual creep can be greatly lessened; it practically stops in about 3 days.

In order to prevent corrosion in unbonded prestressed concrete, wires are sometimes galvanized. When they are galvanized, their tensile strength is slightly reduced, but the other characteristics are similar to those of the time-temperature-treated wire. The latter has practically no creep when it is used within 55 percent of its ultimate strength, but its tendency to creep at stresses above 55 percent cannot be well controlled.

4.9 Embedded Hardware

Embedments are common to all concrete construction. The materials are usually steel and generally consist of fabricated structural steel materials such as plates, angles, bars and sleeves; reinforcing steel of varying grades and strengths; and prestressing tendons such as wires, strands, and bars. Occasionally, other materials such as stainless steel, copper pipe, bronze plates, and Teflon are employed. Structural connections usually require the greatest care during the construction process.

For all embedments, care must be taken to ensure no disintegration of the concrete. Metal hardware must be protected from corroding by sufficient concrete cover, alloying, or coating. Common forms of coating are galvanizing and epoxy. All metal embedments should be protected from chlorides or sulfides or possibly other negative ions occurring in a humid atmosphere if such materials are subjected to any stress in the concrete.

Exposures in industrial atmosphere may lead to H_2S concentrations that can cause brittle fracture in the hardware. Metallizing hardware with highly corrosion-resistant material, by flame or arc application, is an optional way of protecting metal hardware. These failures are generally classified as one form of hydrogen embrittlement which occurs when molecular hydrogen ions are able to enter between the steel molecules. This condition may also develop when dissimilar metals such as aluminum and zinc are used in the vicinity of steel.

4.10   Joint Sealing

Nearly every concrete structure has joints (or cracks) that must be sealed to ensure integrity and serviceability. It is a common experience that satisfactory sealing is not always achieved. The sealant used or its poor installation usually receives the blame, whereas often there have been deficiencies in the location or the design of the joint that would have made it impossible for any sealant to have done a good job.

This section shows that, by combining the right sealant with the right joint design for a particular application and then carefully installing it, there is every prospect of successfully sealing the joint and keeping it sealed. This section is a guide to what can be done rather than a recommended practice because in most instances more than one choice is available. Without specific knowledge of the structure, its design, service use, and environment and the amount to be spent, it is impossible to prescribe a "best joint design" or a "best sealant." The information contained in this guide is, however, based on current practices and experience judged sound by many agencies and organizations. It should therefore be useful in making an enlightened choice of a suitable joint sealing system and ensuring that it is then properly detailed, specified, installed, and maintained.

Prefabricated concrete units require field joining to establish structural continuity of the structure. Grout material proportions, as in conventional concrete practice, are influenced by structural design requirements. Additionally, the grout must be so designed as to flow freely into the joint space or voids without appreciable segregation or water gain so that honeycombing is avoided and intimate bond between the concrete surfaces is assured. Slurries of portland cement and water, with or without sand, have long been used in the construction industry as a grout for filling cracks, voids, and joints. Later development of epoxy formulations, uniquely suited for use as an adhesive with concrete, led to the formulations being commonly accepted for use as concrete grouting and joining material in concrete construction.

4.10.1   Materials

Cement-sand grouts.   Commonly employed grout material proportions of cement to sand range from 1:1 to 1:2, although ratios as lean as 1:3 have been used. When the proportion of cement to sand is 1:3, the water-to-cement ratio, by weight, for grout containing sand of average gradation (fineness modulus of 2.75) may be approximately 1:0.9.

With lower cement-to-sand ratios, the w/c ratio also will be lower, which will increase the strength of the grout. Compressive strength and placability may limit the amount of sand that can be used in a grout because the grout must be sufficiently fluid to penetrate and fill all of the voids and joint space yet be of such consistency that the suspended sand and cement do not filter out. Pozzolan is used to improve the fluid properties of the mix and reduce segregation of solid particles. Usually, the proportions of cement to pozzolan are 2:1, although ratios as low as 1:1 and up to 9:1 have been used. For structural grout, it is usually not desirable to exceed a cement-to-sand ratio of 1:2 by weight, because higher ratios produce lower strengths and excessive segregation of sand in the grout mixture. The ratio of water to cement plus pozzolan ($w/c + p$), by weight, should range from 1:0.45 to 0.50.

Epoxy grout. Epoxy compounds are generally supplied in two or more parts. Almost without exception, epoxy systems must be formulated to make them suitable for specific end uses. There are many reasons why epoxies make good adhesives: They may be in liquid form and contain no volatile solvent; they adhere to most materials used in construction; no by-products are generated during curing; curing shrinkage is low; long-term dimensional stability is good; and they have high tensile and compressive strengths. Appropriate formulations are resistant to the action of weathering, moisture, acids, alkalis, and many other environmental factors.

Epoxy resins find wide application as grouting materials. The filling of cracks, either to seal them from the entrance of moisture or to restore the integrity of a structural member, is one of the more frequent applications. Cracks or joints 6 mm (0.25 in) or less are most effectively filled with pourable epoxy compound, whereas an epoxy resin mortar should be used for wider openings. Surfaces upon which epoxy compounds are to be used must be given careful attention because the bonding capability of a properly selected epoxy compound is primarily dependent on surface preparation. All surfaces must be meticulously cleaned and dried and be at proper surface temperature at the time of epoxy application.

If it is impracticable or impossible to obtain a dry surface, an epoxy system formulated to bond to damp surfaces must be used. For the best performance under each condition of use, the properties of the epoxy resin system should be tailored to meet the specific needs of the application. It is unlikely that a system containing only an epoxy resin and a pure hardening agent will find wide use. That is why epoxy resin systems sold commercially are generally the products of formulators who specialize in modifying the systems with flexibilizers, extenders, diluents, and fillers to meet specific end use requirements and why it is important to adhere to the formulator's recommendation for use.

Accuracy is required in mixing epoxy components; and although a tolerance of ± 5 percent is acceptable, a ± 2 percent tolerance is highly desirable. The mixing of epoxy mortar or grout requires that the epoxy binder thoroughly wet each and every one of the aggregate particles. Although it is difficult, hand mixing of small quantities can be accomplished by using a trowel. The preferred way to mix epoxy is to use some mechanical means. Epoxy concretes are mixed in a manner similar to that used for epoxy mortars except that, in stiff mixes, the large aggregate should be added to the mixed binder first and be followed by the finer aggregate or sand to help prevent the tendency of the mix to "ball." The finer aggregates should be added slowly. Care should be taken to avoid segregation to obtain a uniform epoxy concrete. Epoxy grouts or mortars usually consist of 4 to 7 parts by weight to 1 part of binder.

The ease and effectiveness of epoxy application are greatly influenced by the temperatures of surfaces on which the epoxy compound is applied. Commonly available epoxy compounds in use today react most favorably when temperatures are in the range of 16 to 38°C (60 to 100°F). If the temperature is below 16°C but above 5°C, the application of an epoxy compound can be accomplished provided the compound is formulated for use within the temperature range and an increased hardening period is not objectionable. At temperatures above 32°C (90°F) difficulties in application may be experienced because of acceleration of the reaction and hardening rates.

4.10.2 Joint requirements

Concrete is normally subject to changes in length, plane, or volume caused by changes in its moisture content or temperature, reaction with atmospheric carbon dioxide, or by the imposition or maintenance of loads. The effect may be permanent contractions due to, for example, initial drying shrinkage, carbonation, and irreversible creep. Other effects are cyclical and depend on service conditions such as environmental differences in humidity and temperature or the application of loads, and they may result in either expansions or contractions. In addition, abnormal volume changes, usually permanent expansions, may occur in the concrete because of sulfate attack, reactions between alkali from the cement and certain aggregates, and other causes.

The results of these changes are movements, both permanent and transient, of the extremities of concrete structural units. If for any reason contraction movements are excessively restrained, then cracking within the unit may occur. The restraint of expansion movements may result in distortion and cracking within the unit or crushing of the unit's ends and the transmission of unanticipated forces to abutting units. In most concrete structures these effects are objectionable from a structural or an appearance viewpoint. One means of minimizing these effects is to provide joints at which movement can be accommodated without loss of integrity of the structure.

There may be other reasons for providing joints in concrete structures. In many buildings the concrete serves to support or frame curtain walls, cladding, doors, windows, partitions, and mechanical and other services. To prevent development of distress in these structural units, it is often necessary for the units to move to a limited extent independently of overall expansions, contractions, and deflections occurring in the concrete. Joints may also be required to facilitate construction without serving any structural purpose.

The introduction of joints creates openings which must usually be sealed in order to prevent passage of gases, liquids, or other unwanted substances into or through the openings. In buildings, it is important to prevent intrusion of wind and rain. In tanks, most canals, pipes, and dams, joints must be sealed to prevent loss of the contents. Moreover, in most structures exposed to the weather the concrete itself must be protected against the possibility of damage from freezing and thawing, wetting and drying, leaching or erosion caused by any concentrated or excessive influx of water at joints. Foreign solid matter, including ice, must be prevented from collecting in open joints; otherwise, the joints cannot be closed freely later. Should that happen, high stresses may be generated and damage to the concrete may occur.

In industrial floors the concrete at the edges of joints often needs the protection of a filler or sealant (possibly between armored faces) capable of preventing damage from the impact of concentrated loads such as steel-wheeled traffic. The specific function of sealants is to prevent the intrusion of liquids (sometimes under pressure), solids, or gases and to protect the concrete against damage. In certain applications, secondary functions are to improve thermal and acoustical insulation, damp down vibrations, and prevent unwanted matter collecting in crevices. Sealants must often perform their prime functions while subject to repeated contractions and expansions as the joint opens and closes and while exposed to heat, cold, moisture, sunlight, and, sometimes, aggressive chemicals.

In most concrete structures, all concrete-to-concrete joints (contraction, expansion, and construction) and the peripheries of openings left for other purposes require sealing. One exception is contraction joints (and cracks) that have very narrow openings, e.g., those in certain short plain slab or reinforced pavement designs. Other exceptions are certain construction joints, e.g., monolithic joints, that are not subject to fluid pressure and joints between precast units, used either internally or externally, with intentional open-draining joints.

4.10.3 Types of joints

Several types of joints are used according to the type and size of structure and the adjoining materials.

Contraction (control) joints. Contraction joints are purposely made planes of weakness designed to regulate cracking that might otherwise occur because of the unavoidable and often unpredictable contraction of concrete structural units. They are appropriate only when the net result of the contraction and any subsequent expansion during service is such that the units abutting are always shorter than when the concrete was placed. They are frequently used to divide large, relatively thin structural units, e.g., pavements, floors, canal linings, and retaining and other walls, into smaller panels. Contraction joints in structures are often called control joints because they are intended to control crack location.

Contraction joints may form a complete break dividing the original concrete unit into two or more units. When the joint is not wide, some continuity can be maintained by aggregate interlock. When greater continuity is required without restricting freedom to open and close, dowels and, in certain cases, steps or keyways may be used. When restriction of the joint opening is required for structural stability, appropriate tie bars or continuation of the reinforcing steel across the joint may be provided. The necessary plane of weakness can be formed by either partly or fully reducing the concrete cross section. This may be done by installing thin metallic, plastic, or wooden strips when the concrete is placed or by sawing the concrete soon after it has hardened.

Expansion (isolation) joints. Expansion joints are designed to prevent the crushing and distortion (including displacement, buckling, and warping) of the abutting concrete structural units that might otherwise occur because of the transmission of compressive forces that may be developed by expansion, applied loads, or differential movements arising from the configuration of the structure or its settlement. They are frequently used to isolate walls from floors or roofs, columns from floors or cladding, pavement slabs and decks from bridge abutments or piers, and in other locations where restraint or transmission of secondary forces is not desired. Many designers consider it good practice to place such joints where walls change direction, as in L-, T-, Y-, and U-shaped structures, and where different cross sections develop. Expansion joints in structures are often called isolation joints because they are intended to isolate structural units that behave in different ways.

Expansion joints are made, when the concrete is placed, by providing a space between abutting structural units for the full cross section. The space is formed by the use of filler strips of the required thickness or by leaving a gap when precast units are positioned. Provision for continuity or for restricting undesired lateral displacement can be made by incorporating dowels, steps, or keyways.

Construction joints. Construction joints are made at the surfaces created before and after interruptions in the placement of concrete or through the positioning of precast units. Locations are usually predetermined by agreement between the engineer and the contractor so as to limit the work that must be done at one time with least impairment of the finished structure. Joint location may also be necessitated by unforeseen interruptions in concreting operations. Depending on the structural design, they may be required to function later as expansion or contraction joints having the features already described, or they may be required to be monolithic; i.e., the second placement must be soundly bonded to the first so as to maintain complete structural integrity. Construction joints may run horizontally or vertically depending on the placing sequence prescribed by the design of the structure.

Combined and special-purpose joints. Construction joints at which the concrete in the second placement is intentionally separated from that in the preceding placement by a bond-breaking membrane, but without space to accommodate expansion of the abutting units, also function as contraction joints. Similarly, construction joints in which a filler is placed, or a gap is otherwise formed by bulkheading or the positioning of precast units, function as expansion joints. Conversely, expansion joints are often convenient for forming nonmonolithic construction joints. Expansion joints automatically function as contraction joints, though the converse is true only to an amount limited to the size of gap created by initial shrinkage.

Hinge joints are joints that permit hinge action (rotation) but at which the separation of the abutting units is limited by tie bars or the continuation of reinforcing steel across joints. The term has wide usage in, but is not restricted to, pavements, where longitudinal joints function in this manner to overcome warping effects while resisting deflections due to wheel loads or settlement of the subgrade. Hinge joints in structures are often referred to as articulated joints.

Sliding joints may be required where one unit of a structure must move in a plane at right angles to the plane of another unit; e.g., in certain reservoirs where the walls are permitted to move independently of the floor or roof slab. These joints are usually made with a bond-breaking material such as a bituminous compound, paper, or felt that also facilitates sliding.

Cracks. Although joints are placed in concrete so that cracks do not occur elsewhere, it is seemingly impossible to prevent occasional cracks between joints for a variety of reasons. As far as the problem of sealing is concerned, cracks may be regarded as construction joints of irregular line and form.

4.10.4 Joint configurations

From the standpoint of the functioning of the sealant, two basic configurations of the schematic joint details for various types of concrete structures occur. They are known as butt joints and lap joints.

In butt joints, the structural units being joined abut each other and any movement is largely at right angles to the plane of the joint. In lap joints, the units being joined override each other and any relative movement is one of sliding. Butt joints—and they include most stepped joints—are by far the most common. Lap joints may occur in certain sliding joints, between precast units or panels in curtain walls and at their junctions, and at the junctions of cladding and glazing with their concrete or other framing. The difference in the mode of the relative movement between structural units at butt and lap joints in part controls the functioning of the sealant. In many of the applications of concern, pure lap joints do not occur and the functioning of the lap joint is in practice a combination of butt and lap joint action.

In regard to the sealant, two systems should be recognized. First, there are open surface joints, as in pavements and buildings in which the joint sealant is exposed to outside conditions on at least one face. Second, there are joints, such as those in containers, dams, and pipelines, in which the primary line of defense against the passage of water is a sealant such as a waterstop or gasket buried deeper in the joint. The functioning and type of sealant material that is suitable and the method of installation are affected by these considerations.

In conclusion, two terms that are in wide, though imprecise use should be mentioned. Irrespective of their type of configuration, joints are often called working joints when significant movement occurs and nonworking joints when movement does not occur or is negligible.

4.10.5 Use of sealant materials

Sealant materials come in a variety of shapes, properties, and textures which allow a selection of joint types but require that the right sealant for the joint selected be used.

Performance in open joints. For satisfactory performance a sealant in open surface joints must:

- Be an impermeable material
- Deform to accommodate the movement and rate of movement occurring at the joint
- Sufficiently recover original properties and shape after cyclic deformations
- Remain in contact with the joint faces (This means that for all sealants, except preformed sealants that exert a force against the joint face, the sealant must bond to the joint face and not fail in adhesion or peel at corners or other local areas of stress concentration.)
- Not internally rupture (i.e., fail in cohesion)
- Resist flow due to gravity (or fluid pressure) or unacceptable softening at higher service temperatures
- Not harden or become unacceptably brittle at lower service temperatures
- Not be adversely affected by aging, weathering, or other service factors for a reasonable service life under the range of temperatures and other environmental conditions that occur

Performance buried in joints. Sealants buried in joints, such as waterstops and gaskets, generally require similar properties. The method of installation may, however, require the sealant to be in a different form and, because replacement is usually impossible, exceptional durability is required. In addition, depending on the specific service conditions, the sealant may be required to resist one or more of the following: intrusion of foreign material, wear, indentation, pickup, and attack by chemicals. Further requirements may be that the sealant be a specific color, resist change of color, or be nonstaining. Finally, it must be relatively easy to handle and install and be free of substances harmful to the user and to concrete or other material that may abut.

Materials available. No one material has the perfect properties necessary to fully meet each and every one of the requirements for each and every application. Therefore, it is a matter of selecting, from among a large range of materials, a particular material that has more of the right properties at the right price to do the job. Table 4.16 lists commonly used joint sealant materials.

For many years, oil-based mastics or bituminous compounds and metallic materials were the only sealants available. In many applications these traditional materials do not perform well, and in recent years there has been an active development of elastomeric sealants. These sealants are largely elastic rather than plastic in behavior, and they are flexible rather than rigid at normal service temperatures. Elastomeric materials are available as field-molded and preformed sealants. Though initially more expensive, they may be cheaper in the long run because they usually have a longer service life. Furthermore, they can seal joints where considerable movement occurs that could not possibly be sealed by the traditional materials.

Field-molded sealants. The following types of materials, as listed in Table 4.16, are currently used as field-molded sealants.

Mastics. Mastics are composed of viscous liquid rendered immobile by the addition of fibers and fillers. They do not usually harden, set, or cure after application; instead, they form a skin on the surface exposed to the atmosphere. The vehicle in mastics may be drying or nondrying oils (including oleoresinous compounds), polybutenes, polyisobutylenes, low-melting-point asphalts, or combinations of these materials. With any of them, a wide variety of fillers, including asbestos fiber, fibrous talc, or finely divided calcareous or siliceous materials, is used. The functional extension-compression range for these materials is approximately ± 3 percent.

Thermoplastics, hot-applied. The hot-applied thermoplastics are materials which soften on heating and harden on cooling, usually without chemical change. They are generally black and include asphalts, rubber asphalts, pitches, coal tars, and rubber tars. They are usable over an extension-compression range of ± 5 percent. This limit is directly influenced by service temperatures and aging characteristics of specific materials. Though initially cheaper than some of the other sealants, their effective life is, in practice, shorter. They tend to lose elasticity and plasticity with age, to accept rather than reject foreign materials, and to extrude from joints that close tightly or that have been overfilled. Overheating during the melting process adversely affects the properties of the compounds that contain rubber. Those with an asphaltic base are softened by hydrocarbons such as oil, gasoline, or jet fuel spillage. Tar-based materials are fuel- and oil-resistant; they are preferred for service stations, refueling and vehicle parking areas, airfield aprons, and holding pads.

Use of sealants of this class is restricted to horizontal joints, since they would run out of vertical joints when installed hot or subsequently in warm weather. The sealants have been widely used in pavement joints, but they tend to be superseded by chemically curing thermosetting field-molded sealants or compression seals. They are also used in building roof decks and containers.

TABLE 4.16 Materials Used for Sealants in Joints Open on at Least One Surface

Group	Field-molded					Preformed
	Thermoplastics			Thermosetting		
Type	Mastic	Hot-applied	Cold-applied	Chemically curing	Solvent release	Compression seal
Composition	(A) Drying oils (B) Nondrying oils (C) Low melt point asphalt (D) Polybutenes (E) Polyisobutylenes or combination of D & E. All used with fillers such as asbestos fiber or siliceous materials, all contain 100% solids, except D & E which may contain solvent.	(F) Asphalts (G) Rubber asphalts (H) Pitches (I) Coal tars (J) Rubber coal tars All contain 100% solids (W) Hot applied PVC coal tar	(K) Rubber asphalts (L) Vinyls (M) Acrylics (K) Contains 70 to 80% solids (L) (M) Contain 75 to 90% solids All contain solvent (K) may be an emulsion (60 to 70% solids).	(N) Polysulfide (O) Polysulfide coal tar (P) Polyurethane (Q) Polyurethane coal tar (R) Silicones (S) Epoxy (N),(R) contain 95 to 100% solids (O),(Q),(S) contain 90 to 100% solids (P) contains 75 to 100% solids (N),(P),(R) may be either one- or two component system (O),(Q),(S) two-component system	(T) Neoprene (U) Butadienestyrene (V) Chlorosulfonated polyethylene (T) (V) contain 80 to 90% solids (U) contains 85 to 90% solids (R) Silicones	(3) Neoprene rubber
Colors	(A) (B) Varied (C) Black only (D) (E) Limited	Black only	(K) Black only (L) (M) Varied	(N) (R) (S) Varied (O) (P) Limited (Q) Black only	(T) Limited (V) Varied	Black, exposed surfaces may be treated to give varied colors

TABLE 4.16 Materials Used for Sealants in Joints Open on at Least One Surface (Continued)

| Group | | Field-molded | | | | Preformed |
| | | | Thermoplastics | | Thermosetting | | |
Type	Mastic	Hot-applied	Cold-applied	Chemically curing	Solvent release	Compression seal
Setting or curing	Noncuring, remains viscous, A and B form skin on exposed surface.	Noncuring, sets upon cooling. Softens on warming, hardens on cooling. (W) Resilient	Noncuring, sets on release of solvent or evaporation of water. (M) remains soft except for surface skin.	Two-component-system catalyst. One component; moisture pickup from the air	Release of solvent	—
Aging and weathering resistance	Low	Moderate (W) High resistance to weather	Moderate	High	High	High
Increase in hardness in relation to (1) age	High	High to moderate (W) No hardness	High	(S) High (N) (O) (P) (Q) (R) Moderate	High	Low
or (2) low temp.	High	High to moderate (W) No hardness	High	(S) (N) (O) (P) (Q) (R) Low	High	Low
Recovery	Low	Moderate (W) High	Low	(N) (O) Moderate (P) (Q) (R) High (S) Low	Low	High
Resistance to wear	Low	Moderate	Moderate	(P) (Q) (R) (S) High (N) (O) Moderate	Moderate	High
Resistance to indentation and intrusion of solids	Low	Low at high temperatures (W) High	Low at high temperatures	High	Low	High
Shrinkage after installation	High	Varies (W) None	High	Low	High	None

Resistance to chemicals	High except to solvents and fuels	(F) (G) High except to solvents and fuels (H) (I) (J) High and fuel resistant (W) High	(K) High except to solvents and fuels (L) (M) High except to alkalis and oxidizing acids	(N) (P) Low to solvents, fuels, and oxidizing acids (O) (Q) Low to solvents but moderate fuel resistance (R) Low to alkalis (S) High	Low to solvents, fuels, and oxidizing acids	High
Modulus at 100% elongation	Not applicable	Low	Low	(R) (O) (P) (Q) Low (R) High and low (S) Not applicable	Moderate	
Allowable extension and compression	± 3%	± 5% (W) ± 25% extension	± 7%	± 25% except (S) less	± 7%	Must be compressed at all times to 45 to 85% of its original width
Other properties	(A) (B) (D) (E) Nonstaining (D) (E) Pick up dirt; use in concealed location only.	Due to softening in hot weather usable only in horizontal joints (W) No flow at elevated temperatures	(K) Usable in inclined joints	(N) (P) (R) (S) Nonstaining	(U) (V) Nonstaining (V) Good vapor and dust sealer	
Unit first cost	(A) (B) (C) very low (D) (E) Low	(F) (G) (H) (I) (J) Very low (W) Medium	(K) Very low (L) Low (M) High	(O) (Q) High (N) (P) (R) (S) Very high	(T) (U) (V) Low	(3) High

Thermoplastics, cold-applied (solvent or emulsion type). The cold-applied thermoplastics set either by the release of solvents or by breaking of the emulsion on exposure to air. They are sometimes heated to a temperature not exceeding 49°C (120°F) to facilitate application, but they are usually handled at ambient temperature. Release of solvent or water can cause shrinkage and increased hardness with resulting reductions in the permissible joint movement and serviceability. Products in this category include acrylic, vinyl, and modified butyl types which are available in a variety of colors. Their maximum extension-compression range is ± 7 percent, but heat softening andcold hardening may reduce that range. The materials are restricted in use to joints with small movements. Acrylics and vinyls are used in buildings, mainly for caulking and glazing. Rubber asphalts are used in canal linings, tanks, and fillers for cracks.

Thermosetting, chemically curing sealants. Sealants in the chemically curing thermosetting class are either one- or two-component systems which cure, by chemical reaction, to a solid state from the liquid form in which they are applied. They include polysulfide-, silicone-, urethane-, and epoxy-based materials. The properties that make them suitable as sealants for a wide range of uses are their resistance to weathering ozone, flexibility and resilience at both high and low temperatures, and inertness to a wide range of chemicals, including, for some, solvents and fuels. In addition, the abrasion and indentation resistance of urethane sealants is above average. Thermosetting, chemically curing sealants have expansion-compression ranges up to ± 25 percent, depending on the sealant used, at temperatures from −40° to +82°C (− 40 to 180°F). Silicone sealants remain flexible over an even wider temperature range. They have a wide range of uses in buildings and containers for both vertical and horizontal joints or in pavements. Though initially more expensive, thermosetting, chemically curing sealants can stand greater movements than other field-molded sealants, and they generally have a much greater service life.

Thermosetting solvent-release sealants. The thermosetting sealants of another class cure by the release of solvent. Chlorosulfonated polyethylene and certain butyl and neoprene materials are included in this class, and their performance characteristics generally resemble those of thermoplastic solvent-release materials. They are, however, less sensitive to variations in temperature once they have set up on exposure to the atmosphere. Their maximum extension-compression range does not, however, exceed ± 7 percent. They are mainly used as sealants for caulking and joints in buildings which have small movements both horizontally and vertically. The cost is somewhat less than that of other elastomeric sealants, and the service life is likely to be satisfactory, though for some recent products this has not yet been established by experience.

Rigid materials. When special properties are required and movement is negligible, certain rigid materials can be used as field-molded sealants for joints and cracks. They include lead (cool or molten), sulfur, and modified epoxy resins.

Preformed sealants. Preformed sealants (listed in Table 4.17) can be divided into two classes: rigid and flexible. Most rigid preformed sealants are metallic; examples are metal waterstops and flashings. Flexible sealants are usually made from natural or synthetic rubbers, polyvinyl chloride (PVC), and like materials and are used for waterstops, gaskets, and miscellaneous sealing purposes. Preformed equivalents of certain materials, e.g., rubber asphalts, usually categorized as field-molded, are available as a convenience to handling and installation.

Compression seals should be included with the flexible group of preformed sealants. However, because their functional principle is different and because the compartmentalized neoprene type can be used in almost all joint sealant applications as an alternative to field-molded sealants, compression seals are treated separately. Preformed tension-compression seals also are discussed separately.

4.10.6 Rigid waterstops

Rigid waterstops are made of steel, copper, and occasionally lead. Steel waterstops are primarily used in dams and other heavy-construction projects. Because ordinary steel may require additional protection against corrosion in dam construction, stainless steels are used. They must be low in carbon and be stabilized with columbium or titanium to facilitate welding and retain corrosion resistance after welding. Although annealing is required for improved flexibility, the stiffness of steel waterstops may still lead to cracking in the adjacent concrete.

Copper waterstops are used in dams and general construction. They are highly resistant to corrosion, but they must be handled with care to avoid damage. For that reason, and because of cost, flexible waterstops often are used instead. Copper also is used for flashings.

Use of lead as waterstops and flashings or as protection in industrial floor joints is now very limited. Bronze strips find wide application in dividing, rather than sealing, terrazzo and other floor toppings into smaller panels.

TABLE 4.17   Preformed Materials for Waterstops, Gaskets, and Sealing Purposes

Composition and type	Properties significant to application	Available in	Uses
(1) Butyl—Conventional rubber-cured	High resistance to water, vapor, and weathering. Low permanent set and modulus of elasticity formulations possible, giving high cohesion and recovery. Tough. Color black; can be painted.	Beads, rods, tubes, flat sheets, tapes, and purpose-made shapes.	Waterstops, combined crack inducer and seal, pressure-sensitive dust and water sealing tapes for glazing and curtain walls.
(2) Butyl—raw; polymer modified with resins and plasticizers	High resistance to water, vapor, and weathering. Good adhesion to metals, glass, plastics. Moldable into place but resists displacement, tough and cohesive. Color black; can be painted.	Beads, tapes, gaskets, grommets.	Glazing seals, lap seams in metal cladding, curtain-wall panels.
(3) Neoprene—conventional rubber cured	High resistance to oil, water, vapor, and weathering. Low permanent set. Color basically black, but other surface colors can be incorporated.	Beads, rods, tubes, flat sheets, tapes, purpose-made shapes. Either solid or open or closed-cell sponges.	Waterstops, glazing seals, insulation and isolation of service lines. Tension-compression seals. Compression seals, gaskets.
(4) Polyvinylchloride (PVC), thermoplastic, extrusions, or moldings	High water and vapor but only moderate chemical resistance. Low permanent set and modulus of elasticity formulations possible, giving high cohesion and recovery. Tough. Can be softened by heating for splicing. Color pigmented black, brown, green, etc.	Beads, rods, tubes, flat sheets, tapes, gaskets, purpose-made shapes.	Waterstops, gaskets, combined crack inducer and seal.

TABLE 4.17   Preformed Materials for Waterstops, Gaskets, and Sealing Purposes (Continued)

Composition and type	Properties significant to application	Available in	Uses
(5) Polyisobutylene, noncuring	High water, vapor resistance. High flexibility at low temperature. Flows under pressure, surface; pressure-sensitive, high adhesion. Sometimes used with butyl compounds to control degree of cure. Color black, gray, white.	Beads, tapes, grommets, gaskets.	Gaskets, glazing seals, curtain-wall panels, acoustical partitions.
(6a) SBR (styrene butadiene rubber) (6b) Nitrile butadiene rubber (NBR) Polyisoprene–polydiene—conventional rubber cure	High water resistance; NBR has high oil resistance.	Beads, rods, flat sheets, tapes, gaskets, grommets, purpose-made shapes. Either solid or cellular sponges.	Waterstops, gaskets for pipes, insulation and isolation of service lines.
(7) Polyurethane foam impregnated with polybutylene	Low recovery at low temperature, can be installed in damp joints. Color black, gray.	Rods, flat sheets (strips), open-cell sponges.	Gaskets, compression seals.
(8) Natural rubber cured (vulcanized)	High water resistance but deteriorates when exposed to air and sun. Low resistance to oils and solvents. Now largely superseded by synthetic materials. Color black.	Purpose-made shapes.	Waterstops, gasket for pipes.

TABLE 4.17   Preformed Materials for Waterstops, Gaskets, and Sealing Purposes (*Continued*)

Composition and type	Properties significant to application	Available in	Uses
(9) Metals (*a*) Copper (*b*) Steel (stainless) (*c*) Lead (*d*) Bronze	For waterstops: (*a*) Ductile and flexible, but work-hardens under flexing and fracturess. (*b*) Rigid; must be V- or U-corrugated to accommodate any movement and anchored. (*c*) Deforms readily but inelastic to deformation under movement.	Flat and preshaped strips. Lead also molten or yarn.	(a) (b) Waterstops. (c) Protection for joint edges in floor. (d) Panel dividers in floor toppings.
(10) Rubber asphalts	Natural rubber 8, butyl 1, or neoprene 3 digested in asphalt. High viscosity, some elasticity. Moldable into place.	Gaskets for beads, rods, flat sheets (strips). (IIG IIIK)	As alternative to hot- or cold-applied rubber asphalts (IIG IIK). Gasket for pipes.

SOURCE: American Concrete Institute, 1979, part 3.

4.10.7   Flexible Waterstops

The types of materials that are suitable and are in use as flexible waterstops are listed in Table 4.17. Butyl, neoprene, and natural rubbers have good extensibility and resistance to water or chemicals and may be formulated to give good recovery and fatigue resistance. PVC compounds are, however, probably now most widely used. Although PVC is not quite as elastic as the rubbers, recovers more slowly from deformation, and is susceptible to degradation by oils, grades of PVC with sufficient flexibility (especially important at low temperatures) can be formulated. PVC has the great advantage of being thermoplastic; hence, it can easily be spliced on the job or special configurations can be made for joint intersections. Flexible waterstops are widely used as the primary sealing systems in water-containing projects such as dams, tanks, monolithic pipelines, floodwalls, and swimming pools to keep water in and in buildings below grade or in earth-retaining walls to keep water out.

4.10.8 Gaskets

Gaskets and tapes are widely used sealants between glazed surfaces, around windows and other openings in buildings, and at joints between metal or precast concrete panels in curtain walls. Gaskets are also extensively used at joints between precast pipes and where mechanical joints are needed in service lines. The sealing action is obtained either because the sealant is compressed between the joint faces (gaskets) or because the surface of the sealant, as in the case of polyisobutylene, is pressure-sensitive and thus adheres.

4.10.9 Compression seals

Compression seals are preformed, compartmentalized, or cellular elastomeric devices which, when in compression between the joint faces, function as sealants.

Compartmentalized seals. Neoprene extruded to the required configuration is currently used for most compression seals. The neoprene formulation used must have special properties for this application. To seal effectively, sufficient contact pressure must be maintained at the joint face. That requires the seal to be in some degree of compression, and therefore good resistance to compression set is required (i.e., the material must recover sufficiently when released). In addition, the neoprene must be crystallization-resistant at low temperatures (the resultant stiffening may make the seal temporarily ineffective, though recovery will occur on warming). If during the manufacturing process the neoprene is not fully cured, the interior webs may adhere during service (often permanently) when the seal is compressed.

To facilitate installation of compression seals, liquid neoprene-based lubricants are used. For machine installation, additives to make the lubricant thixotropic have been found necessary. Special lubricant adhesives, which both prime and bond, have been formulated for use when improved seal–to–joint face contact is required.

Neoprene compression seals are effective joint sealants over a wide range of temperatures in most applications. Seals may be used individually or as components for modular systems.

Modular systems. In modular systems designed to accommodate larger movements, standard compartmentalized compression seals or rubber tubes are placed between vertical steel I-sections to form modules each of which can accommodate about 38.1-mm (1.5-in) movement. The complete unit of modules in a series to take the total anticipated movement is supplied prefabricated and ready for installation at the appropriate precompression to suit conditions. A certain amount of out-of-plane movement arising from skew or other causes

can be accommodated, and field modifications to allow for unanticipated irreversible opening or closing of the joint can be made by adding or removing seals and separation plates as required.

Individual seals must remain in at least 15 percent compression at
the widest opening. The allowable movement is approximately 40 percent of the uncompressed seal width.

4.10.10   Tension-compression seals

One device of the tension-compression type currently in use consists of
neoprene expansive elements combined with encased steel bearing
plates and anchorage angles to form a single unit that can be extended
or compressed without buckling. Such a unit can support traffic or
other loads on its upper surface. Individual units are available in
varying lengths up to 1.83 m (6 ft) and may be butted side by side
along the length of the joint opening. The device is bolted directly to
the concrete surface at each side of the joint, and this mechanical anchoring permits the device to function in tension or compression in response to the movement of the joint.

To date, tension-compression devices have been used exclusively on
bridge decks, and special sections have been developed to fit curb contours. They may also have application on dam faces or other locations
where sealing against considerable pressure and movement is required.

The long-term performance of tension-compression seals remains to
be evaluated. Observations of performance indicate that careful installation is important.

4.11   Repair of Concrete

To objectively evaluate the damage to a structure, it is necessary to
determine what caused the damage in the first place. The damage
may be the result of poor design, faulty workmanship, mechanical
abrasive action, cavitation or erosion from hydraulic action, leaching,
chemical attack, chemical reaction inherent in the concrete mixture,
exposure to deicing agents, corrosion of embedded metal, or other
lengthy exposure to an unfavorable environment. Figures 4.4 and 4.5
show damage from various sources.

Whatever may have been the cause, it is essential to establish the
extent of the damage and determine if the major part of the structure
is of quality suitable to the building of a sound repair. Based on that
information, the type and the extent of the repair are chosen. This is
the most difficult step, one which requires a thorough knowledge of
the subject and mature judgment by the engineer. If the damage is the
result of moderate exposure of what was an inferior concrete in the

Figure 4.4 Piles damaged by corrosion of steel reinforcement. *(Courtesy of Los Angeles Harbor Department)*

first place, then replacement by good-quality concrete should ensure lasting results. On the other hand, if good-quality concrete was destroyed, the problem becomes more complex. In that case, a very superior quality of concrete is required or the exposure conditions must be altered.

The repair of spalls from reinforcing bar corrosion requires a more

Figure 4.5 Concrete pile damaged by overloading. *(Courtesy of Los Angeles Harbor Department)*

detailed study. Simply replacing the deteriorated concrete and restoring the original cover over the steel will not solve the problem. Also, if the structure is salt-contaminated, the electrolytic conditions will be changed by the application of new concrete, and the consequences of the changed conditions must be considered before any repairs are undertaken.

Basic requirements for achieving a durable repair are:

- The repair material must be thoroughly bonded to the sound concrete of the cavity.

- The shrinkage of the patch should be small enough not to jeopardize the bond.

- The patch and its substrate should be free of cracks.

- The response of the patch and the old concrete to changes in temperature, moisture, and load should be similar enough to avoid gross differences in movement.

- The patch should be low enough in permeability that moisture will not migrate through it to the old concrete underneath.

- The patch should be resistant to weathering and be durable in the environment in which it is exposed.

4.11.1 Replacement concrete

The concrete replacement method consists of replacing defective concrete with machine-mixed concrete of suitable proportions and consistency so the latter will become integral with the base concrete. Concrete replacement is the desired method if there is honeycomb in new construction or deterioration of old concrete which goes entirely through the wall or beyond the reinforcement or if the quantity is large. For new work, the repairs should be made immediately after stripping the forms (Tuthill, 1960). Considerable concrete removal is always required for this type of repair. Excavation of affected areas should continue until there is no question that sound concrete has been reached. Additional chipping may be necessary to accommodate the repair method and shape the cavity properly. Although opinions differ on the value of wetting the cavity before placing plastic mortar, most authorities believe it is advisable to keep the faces of the cavity wet for several hours before placing operations are begun. No standing water should be present, however, at the time of placement. Concrete for the repair should generally be similar to the old concrete in maximum size of aggregate and water-cement ratio.

4.11.2 Dry Pack

The dry-pack method consists of ramming a very stiff mix into place in thin layers. It is suitable for filling form tie rod holes and narrow slots and for repairing any cavity which has a relatively high ratio of depth to area. Practically no shrinkage of this mix will occur, and the mix develops a strength equaling or exceeding that of the parent concrete.

4.11.3 Preplaced-aggregate concrete

Preplaced-aggregate concrete can be used advantageously for certain types of repairs. It bonds well to concrete and has low drying shrinkage. It is also well adapted to underwater repairs.

4.11.4 Shotcrete

Shotcrete, or gunite, has excellent bond with new or old concrete and is frequently the most satisfactory and economical method of making shallow repairs. It is particularly adapted to vertical or overhead surfaces, where it is capable of supporting itself (without a form) and not sagging or sloughing. Shotcrete repairs generally perform satisfactorily where recommended procedures are followed.

4.11.5 Repair of scaled areas and spalls in slabs

Scaling of concrete pavement surfaces is not unusual when the surfaces are subject to deicing salts, particularly if the concrete is inadequately air-entrained. Such areas may be satisfactorily repaired by a thin concrete overlay provided the surface of the old concrete is sound, durable, and clean. A minimum overlay thickness of about 38 mm (1.5 in) is needed for good performance. The temperature of the underlying slab should be as close as possible to that of the new concrete.

Spalls may occur adjacent to pavement joints or cracks. Spalls usually are several inches in depth, and even deeper excavation may be required to remove all concrete which has undergone some degree of deterioration. Numerous quick-setting patching materials, some of which are proprietary, are available. Information on the field performance of these materials is given by the Federal Highway Administration (1975).

4.11.6 Bonding agents

Bonding agents are used to establish unity between fresh concrete or mortar and the parent concrete. An enriched sand-cement mortar or neat cement paste has generally been used in the past. Epoxy resin is now used frequently as a bonding agent, with the expectation of durable results. This material develops a bond having greater tensile, compressive, and shear strength than concrete has. It is waterproof and highly resistant to chemical and solvent action. It is possible to have acceptable results when the concrete is brought to a featheredge; however, better results are obtained if a 25-mm (1-in) minimum thickness is maintained.

Other types of bonding agents have recently become available. Certain latexes, supplied as emulsions or dispersions, improve the bond and have good crack resistance. Polyvinyl acetates, styrenebutadiene, and acrylic are among those used. These materials, particularly the polyvinyl acetates, must be properly compounded if the dried film is to be resistant to moisture. They may be used either as a bonding layer or added to the concrete or mortar mix.

4.11.7 Appearance

Unless proper attention is given to all the factors influencing their appearance, concrete repairs are likely to be unsightly. When appearance is important, particular care should be taken to ensure that the texture and color of the repair will match the surrounding concrete. A proper blend of white cement with the job cement is important to come close to matching the color of the original concrete. A patch on a formed concrete surface should never be finished with a steel trowel, since the troweling produces a dark color which is impossible to remove.

4.11.8 Curing

Except when epoxy mortar or epoxy concrete is used, all patches must be properly cured to assure proper hydration of the cement and durable concrete or mortar.

4.11.9 Treatment of cracks

The decision whether a crack should be repaired to restore structural integrity or merely sealed is dependent on the nature of the structure and the cause, location, and extent of the crack. If the stresses which caused the crack have been relieved by its occurrence, the structural integrity can be restored with some expectation of permanency. However, in the case of a working crack (such as a crack caused by foundation movements or a crack which opens and closes from temperature changes), the only satisfactory solution is to seal the crack with a flexible or extensible material.

Thorough cleaning of the crack is essential before any treatment takes place. All loose concrete, oil-based joint sealant, and other foreign material must be removed. The method of cleaning depends on the size of the crack and the nature of the contaminants. It may include any combination of the following: compressed air, wire brushing, sandblasting, routing, or the use of picks or similar tools.

Structural integrity across a crack has been successfully restored by using pressure and vacuum injection of low-viscosity epoxies and other monomers which polymerize in situ and rebond the parent concrete. Crack sealing without restoration of structural integrity requires the use of materials and techniques similar to those used in sealing joints.

Epoxy resin has become a common and satisfactory material for sealing cracks. The U.S. Navy Civil Engineering Laboratory has developed the following information on such resins. Epoxy, when mixed with a curing agent, becomes epoxy resin, which is a thermosetting plastic that rapidly develops adhesive strength. This synthetic organic compound is stable chemically and physically; it is durable and crack-resistant, and it undergoes little reduction in volume (2 to 3 percent) as the result of curing. Adhesives of this type become irreversibly set as the result of exothermic chemical changes initiated by the chemical changes initiated by the chemical curing agent. Epoxy resins can, by means of various hardeners, fillers, flexibilizers, and plasticizers, be formulated to have specific values of mechanical and physical characteristics.

Epoxy resin pressure-injected into the cracks of concrete can restore the structure to its original strength. Cracks as narrow as 0.13 mm (5 mils) and as wide as 6.35 mm (250 mils) can be repaired by injecting epoxy resin. The type of resin needed depends on the width of the crack, on whether the crack is working or stationary, and on the particular method chosen for applying the resin. Repair of a working crack requires a formulation that will set up rather rapidly so the bond is not broken before the resin has developed sufficient strength. Narrow cracks require a low-viscosity system to ensure complete penetration of the crack. However, a more viscous system with high-pressure injection methods can be used. The following is a recommended method for repairing cracked concrete by injecting epoxy resin:

- Clean the crack with compressed air; remove any salt, oil, or grease deposits from the adjacent concrete surface, and, if possible, from the crack itself.

- Seal the exposed crack along its entire length. The sealant, which may be an epoxy resin, must be able to withstand internal pressures of at least 862 kPa (125 lb/in^2). If the sealant is applied to either vertical or overhead cracks, it should be stiff enough that it will not slough off or sag before hardening. It should be able to bridge cracks as wide as 6.35 mm (0.25 in). Alternatively, a special thermoplastic sealant tape can be applied directly to the concrete surface and will not deface the concrete when removed later.

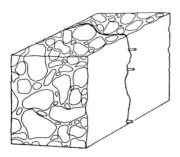

Figure 4.6 Concrete structure showing nipples through which epoxy resin is injected.

- If the sealant is an epoxy resin, drill holes about 6.35 mm (0.25 in) in diameter and about 25.4 mm (1 in) deep along the crack, with spacing of the holes generally not less than the thickness of the concrete member being repaired.

- Insert metallic nipples, secured in place with a puttylike epoxy resin sealant, in the holes to serve as ports for the epoxy resin.

- Inject the epoxy resin adhesive under pressure through the first nipple (lowest in the case of a vertical or diagonal crack) until the level of adhesive reaches the next nipple (Fig. 4.6).

- Use an inert gas to maintain a pressure of 620 kPa (90 lb/in^2) for about 1 min to force the adhesive into any interior microcracks adjoining the crack under repair.

- Release the pressure and then pump more adhesive through the same nipple until the next nipple overflows; then disconnect the hose and cap the nipple.

- Follow the procedure until the entire crack is filled and all nipples are capped.

- Cut off the protruding ends of the capped nipples flush with the concrete surface, and plug the resultant exposed openings with epoxy resin sealant.

4.11.10 Repair of joints

There has been much experience with poor sealant performance and resulting damage to a wide variety of structures. Concern with such problems spurred the development and introduction of higher-class sealants, both field-molded and preformed. Failures have continued to occur, however, often within days and weeks rather than months or years, for five main reasons:

- The joint as designed was of impossible width, shape, and potential movement to seal successfully, yet somebody went ahead to construct and seal it.

- Unanticipated service conditions have resulted in greater joint movements than those allowed for when the joint design and type of sealant were determined.

- The wrong type of sealant for the particular conditions was selected, often on the false grounds of economy in first cost.

- New sealants have sometimes been initially overpromoted and used before their limitations were realized.

- Poor workmanship occurred when constructing the joint, in preparing it to receive the sealant, or during sealant installation.

At joints, minor touch-up of small gaps and soft or hard spots in field-molded sealants can usually be made with the same sealant; but when the failure is extensive, it is usually necessary to remove the sealant and replace it. When the sealant has generally failed but has not come out of the sealing groove, it can be removed with hand tools; on larger projects, such as pavements, it can be removed by routing or plowing with suitable tools. When widening is required to improve the shape factor, the sealant reservoir can be enlarged by sawing.

After proper preparation to ensure clean joint faces and additional measures designed to improve sealant performance, such as the improvement of shape factor, provision of backup material, and possible selection of a better type of sealant, the joint can be resealed.

Minor edge spalls to concrete joint faces can be repaired with an epoxy resin mortar, an essential operation if a compression seal is being used. Otherwise, most repairs to correct defects in the original construction of the joint involve major, exacting, and often expensive work. The reason for the failure must be identified and, depending on the cause, continuity must be restored in the joint system either by removing whatever is blocking the free working of the joint or by cutting out the whole joint and rebuilding it.

When cracks have taken over from a nonworking or absent joint, they can be routed out and sealed with a suitable field-molded sealant to prevent damage to the structure. Selections of a suitable sealant and an installation method follow those for the equivalent joint. An additional problem occurs when water is flowing through the crack and the upstream face cannot be reached for sealing. Before sealing can be successfully undertaken, the water flow must be stopped. If the source of water cannot be cut off by dewatering, then (depending on the circumstances) one of the many alternatives such as cutting back the crack and plugging it with a quick-setting or dry-pack mortar or cement or grouting with a chemical or epoxy resin can be tried. Successful execution of any of these operations usually requires specialized knowledge, experience, and workmanship.

Few exposed sealants have a life as long as that of the structure whose joints they are intended to seal. Fortunately, buried sealants such as waterstops and gaskets have a long life because they are not exposed to weathering and other deteriorating influences.

Most field-molded sealants will, however, require renewal sooner or later if an effective seal is to be maintained and deterioration of the structure is to be avoided. The time at which this becomes necessary is determined by service conditions, by the type of material used, and by whether any defects of the kind already enumerated were built in at the time of the original sealing.

4.12 Delivery and Placement

Delivery and placement are critical to obtaining satisfactory concrete structures and are a significant part of any specification for concrete.

4.12.1 Batching

During measurement operations, aggregates should be handled in a manner that will maintain their desired grading, and all materials should be weighed to the tolerances required for desired reproducibility of the concrete mix selected. In addition to accurate weighing, an important objective of successful batching is the proper sequencing and blending of the ingredients during charging of the mixers. The final objective is to obtain uniformity and homogeneity in the concrete produced as indicated by such physical properties as unit weight, slump, air content, strength, and air-free mortar content in successive batches of the same mix proportions.

Tolerances. Most engineering organizations, both public and private, issue detailed specifications of manual, semiautomatic, and automatic batching of concrete.

Plant type. Factors affecting the choice of the proper batching systems are (1) size of job, (2) required production rate, and (3) required standards of batching performance. The productive capacity of a plant is determined by a combination of such items as the materials handling system, bin size, batcher size, and plant mixer size and number. The available batching equipment falls into three general categories: manual, semiautomatic, and automatic.

Manual batching. As the term implies, all the operations in weighing and batching the concrete ingredients are done manually. Manual plants are acceptable for small jobs having low-batching-rate requirements, generally for jobs up to 3800 m^3 (5000 yd^3) and rates up to 20 cubic meters per hour (m^3/h) [25 cubic yards per hour (yd^3/h)]. As the job size increases, automation of batching operations is rapidly justified. Attempts to increase the capacity of manual plants by rapid batching invariably result in excessive weighing inaccuracies.

Semiautomatic batching. In semiautomatic batching, aggregate bin gates for charging batchers are opened by manually operated push buttons or switches. Gates are closed automatically when the designated weight of material has been delivered.

Automatic batching. Automatic batching of all materials is electrically activated by a single starter switch. Interlocks interrupt the batching cycle when the scale has not returned to ± 3 percent of zero balance or when preset weighing tolerances are exceeded. An individual automatic batching system provides separate scales and batchers for each aggregate size and for each of the other materials batched. The weighing cycle is started by a single starter switch, and individual batchers are charged simultaneously.

4.12.2 Mixing

However the batching is accomplished, mixing is the next step. It must be carefully controlled, especially if the delivery time is longer than the ideal mixing time.

Total mixing water. Uniformity in the measurement of total mixing water, in addition to the accurate weighing of added water, involves control of such additional water sources as mixer wash water, ice, and free moisture in aggregates. One specified tolerance (ASTM Standard C94) for accuracy in measurement of total mixing water, from all sources, is ± 3 percent.

Measurement of admixtures. Use of admixtures in concrete, particularly air-entraining agents, is a universally accepted practice. Batching tolerance and charging-discharge interlocks should be provided for admixtures also.

Measurement of materials for small jobs. Occasionally the concrete volume on a job is so small—76 m³ (100 yd³) or less—that it is not practical to establish and maintain a batch plant and a mixer at the construction site. In that case it is preferable to use ready-mixed concrete. If centrally dry-batched concrete is not available, proper precautions should still be taken to measure and mix concrete materials properly. Thorough mixing is essential to the production of uniform concrete. Therefore, equipment and methods used should be capable of effectively mixing concrete materials containing the largest specified aggregate to produce uniform mixes of the lowest slump practical for the work.

4.12.3 Transporting

Concrete can be transported by a variety of methods and equipment: truck mixers, stationary truck bodies with and without agitators, buckets hauled by truck or railroad car, pipeline or hose, or conveyor belts. Each type of transportation has specific advantages and disadvantages depending on the conditions of use, mix ingredients, accessibility and location of the placing site, required capacity and time for delivery, and weather conditions.

Delivery by mixing truck. Some specifications limit the total number of drum revolutions that can be used for charging, mixing, agitating, and discharging concrete in revolving-drum trucks (Fig. 4.7). Others place limits on the number of revolutions at mixing speed only. A maximum elapsed time of 1.5 h after the cement has entered the drum until completion of discharge is also frequently specified. Also, provision is made for reduction of the maximum elapsed time in warm weather (ASTM C94-60). Another specification method used is to place no limits on revolutions or elapsed time as long as the specified mixing water is not exceeded, no retempering water is added, the concrete has proper plastic physical properties, and the concrete is of adequate consistency and homogeneity for satisfactory placement and consolidation. This latter approach is favored specifically with regard to maximum allowable time for discharge, and it is particularly applicable when low concrete temperatures are used or when cooler weather prevails. Final determination of whether mixing is being accomplished

Figure 4.7 Concrete delivered by revolving-drum truck. *(Courtesy of Los Angeles Harbor Department)*

satisfactorily should be based on standard mixer uniformity tests (ASTM Standard C94-81).

Revolving drum. In the revolving-drum method, the mixer serves as an agitating transportation unit. The drum is rotated at charging speed during loading and either reduced to agitating speed or stopped after loading is complete. Elapsed time for discharge of the concrete can be the same as for truck mixing.

Final objective. The method of transportation used should be such as to deliver the concrete to the point of placement efficiently without significantly altering the concrete's desired properties with regard to water-cement ratio, slump, air content, and homogeneity. Each method of transportation has advantages under particular conditions of use pertaining to such items as mix materials and design, type and accessibility of placement, required delivery capacity, and location of batch plant. These various conditions should be carefully reviewed during selection of the type of transportation best suited for obtaining quality in-place concrete economically.

4.12.4 Placing concrete

A basic requirement for placing equipment and methods, as for all other handling equipment and methods, is that the quality of the concrete, in terms of water-cement ratio, slump, air content, and homogeneity, must be preserved. Selection of equipment should be based on capability for efficiently handling concrete of the most advantageous proportions that can be readily consolidated in place with vibration.

Sufficient placing capacity, as well as mixing and transporting capacity, should be provided so that the concrete can be kept plastic and free of cold joints while it is being placed. It should be placed in horizontal layers not exceeding 0.6 m (24 in) in depth, and inclined layers and cold joints should be avoided. For monolithic construction, each concrete layer should be placed while the underlying layer is still responsive to vibration, and layers should be sufficiently shallow to permit knitting the two together by proper vibration. Concrete should be deposited at or near its final position in the placement, thereby eliminating the tendency to segregate when it has to be flowed laterally into place. On sloping surfaces, concrete should be placed at the lower part of the slope first. Placement should progress upward and thereby increase natural compaction of the concrete. High-velocity discharge, which will cause segregation of the concrete, should be avoided.

The equipment and method used for placing concrete should avoid separating the coarse aggregate from the concrete. Clusters and pockets of coarse aggregate should be scattered before placing concrete over them to prevent rock pockets and honeycomb in the completed work.

Requests for increases in mixing water are frequently made on the job when concrete of relatively stiff consistency will not flow down chutes, drop out of buckets or hoppers, or discharge through gates or trucks. If the concrete is readily workable and satisfactorily consolidated in place with proper vibration, these requests for additional water are not valid. A limitation on the use of reasonable mix proportions and slump should not be imposed because inadequate placing equipment is being used.

Preplaced-aggregate concrete. In the preplaced-aggregate method of construction, forms are first filled with clean, well-graded coarse aggregate and then structural quality grout is injected into the voids of the aggregate mass to produce concrete. The method is especially adaptable to underwater construction, to concrete and masonry repairs, and, in general, to new structures, where placement by conventional means is usually difficult or where concrete of low volume change is required. Since preplaced-aggregate concrete construction is of a specialized nature, it is advisable that the work be undertaken by qualified personnel experienced in the method. The physical properties of preplaced-aggregate concrete are similar to those of conventional concrete; therefore, the allowable working stresses used for conventional concrete structural design can be used (U.S. Army Engineer, Waterways Experiment Station, 1954).

Tremie concrete. The tremie method is frequently used to place concrete underwater. By this method the concrete is deposited under the surface of fresh concrete previously placed. Placement is usually by gravity feed from above the water surface through a vertical pipe connected to a funnel-shaped hopper at the top (Fig. 4.8). Tremie concrete flows outward from the bottom of the pipe and pushes the existing surface of the concrete outward and upward. As long as flow is smooth so that the concrete surface adjacent to the water is not physically agitated, high-quality concrete will result. Placement can also be carried out through other liquids lighter than concrete, such as bentonite slurry, to suit special conditions. Tremie concrete is used primarily for cofferdam or caisson seal, underwater structural sections such as bridge piers, drydock walls, and floors and as a seal for precast tunnel sections.

The concrete mix proportions for tremie placement differ from ordinary structural mix proportions because of the need to have the mix flow into place slowly by gravity without vibration or mechanical help. The mix should be proportioned for a slump 15 to 23 cm (6 to 9

Figure 4.8 Tremies in place before pouring.

in). It is generally preferable to use a natural round gravel rather than crushed rock because of flow requirements. The maximum size of aggregate is usually 38.1 mm (1.5 in). However, a nominal size of 19.0 or 9.51 mm (¾ or ⅜ in) can be used for complex sections and critical flow conditions. The proportion of fine aggregate (sand) is usually in the range of 40 to 50 percent of the total weight of aggregate. Water-reducing retarding admixtures conforming to ASTM Standard C494 have been found to be an aid in placement of the concrete, and the retarding effect slows the setting (Williams, 1959). Air-entraining admixtures and pozzolans also are beneficial to flow characteristics. The concrete temperatures should be kept as low as practical, usually below 21.1°C (70°F) to improve placement and structural qualities. The recommended maximum water-cement ratio for concrete deposited by tremie under water is 0.44 by weight.

The compressive strength of rich, high-slump tremie concrete mixes will often be approximately 28 to 56 MPa/m^2 (4000 to 8000 lb/in^2) at 28 days. Curing conditions are excellent, and shrinkage is low. The surfaces that will be in contact with the concrete should be free of mud, marine growth, sewage, and other matter. The bond to clean surfaces of steel, rock, and timber is generally excellent. The heat of hydration developed in rich mixes produces high early strength even when water temperature is as low as 4.4°C (40°F). When large masses of tremie concrete are placed, volume change due to heat development may warrant special consideration. The use of suitable instrumentation to monitor the temperature rise in these structures may be required.

Pumping. Pumped concrete can be defined as concrete conveyed by pressures through either rigid pipe or flexible hose and discharged directly into the desired area (Fig. 4.9). Pumping can be used for most concrete construction, but it is especially useful where space or access of construction equipment is limited. Most concrete transported to the placement areas by pumping methods is pumped through a rigid pipe or a combination of rigid pipe and heavy-duty flexible hose. Effectivepumping range will vary from 90 to 305 m (300 to 1000 ft) horizontally or 30 to 90 m (100 to 300 ft) vertically.

To establish the optimum slump for a pump mix and to maintain control of that particular slump through the course of the job are both extremely important factors. Experience indicates that slumps below 50 mm (2 in) are impractical for pumping, and slumps above 152 mm (6 in) should be avoided. In mixtures with high slump, the aggregate will separate from the mortar and paste and may cause blocking in the pump line. Overly wet mixes also bleed and increase shrinkage. It is important to obtain a truly plastic mix through proper proportioning rather than to try to overcome deficiencies by adding more mortar.

Figure 4.9 Pumped concrete being applied with a pressure nozzle.

4.13   Environmental Effects

The environmental effects discussed in the following subsections are generally to be considered in the use of concrete in coastal structures. Portland cement concrete is a durable material and is well suited for use in the coastal environment. When properly designed, placed and cured, it will resist most coastal environments for many years.

4.13.1   Corrosive and pollutant attacks

Concrete is rarely attacked by solid, dry chemicals. In order to significantly attack concrete, corrosive chemicals must be in solution form and be above some minimum concentration. Chemical attack on concrete is generally the result of exposure to sulfates or acids. Naturally occurring sulfates of sodium, potassium, calcium, and magnesium, which are sometimes in soil or dissolved in ground water adjacent to concrete structures, can attack concrete.

Two chemical reactions are likely to be involved in sulfate attack on concrete: (1) the combination of sulfate with free calcium hydroxide liberated during the hydration of the cement to form calcium sulfate (gypsum) and (2) the combination of gypsum and hydrated calcium aluminate to form calcium sulfoaluminate.

The effects of some of the more common chemicals on the deterioration of concrete are indicated in Table 4.18. Many pollutants contain the chemicals indicated in the table.

4.13.2   Sunlight exposure effects

Sunlight has virtually no effect on the deterioration of concrete.

4.13.3   Water penetration effects

Pure water does not attack concrete, but it can be the medium for dissolving most chemicals, which in solution may cause concrete deterioration. Seawater has a high sulfate and chloride content which may be only moderately aggressive to concrete; but if those chemicals can penetrate the concrete to the steel reinforcing, then rapid deterioration will occur.

4.13.4   Wave and current effects

Waves and currents have no direct effect on concrete or are not the direct cause of its deterioration. Concrete destruction occurs only when the structure is not adequately designed. Wear to concrete structures does occur as a result of cavitation caused by the collapse of bubbles of water vapor. Concrete wear by abrasion may also occur as a result of

TABLE 4.18   Effect of Commonly Used Chemicals on Concrete

Rate of attack at ambient temperature	Inorganic acids	Organic acids	Alkaline solutions	Salt solutions	Miscellaneous
Rapid	Hydrochloric Nitric Sulfuric	Acetic Formic	—	Aluminum chloride Ammonium nitrate Ammonium sulfate Sodium sulfate	— Bromine (gas) Sulfite liquor
Moderate	Phosphoric	Tannic	Sodium hydroxide- 20%^{a} Sodium hydroxide 10 to 20%^{a}	Magnesium sulfate Calcium sulfate Ammonium chloride Magnesium chloride	Chlorine (gas) Seawater Soft water
Slow	Carbonic	—	Sodium hypochlorite	Sodium cyanide Calcium chloride	Ammonia (liquid)
Negligible	—	Oxalic Tartaric	Sodium hydroxide 10%^{a} Sodium hypochlorite Ammonium hydroxide	Sodium chloride Zinc nitrate Sodium chromate	

^{a}Avoid siliceous aggregates because they are attacked by strong solutions of hydroxide.

solid particles (such as sand) that are transported by waves and currents and impinge on a concrete surface.

4.13.5   Effect of severe temperature and ice

Resistance to severe temperature changes in concrete is more a function of proper mix design with good aggregate and proper curing than any other factors. Generally, high temperatures do not affect well-cured concrete. Building codes generally require concrete to resist temperatures of 538°C (1000°F) for 5 min to more than 1038°C (1900°F) for 3 h, depending on the thickness of concrete tested. Sustained high-ambient temperatures [above 200°C (392°F)] will stop normal crystal growth in concrete and normal strength gain with aging.

To provide a high degree of resistance to the disruptive action of freezing and thawing and of deicing chemicals, air-entraining admixtures are used. Unless low temperatures are very extreme, properly designed concrete will not deteriorate or spall in freezing conditions.

4.13.6 Marine organisms

Marine organisms do not injure good concrete that contains sound aggregate. For example, concrete which is composed of siliceous aggregates is resistant to marine borer activity because the material is extremely abrasive to the lime shells of boring organisms. However, in tropical and semitropical water there have been instances of borer damage in concretes in which limestone or similar sand has been used.

The first record of marine animals entering concrete is found in Hill and Kofoid (1927). This record indicated that Pholadidae were found drilling into concrete jackets used to protect wood piles from attacks by Limnoria and Teredo. Subsequent tests showed that the mortar jackets bore no resemblance to even a poor grade of structural concrete and offered no resistance to the boring mechanisms of pholads. Although mollusks have been found in lightweight mortar pontoons of dubious quality, additional tests prove that pholads could not enter any material harder than their shells (about 2.5 on the diamond scale).

4.13.7 Periodic wetting and drying

Periodic wetting and drying may cause the formation on the concrete surface of D cracks (the progressive formation of fine cracks, often in random pattern). Such cracks may enlarge in time; and if they are exposed to freezing and thawing, they can result in concrete spalling.

4.13.8 Wind erosion

Concrete resistance to wind is usually not a serious problem in coastal structures. Where strong winds may pick up sand particles, which cause some etching of concrete surfaces similar to surf zone abrasion (usually near the ground line), it would take many years of exposure of structural grade concrete for wind erosion to become a problem.

4.13.9 Burrowing animal effects

Marine animals do not penetrate good concrete, as indicated in Sec. 4.13.6, and the larger dry land animals do not attack concrete. Concrete is one of the hardest materials in the coastal environment, and it contains nothing of food value for such animals.

4.13.10 Effects of flora

There are no reported effects of flora growth on concrete.

4.13.11   Fire

Concrete has resistance to fire or extreme high temperatures as stated above.

4.13.12   Abrasion

Abrasion is defined as the ability of a surface to be worn away by rubbing and friction. Wind- or waterborne particles can abrade or etchconcrete surfaces. If windborne particles do cause abrasion, some dusting problem can develop; however, the slow rate of the abrasion process in the coastal zone is usually unnoticeable. Wear on concrete structures exposed to high velocities and negative pressures is generally known as cavitation erosion. Precise limits for abrasion resistance of concrete are not possible. It is necessary to rely on relative values based on weight or volume loss, depth of wear, or visual inspection.

4.13.13   Seismic effect

Severe seismic forces can cause direct failure of a concrete structure or indirect failure by altering the foundation on which the structure rests; subsequent settlement can then result in structural failure or deterioration. With proper design, concrete can be made to resist seismic effects.

4.13.14   Human activity

Human activity has very little effect on concrete structures except the visual impact of graffiti or other defacing actions.

4.14   Use in Coastal Structures

Concrete is easily adapted to coastal construction in that aggregates are normally available at or near the site so that the only imported materials are cement and steel reinforcing. It can be cast in almost any shape or size to fit site requirements, and the structures can be built in sections either by casting separate members and assembling them in place to create a large structure or by casting mass concrete a section at a time to produce a large continuous structure. This characteristic allows the design engineer a wide selection of type, size, and configuration of structure design. Excellent physical and strength properties, as well as stability and resistance to the environment, make concrete an ideal coastal zone construction material. However, because it is a relatively heavy material, concrete is limited to uses in which its heavy weight is not a deterrent.

In addition to structures constructed totally of concrete, many concrete structural elements are used in a variety of coastal projects such as armor units of various shapes and sizes, concrete caissons, solid and perforated blocks, sheet and bearing piles, and beams and slabs. Floating structures such as caissons, barges, and pontoons have been successfully built and used. Concrete is also used unreinforced in mass structures. Reinforced and prestressed units are usually precast structure elements.

4.14.1 Seawalls, bulkheads, and revetments

Seawalls, bulkheads, and revetments are distinguished by purpose. In general, seawalls are the most massive of the three structures, because they resist the force of the waves. Bulkheads are next in size. Their function is to retain fill; in general, they are not exposed to severe wave action. Revetments are the lightest, because they are designed to protect shorelines against erosion by currents or light wave action.

Curved-face seawalls and combination stepped and curved-face seawalls are usually massive structures which are built to resist high wave action and reduce scour. Figure 4.10 shows an example of reinforced-concrete combination stepped and curved-face seawall. It was designed for stability against moderate waves.

Bulkheads.   Concrete bulkheads can take virtually any form or configuration required for the intended use and location.

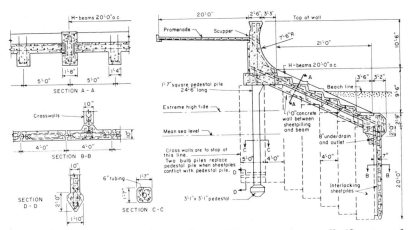

Figure 4.10 Combination stepped and curved-face concrete seawall. (*Courtesy of U.S. Army Corps of Engineers, CERC, 1977*)

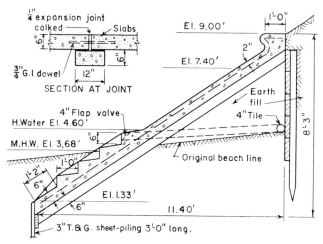

Figure 4.11   Concrete revetment. *(Courtesy of Portland Cement Association)*

Revetments. Structural types of revetments used for coastal protection in exposed and sheltered areas are illustrated in Figs. 4.11 to 4.13. There are two types of revetments: the rigid, cast-in-place concrete type (Fig. 4.11) and the flexible or articulated armor unit type (Figs. 4.12 and 4.13). A rigid concrete revetment provides excellent bank protection, but the site must be dewatered during construction to pour the concrete. A flexible structure also provides excellent bank protection, and it can tolerate minor consolidation or settlement without structural failure. The articulated block structure shown in Fig.

Figure 4.12   Articulated armor unit revetment. *(Courtesy of Marine Modules Inc.)*

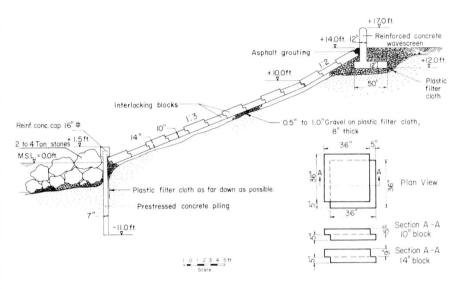

Figure 4.13 Interlocking concrete block revetment. *(Courtesy of Carthage Mills Inc, Erosion Control Division)*

4.13 allows for the relief of hydrostatic uplift pressure generated by wave action.

Interlocking concrete blocks have been used extensively for shore protection in the Netherlands and England, and they have become popular in the United States. Typical blocks are generally square slabs with shiplap-type interlocking joints (Fig. 4.13). A joint of the shiplap type provides a mechanical interlock with adjacent blocks. Stability of an interlocking concrete block depends largely on the type of mechanical interlock. Concrete piles are sometimes used as cutoff walls for revetments and seawalls. Concrete foundation piles are sometimes used to support seawalls and other massive concrete structures, such as caisson breakwaters.

4.14.2   Groins

Concrete groins are built of concrete sheet piles or king piles and panels if they are impermeable. The piles are usually prestressed units tied together with a cast-in-place concrete cap. If greater pile flexibility is required, timber wales have been used. Permeable concrete groins that permitted the passage of sand through the structure have been built in the past; they are not used at present. In low-wave climates, grout-filled bags are also used as an installation convenience; the bags, usually plastic, deteriorate and leave the small concrete shapes as protection of the groins.

4.14.3   Jetties and breakwaters

In exposed locations, jetties and breakwaters are generally composed of some variation of a rubble mound structure containing concrete either as a binding material to hold rock together or as a separate element. When used as a separate element, concrete takes the form of precast armor units having heavy weight as well as energy absorption characteristics. Some types of jetties are illustrated in Figs. 4.14 and 4.15. In less severe exposures, both cellular steel and concrete caissons have been used. In low-wave climates grout-filled bags are used.

When rock armor units in adequate quantities or size are not economically available, concrete armor units are used. Also, concrete sheet piles are sometimes used as cores for jetties. Figure 4.16 illustrates the use of quadripod armor units on the rubble mound jetty at Santa Cruz, California. Figure 4.15 illustrates the use of the more recently developed dolos armor unit in which 374- and 383-kg (42- and 43-ton) dolosse were used to rehabilitate the seaward end of the Humboldt Bay jetties against 12-m (40-ft) breaking waves (Magoon and Shimizu, 1971).

Concrete caisson breakwater.   Breakwaters of the concrete caisson type are built of reinforced concrete shells that are floated into position, settled on a prepared foundation, filled with stone or sand for stability, and then capped with concrete or stones. These structures may be constructed with or without parapet walls for protection against wave overtopping. In general, concrete caissons have a reinforced concrete

Figure 4.14   Fabric tubes filled with concrete form a jetty. *(Courtesy of Fabriform)*

bottom, although open-bottom concrete caissons have been used. The open-bottom type is closed with a temporary wooden bottom that is removed after the caisson is placed on the foundation. The stone used to fill the compartments combines with the foundation material to provide additional resistance against horizontal movement.

Figure 4.17 illustrates a perforated type of caisson breakwater (Jarlan, 1961). The installation at Baie Comeau, Quebec (Stevenston, 1963) utilized the caisson as a wharf on the harborside. The holes or perforations on the seaward side reduce the undesirable conditions of a smooth vertical face wall and are an illustration of complex structural shapes that are possible because of the way concrete is cast.

Concrete armor units.   Many different concrete shapes have been developed as armor units for rubble structures. The major advantage of concrete armor units is that the units usually have higher stability coefficient values, thus permitting the use of steeper structure side slopes or lighter weights of armor units. This property is especially valuable when quarrystone of the required size is not available.

The unit weight of concrete containing normal aggregates will range from 22 to 24.3 kt/m^3 (140 to 155 lb/ft^3) but can be increased by using heavy aggregate to 28.3 kt/m^3 (180 lb/ft^3) usually at some additional cost. The technique of placement and the size of the armor unit

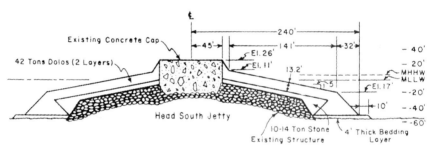

(after Magoon and Shimizu, 1971)

Figure 4.15 Dolos rubble mound jetty. *(Courtesy of U.S. Army Corps of Engineers, CERC, 1977)*

will determine if reinforcing is required in dolos or tribar units. Heavy units, exceeding about 179 kg (20 tons) will require reinforcing if placed from a landside unit. Placing armor units from floating equipment where the wave action may cause bumping of the units may require reinforcing in armor units as little as 9.0 kt (10 tons) in weight.

Table 4.19 lists the concrete armor units that have been cited in literature and shows where and when the units were developed. Commonly used types of units are illustrated in Fig. 4.18. Projects using

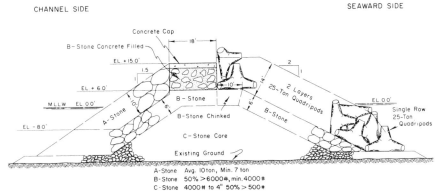

Figure 4.16 Quadripod rubble mound jetty. *(Courtesy of U.S. Army Corps of Engineers, CERC, 1977)*

tetrapods, tribars, quadripods, and dolosse in the United States are listed in Table 4.20.

4.14.4 Other structures

Concrete has been adapted to many kinds of marine structures as monolithic or cast-in-place structures as well as precast or prestressed

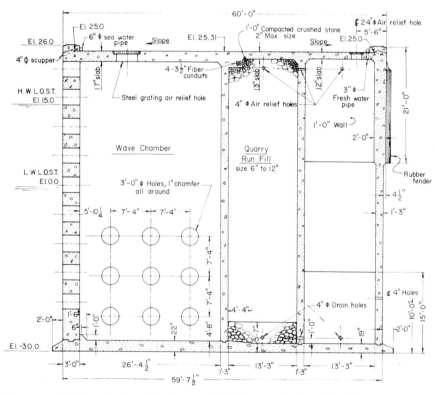

Figure 4.17   Perforated caisson breakwater, Baie Comeau, Quebec, Canada.

TABLE 4.19   Typical Concrete Armor Units

Name of unit	Country of development	Year
Accropode	France	1978
Antifer[a,b]	Netherlands	1978
Cube	—	—
Cube (modified)[a]	United States	1959
Dolos[a]	South Africa	1963
Handbar	Australia	1979
Hexapod[a]	United States	1959
Quadripod[a]	United States	1959
Rectangular block[a,b]	—	—
Stabit	England	1961
Tetrahedron (perforated)[a,c]	United States	1959
Tetrapod	France	1950
Tribar[a]	United States	1958

[a]The units have been tested, some of them extensively, at the Waterways Experiment Station (WES).

[b]Cubes and rectangular blocks are known to have been used in masonry-type breakwaters since early Roman times and in rubble mound breakwaters during the last two centuries. The cube was tested at WES as early as 1943.

[c]Solid tetrahedrons are known to have been used in hydraulic works for many years. This unit was tested in WES in 1959.

units. Concrete is an optimum material for marine structures because it combines durability, strength, and economy. The ability to produce it in almost any geometric form gives concrete a high adaptability to almost any location and condition of use required.

Navigation structures.   Prestressed concrete piles are used for navigation light standards. Navigation aids located on breakwaters or along the shoreline as lighthouses and radio signal towers are usually of

TABLE 4.20   Concrete Armor Projects in the United States

Date	Location	Structure	Armor Unit
1956	Crescent City, Calif.	Breakwater	25-ton tetrapods
1957	Kahului, Hawaii	Breakwater	33-ton tetrapods
1958	Nawiliwili, Hawaii	Breakwater	18-ton tribars
1958	Rincon Islands, Calif.	Revetment	31-ton tetrapods
1963	Kahului, Hawaii	Breakwater	19- to 50-ton tribars
1963	Santa Cruz, Calif.	Breakwater	28-ton quadripods
1963	Ventura, Calif.	Jetty	10.7-ton tribars
1971	Diablo Canyon, Calif.	Breakwater	21.5- to 36.5-ton tribars
1971	Humboldt Bay, Calif.	Jetty	42- to 43-ton dolosse
1973	Crescent City, Calif.	Breakwater	40-ton dolosse
1980	Cleveland Harbor, Ohio	Breakwater	2-ton dolosse
1982	Manasquan Inlet, N.J.	Jetty	16-ton dolosse

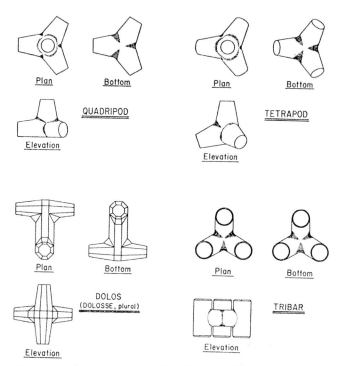

Figure 4.18 Concrete armor units. *(Courtesy of U.S. Army Corps of Engineers, CERC, 1977)*

concrete construction. Also included as navigation structures are mooring anchors for buoys of all kinds.

Piers and wharves. Concrete is the construction material most used in building piers and wharves either on the coastline or in protected harbors. All the elements of pier construction such as piles, dock units, pier girders, substructures, and bulkheads are built of concrete or a combination of concrete and wood or steel. Even then, concrete can be used to protect the wood or steel from erosion, corrosion, dryrot, or attack by marine organisms.

Figures 4.19 to 4.22 are examples of commercial concrete structures located on the coastline. Special piers have been constructed for product loading lines, wastewater disposal, and other discharge lines. Figure 4.23 shows the piling and deck structure of a recreation pier extending from shore to the open ocean. Innumerable concrete piers and wharves are, of course, constructed in bays and protected harbors along the world's shorelines.

The use of concrete in these structures is most feasible not only because of its durability but because it is readily available in most locations, can be

Figure 4.19   Pier 5, Port of Callao, Peru. *(Courtesy of American Association of Port Authorities)*

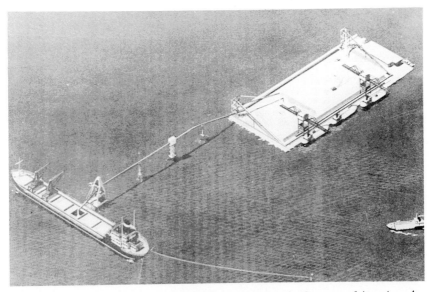

Figure 4.20   Artificial island on continental shelf off Brazil. *(Courtesy of American Association of Port Authorities)*

Figure 4.21 Fixed single-point mooring at Marsa El Brega, Libya. *(Courtesy of American Association of Port Authorities)*

Figure 4.22 Port Lotta, Tasmania, ore terminal. *(Courtesy of American Association of Port Authorities)*

Figure 4.23   Hermosa Beach recreation pier.

produced in virtually any size or shape, and is economical. Recent developments in precast and prestressed concrete units provide the means of using fast and simplified construction procedures to produce final structures that are both stable and durable.

Concrete piling is easily manufactured in almost any length to about 36 m (118 ft)—although longer piles have been made—and commonly in round, square, octagonal, or hollow core cross sections. Either reinforced or prestressed concrete piles can be designed to support very heavy loads. These piles may vary in cross section from 0.15-m- (6-in-) diameter round piles to 0.76-m- (30-in-) solid square or octagonal piles and 1.2-m- (4 ft-) round hollow piles. Hollow piles usually have a 0.15- to 0.30-m (0.5- to 1-ft) wall. The restrictions on concrete pile size are determined by the equipment required to manufacture them, e.g., the pile bed, forms, and, if required, prestressing equipment, as well as the pile-handling equipment such as cranes, barges, and piledrivers.

Submerged structures.   Concrete is an ideal material for the construction of submerged structures. It can be used for both monolithic submerged structures, such as structural elements of bridges, or precast and prestressed for pipelines, intake and outfall structures, and bridge piers. Other submerged structures usually built of concrete include tunnels for vehicular and railroad traffic and utilities.

Floating structures.   Many concrete pontoons of various shapes and sizes have been constructed as parts of pontoon bridges, quays, wharves, and floating facilities for small boats and seaplanes. Floating breakwaters of precast reinforced-concrete pontoons have been installed in Tenakee Springs, Sitka, and Ketchikan, Alaska, and Blaine, Everett, and Port Orchard, Washington. Other structures such as skiffs, launches, scows, barges, floating drydocks, and permanent offshore structures have been constructed of concrete. The construction of fixed breakwaters by floating precast units into place is not uncommon. Caisson units are constructed on a land site, floated into position, sometimes thousands of miles from the construction site, and then submerged on the ocean bottom and filled with sand or dredged material. Almost every type of floating or submersible structure has at one time or another been built of concrete.

Access and roadway structures. Concrete is used for roadways, bridges, structural anchors, and foundations as parts of coastal structures. Overpasses, footbridges, drainage facilities, and concrete pipe also are ancillary to many coastal structures.

Ocean outfall and discharge structures. Because of the excellent durability of concrete in the coastal environment, many ocean outfall and discharge structures in the coastal zone are made of concrete. These structures accommodate rain and flood water runoff as well as industrial and domestic wastes. Ancillary facilities, such as settling ponds and pumping plants required to make these systems work, also are frequently constructed of concrete.

5

Other Types of Concrete
and Grout

5.1 Bituminous Concrete

Asphalt is a primary ingredient of all bituminous concretes. It is a natural constituent of many petroleums in which it exists in solution. If the solvent oils are removed by evaporation or distillation from crude petroleum, an asphalt residue remains. Asphalt is a cement; it is readily adhesive, highly waterproof, and durable. As a plastic substance, it imparts controllable flexibility to mixtures of mineral aggregates with which it is usually combined.

The three general categories of bituminous concrete used in coastal structures are: asphalt concrete, a mixture of asphalt cement and both fine and coarse aggregate placed and compacted to form a monolithic structure; sand asphalt, essentially a type of asphalt concrete with coarse aggregate omitted; and asphalt mastic, basically a sand asphalt having a fluid consistency during placement sufficient to allow it to flow into voids of a rock structure such as a breakwater or jetty.

5.1.1 Types of asphaltic materials

The following terms relating to asphalt are taken from *Asphalt in Hydraulics* (The Asphalt Institute, 1976).

Asphalt cement. Asphalt that is refined to meet specifications for paving, industrial, or special purposes.

Asphalt concrete, hydraulic type. Similar to asphalt concrete for roadway paving except that, to ensure an essentially voidless mix

---

This chapter was written by Lawrence L. Whiteneck.

after compaction, higher mineral filler and asphalt contents are used.

Asphalt facing.   An asphalt surface designed to resist erosion, abrasion, water pressure, and, in some instances, ice pressure. A facing may, in addition, also act as an impermeable layer to prevent leakage through the structure. It may also be termed an asphalt lining or asphalt revetment (see below).

Asphalt grout.   A mixture of asphalt, sand, and mineral filler which, when heated and mixed, will flow into place without mechanical manipulation. It is used to bind together a layer of coarse stone of more or less uniform size. It may also be termed asphalt mastic (see below).

Asphalt injection.   A pressurized subsurface application of asphaltic material. Usually, injections are made for the purpose of filling subsurface cavities or crevices in the foundation soil or voids beneath an existing pavement layer, primarily for controlling water seepage.

Asphalt lining.   That part of a hydraulic structure that functions as a durable, erosion-resistant surface. Usually, its most important function is a waterproof barrier holding water or other liquid inside the structure.

Asphalt mastic.   A mixture of mineral aggregate, mineral filler, and asphalt in such proportions that the mix can be applied hot by pouring or by mechanical manipulation; it forms a voidless mass without being compacted.

Asphalt mat.   A felt or fabric sheet impregnated or coated with asphalt to form a watertight lining or membrane usually 6 mm (0.25 in) or less in thickness. It may be a sheet, first installed in place, to which the asphalt is applied, or it may be a finished material that is watertight and ready for installation.

Asphalt mattress, slab.   According to size, prefabricated flexible units composed of an asphalt mastic mixture reinforced with mesh, netting, lines, or cables as required.

Asphalt membrane.   A relatively thin layer of asphalt formed by spraying a high-viscosity, high-softening-point asphalt cement in two or more applications over the surface to be covered. It is normally about 6 mm (¼ in) thick. It is used for waterproofing or sealing, and it is buried to protect it from weathering and physical damage.

Asphalt revetment.   A protective asphalt facing on a sloped surface, usually placed for the purpose of protecting an embankment from erosion. Revetments may or may not extend all the way to either the toe or crest of the sloped embankment. The term "subaqueous" refers to that part of a revetment placed under the surface of the water. Upper-bank paving is that part placed above the surface of the water.

5.1.2 Asphalt mixes

Various mixes designed for stiffness, durability, permeability, and placeability are available for coastal uses:

Impermeable mixes. Asphalt mixes having low voids (usually less than 4 percent) after installation, designed to prevent the passage of water.

Porous mixes. Asphalt mixes that permit the free flow of water through the mix. Porous asphalt mixes are divided into two general classifications: permeable asphalt mixes and open-graded asphalt mixes (see below).

Permeable mixes. Asphalt mixes that have medium voids after installation and are designed to permit the free passage of water through the lining to and from the supporting layer or embankment.

Open-graded mixes. Asphalt mixes that have high voids and are designed to provide a free drainage layer underneath an impermeable lining.

Prefabricated panels. Layers of a very dense mixture of asphalt and filler sandwiched between two layers of some tough, asphalt-impregnated material and usually coated with waterproofing asphalt.

5.1.3 Properties of asphalt materials

Asphalt has many properties that make it particularly suitable for use in hydraulic and coastal zone structures. It is versatile in form and application. Asphalt can be used alone (as in an asphalt membrane), or it can be mixed with other materials to produce mixes for a variety of purposes. It can be combined with graded aggregate to form a voidless and impermeable mix. On the other hand, it can be combined with an open-graded aggregate to form a porous mixture allowing free passage of water.

Asphalt is stable in the presence of nearly all chemically laden substances. It is normally unaffected by the usual concentrations of acid, salt, and other waste solutions. This important characteristic makes it useful for waterproofing reservoirs. However, since asphalt is refined from petroleum, other petroleum-based products (which are solvents of asphalt) cannot be stored in asphalt-lined structures.

An important property of asphalt is flexibility. It allows asphalt structures to conform to slight irregularities in the subgrade and to adjust to small differential settlements that inevitably occur after the completion of a structure.

The physical properties of asphalt mixes generally depend on stress conditions and temperature. The ingredients that comprise asphalt mixes have completely different characteristics. The mineral aggregate that makes up the major part of the mix is mainly elastic. The asphalt part, on the other hand, behaves as a viscous liquid at high temperature and under impact load; consequently, asphalt mixtures have both plastic and elastic properties.

For many years, asphalt cement has been graded on the basis of the penetration test, an empirical measure of consistency. Recently, however, the penetration grading of asphalt cements has been replaced by the more fundamental viscosity grading. Two systems of viscosity grading are currently used. The AC system is based on the viscosity of the original asphalt cement. The AR system, used mostly on the Pacific coast of the United States, is based on the viscosity of the residue of the asphalt cement after the cement has been subjected to hardening conditions approximating those occurring in normal hot-mix plant operations.

The relations between the various grading systems are shown in Fig. 5.1. The comparisons are based on the rolling thin film oven test (RTFOT) of residues for AR grades and penetration grades and the thin film oven test (TFOT) of residues for AC grades.

5.1.4 Designing asphalt mixes

The design of asphalt mixes, like any other engineering material design, is largely a matter of selecting and proportioning materials to obtain the desired properties in the finished construction. The overall objective in the design of asphalt mixes is to determine an economical blend and gradation of aggregates (within the limits of the project specifications) and asphalt that yields a mix having:

- Sufficient asphalt to ensure durability
- Sufficient mix stability to satisfy the demands of design use without distortion or displacement
- Sufficient voids in the total compacted mix to allow for a slight amount of additional compaction under loading without flushing,

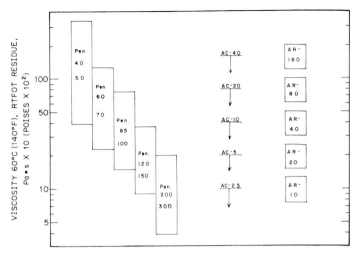

Figure 5.1 Comparison of penetration grade and viscosity grades of asphalt cement.

bleeding, and loss of stability, yet low enough to keep out harmful air and moisture

- Sufficient workability to permit efficient placement of the mix without segregation

Evaluation and adjustment of mix designs. Often, in the process of developing a specific mix design, it is necessary to make several trial mixes to find one that meets the criteria of the design method used. Each trial mix design, therefore, serves as a guide for evaluating and adjusting the trials that follow. For preliminary or exploratory mix designs, it is advisable to start with an aggregate of a gradation that approaches the median of the specification limits. Initial trial mixes for establishing the job mix formula, however, must have an aggregate gradation within the specification limits that the central mixing plant is meeting or is capable of meeting.

When the initial trial mixes fail to meet the design criteria, it will be necessary to modify or, in some cases, redesign the mix. Adjustments in the grading of the original aggregate blend will be required to correct the deficiency.

The strength of many engineering materials is frequently thought to denote quality, but that is not necessarily true of hot-mixed asphalt paving. Extremely high stability is often obtained at the expense of lowered durability, and vice versa. Therefore, when evaluating and adjusting mix designs, always keep in mind that the aggregate gradation and asphalt content in the final mix design must strike a favorable balance between the stability and durability requirements for the use intended. Moreover, the mix must be produced as a practical and economical construction operation.

Grading curves are helpful in making necessary adjustments in mix designs. For example, curves determined from the Fuller equation, a version of the maximum density equation using the power 0.5, represent maximum density and minimum voids in mineral aggregate (VMA) conditions. The Fuller equation is:

$$p = 100(d/D)^{0.5}$$

where p is the total percent passing a given sieve, d is the size of sieve opening, and D is the largest size (sieve opening) in the gradation. Mixtures described by such curves tend to be workable and readily compacted, but their void contents may be too low. Usually, deviations from the curves will result in lower densities and higher VMA. The extent of change in density and VMA depends on the amount of adjustment in fine or coarse aggregate. Figure 5.2 shows a series of Fuller maximum density curves plotted on a conventional semilog grading chart.

Figure 6.3 shows maximum-density curves determined from the maximum-density equation raised to the 0.45 power, $p = 100(d/D)^{0.45}$. It is plotted on the Federal Highway Administration grading chart (based on a scale raising sieve openings to the 0.45 power), which many designers find convenient to use for adjusting aggregate gradings. The curves on the chart need not, however, be determined from the maximum density equation. They can be obtained by drawing a straight line from the origin at the lower left of the chart to the desired nominal maximum particle size at the top. For processed aggregate, the nominal maximum particle size is the largest sieve size listed in the applicable specification upon which any material is permitted to be retained. Gradings that closely approach this straight line usually must be adjusted away from it within acceptable limits to increase the VMA values. This allows enough asphalt to be used to obtain maximum durability without the mixture flushing. The following is a general guide for adjusting the trial mix, but the suggestions outlined may not apply in all cases.

Voids low, stability low.   Voids may be increased in a number of ways. As a general approach to obtaining higher voids in the mineral aggregate (and therefore providing sufficient void space for an adequate amount of asphalt and air voids) the aggregate grading should be adjusted by adding more coarse or more fine aggregate.

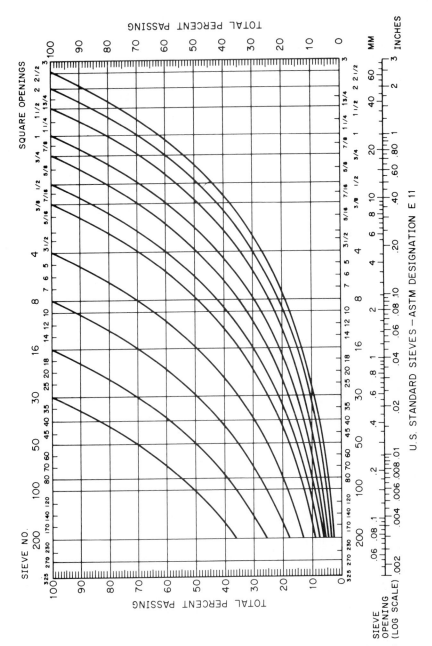

Figure 5.2 Fuller maximum density curves on standard semilog grading chart. *(Courtesy of American Concrete Institute)*

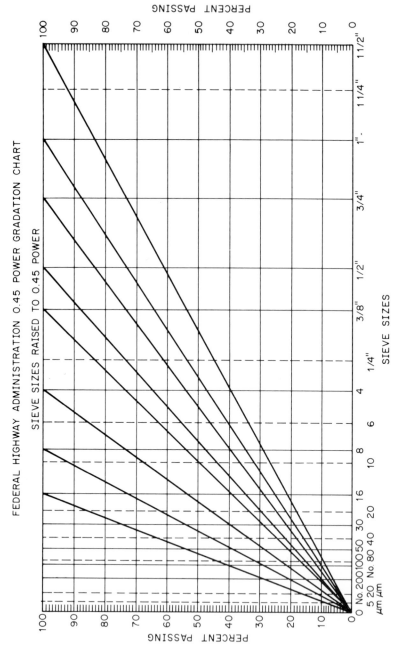

Figure 5.3   Maximum-density curves on Federal Highway Administration 0.45
power gradation chart. *(Courtesy of Federal Highway Administration)*

If the asphalt content is higher than normal and the excess is not required to replace that absorbed by the aggregate, it can be lowered to increase the voids. It must be remembered, however, that lowering the asphalt content increases the void content and reduces the film thickness, which decreases the durability of the pavement. Also, too great a reduction in film thickness may lead to brittleness, accelerated oxidation, and increased permeability. If the above adjustments do not produce a stable mix, the aggregate may have to be changed. It usually is possible to improve the stability and increase the aggregate void content of the mix by increasing the amount of crushed materials. With some aggregates, however, the freshly fractured faces are as smooth as the waterworn faces and an appreciable increase in stability is not possible. That is generally true of quartz or similar rock types.

Voids low, stability satisfactory. Low void content may result in instability or flushing after the mix has been exposed to design loads for a period of time because of reorientation of particles and additional compaction. It may also result in insufficient void space for the amount of asphalt required for high durability, even though stability is satisfactory. Degradation of the aggregate under the action of use may also lead to instability and flushing if the void content of the mix is not sufficient. For these reasons, mixes low in voids should be adjusted by one of the methods given above, even though the stability appears satisfactory.

Voids satisfactory, stability low. Low stability when voids and aggregate grading are satisfactory may indicate some deficiencies in the aggregate. Consideration should be given to improving the quality as discussed above.

Voids high, stability satisfactory. High voids are frequently, although not always, associated with high permeability. High permeability caused by the circulation of air and water through the asphalt cement may lead to premature hardening of the asphalt. Even though stabilities are satisfactory, adjustments should be made to reduce the voids. That can usually be accomplished by increasing the mineral dust content of the mix. In some cases, however, it may be necessary to select or combine aggregates to more closely approximate the gradation of a maximum density grading curve.

Voids high, stability low. Two steps may be necessary when the voids are high and the stability is low. First the voids are adjusted by the methods discussed above. If the adjustment does not also improve the stability, the second step should be an improvement of aggregate quality as discussed above.

Aggregate gradations and fractions. For the purpose of specifications and test reporting it is almost universal practice to specify the gradation of aggregates on the basis of the total aggregate gradation, i.e., the total percent by weight passing the designated sieve sizes. The individual fractions of the total aggregate gradation, however, are designated as follows:

- Coarse aggregate (retained No. 8 sieve)
- Fine aggregate (passing No. 8 sieve)
- Mineral dust (passing No. 200 sieve)

It is also important to note that the aggregate gradations as well as the individual fractions are specified independently of the total mix; i.e., the total aggregate equals 100 percent.

Aggregate materials often are identified in broader terms as rock, sand, and filler. The terms usually are applied to the stockpiled materials supplied to the job site. The following definitions appear to have the greatest usage:

- Rock: material that is predominantly coarse aggregate (retained on No. 8)
- Sand: material that is predominantly fine aggregate (passing No. 8)
- Filler: material that is predominantly mineral dust (passing No. 200)

5.1.5 Functions in coastal structures

Asphalt has many properties that make it particularly suitable for use in hydraulic structures. It is versatile in form and application. It can be used alone (as in an asphalt membrane), or it can be mixed with other materials to produce mixes for a variety of purposes. It can be combined with graded aggregate to form a voidless and impermeable mix, or it can be combined with an open-graded aggregate to form a porous mixture that allows free passage of water.

Many types of asphaltic materials are used in hydraulic applications. Each type can be classified in one of the following distinct categories:

- Impermeable asphalt mixes
- Porous asphalt mixes

- Asphalt mastics
- Asphalt cement
- Prefabricated asphalt materials

These materials can be used in various forms to waterproof, protect, or reinforce a structure. Table 5.1 shows how each type of asphalt material can be used to perform these various functions.

Impermeable asphalt concrete linings. Impermeable asphalt mixes are similar to asphalt mixes for highway paving except that, since a low-void-content mix is required to ensure impermeability, they usually have higher mineral filler and asphalt-cement contents. Also, a harder, or more viscous, grade of asphalt cement normally is used. Mixes are prepared in an asphalt mixing plant and placed with conventional or special paving equipment. Compaction during paving is necessary to produce the required impermeability.

The primary purpose of impermeable asphalt mixes is to waterproof hydraulic structures. Watertight linings are used to impound water in reservoirs, ponds, and lagoons; to waterproof dams, dikes, and embankments; and to prevent seepage losses in canals and channels. They are most often used as surface linings, since they are resistant to wave action and the erosive effects of water currents.

TABLE 5.1   Application Forms of Asphaltic Materials for Hydraulic Structures

Material	Function		
	Waterproof	Protect	Reinforce
Impermeable asphalt mixes	Surface or exposed linings	Surface or exposed linings	
Porous asphalt mixes		Permeable surface linings, open-graded subsurface drainage layers	
Asphalt mastics	Surface linings, seal coats, cutoff walls, dam cores, subsurface injection	Surface linings, seal coats	Grouting, penetration treatments
Asphalt cement	Membranes, seal coats, subsurface pressure injections		Grouting, penetration treatments
Prefabricated asphalt materials	Panels, sheets, mattresses, slabs	Panels, mattresses, slabs	

Revetments constructed with impermeable asphalt mixes are used for bank protection on streams, reservoirs, lakes, and shorelines. Waterproofing properties are not necessarily required in these instances, but quality asphalt concrete linings having low voids effectively resist the destructive effects of wave and current action as well as their abrasive effects (Fig. 5.4). Impermeable asphalt mixes can be used for the entire lining of the structure. They may also constitute a part of a more complex lining. They can, for example, be placed as the surface of a composite section made up of different asphalt layers.

Porous hot-mix asphalt linings. Porous asphalt mixes for hydraulic structures are characterized by the absence or reduced amount of fine aggregate or sand in the mix. As a consequence, the asphalt content also is reduced. The mixes have interconnected pores that permit passage of water. A harder, or more viscous, grade of asphalt cement is desirable in these mixes to allow sufficient film thickness and to prevent drainage from the aggregate. This choice of asphalt also provides additional cohesion in the mix between the aggregate particles.

There are two types of porous asphalt linings: permeable and open-graded. A permeable hot-mix asphalt lining serves as a cover over an earth embankment to protect it from erosion by wave action or surface runoff. An open-graded asphalt lining with higher void content than a permeable lining serves as a drainage layer under an impermeable lining while contributing to the structural strength of the lining. In either case, the purpose is to provide free drainage to prevent hydrostatic pressure from building up in the embankment or within the lin-

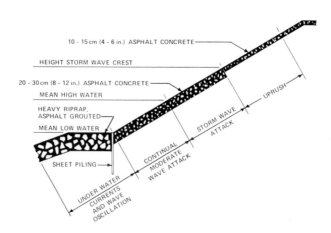

Figure 5.4   Seawall slope revetment.

ing itself. As a surface lining, asphalt allows water to flow to and from the embankment through the lining; as a drainage layer, it collects the subsurface water channeling to drains for removal.

Asphalt mastic mixes. Asphalt mastic mixes for hydraulic structures are essentially mixtures of mineral aggregate and filler wherein the voids in the mineral matrix are overfilled with asphalt cement. The result is an asphalt mix that can be applied by pouring or by hand-floating into place. Asphalt mixes require little or no compaction after placing because void spaces in the aggregate matrix are filled or slightly overfilled with asphalt. Asphalt mastics can be made from a variety of aggregate materials ranging from well-graded coarse and fine aggregates and mineral filler to essentially mineral filler alone with or without an additive such as asbestos fibers. The mastic is voidless except for air bubbles that may be trapped during the manufacture and placing.

Asphalt mastics can be used in several ways to waterproof, protect, or reinforce a hydraulic structure. To waterproof, asphalt mastics have been used for cutoff walls for dams as well as for the central core of the dam itself. They are also used as exposed watertight surface linings.

Asphalt mastic mixes are erosion-resistant, and therefore they can be exposed to waves and abrasive water action. They are also used to form protective covers on embankments or over the floors of channels or estuaries that are subject to erosion. Hot mastic mixes can be placed underwater through tremies or chutes or by simply dumping in masses. They are also used for constructing flexible slabs or mattresses that are lifted into place to form protective blankets or covers.

To reinforce, asphalt mastics are used as grouts to fill and plug the voids in stone structures such as jetties and revetments (Figs. 5.5 and 5.6). The binding action of the mastic tends to make one firm mass, yet mastics are flexible enough to conform to some differential settlement in the structure. Asphalt mastics are also used as joint fillers to

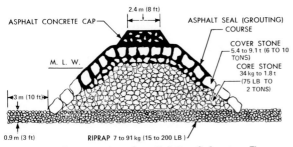

Figure 5.5 Cross section of south jetty, Galveston, Texas.

Figure 5.6   Galveston jetty with asphalt seal. *(Courtesy of Asphalt Institute)*

bind stone blocks together on coastal structures, particularly in European construction.

5.1.6 Surface treatments

A surface treatment can be applied to an asphalt surface for a number of reasons. It may be designed to make the surface more watertight or to protect it from abrasion by waves, water currents, or even by ice. A layer of mud, algae, or other sediment deposited on an asphalt surface and allowed to dry will shrink as it dries and will set up surprisingly large tensile stresses at the surface. The result will be surface curling or cracking. A surface treatment may also be used to protect the surface from mud curl or curling of drying algae along the waterline, to give surface a lighter color in order to reduce temperature extremes, or to reduce the rate of oxidation of the exposed asphalt surface.

Sprayed asphalt seals.   Asphalt cement or emulsified asphalt sprayed over the surface of an asphalt lining at the rate of 1 m^2/L [4 square yards per gallon (yd^2/gal)] will provide a film coating as much as 1 mm (0.04 in) thick. A continuous film coating will fill and seal any exposed pores and increase the watertightness of the asphalt lining. It will also tend to fill and seal small cracks in the surface that may have been caused by improper rolling procedures in compacting the lining. The surface should be clean, dry, and free from loose material. Its temperature should preferably exceed 38°C (100°F). A sloped surface usually necessitates hand spraying. It should be done in a back-and-forth sweeping motion to build up the film and to keep the asphalt from flowing down the slope.

Asphalt mastic seals. In addition to providing a seal, asphalt mastics applied to the surface of asphalt linings provide protection from mechanical abuse. Asphalt mastics, generally placed on an asphalt lining with a screed, permit a coating heavier than that obtained in sprayed applications, and well-designed mixtures can make the surface resistant to abrasion by waves or by scouring by waterborne sands. Asphalt mastic mixtures for this purpose are essentially blends of mineral filler and asphalt cement.

Prefabricated asphalt panels. The typical prefabricated asphalt panel consists of a core of ductile, blown, or oxidized asphalt (asphalt which has certain natural characteristics changed by blowing air through it at elevated temperature) fortified with mineral fillers and reinforcing fibers. The ingredients are blended and molded under heat and pressure. The core is then sandwiched between protective sheets and a protective coating of hot-applied asphalt cement. The protective sheets may be asphalt-impregnated felt or plasticized or flexible glass fabric.

Asphalt panels are usually about 13 mm (½ in) thick, but they are available as thin as 3 mm (⅛ in) thick. They are usually 1.0 to 1.2 m (3 to 4 ft) wide and 3 to 6 m (10 to 20 ft) long for handling and placing.

The most extensive use of prefabricated asphalt panels has been in lining and waterproofing all types of water storage reservoirs, including domestic water reservoirs, sewage lagoons, industrial waste-treatment reservoirs, evaporation ponds, and reflecting pools. They are also used for lining canals and ditches and for bank protection. Prefabricated asphalt panels have been used underneath riprap or rock-reveted embankments to prevent leaching of sand and earth through the rock revetment usually caused by action of waves and tides. Geotechnical fabrics have largely replaced asphalt panels in recent years. Asphalt panels have the advantage of providing a relatively thin watertight barrier that can be used as a surface lining. In addition, their installation does not require heavy machinery. They are useful for relining reservoirs where the concrete lining has cracked badly and where leaking has been excessive. Prefabricated asphalt panel linings are also used as elements of composite lining structures; most frequently they serve as the watertight surface of a built-up lining.

5.1.7 Miscellaneous Uses

Sand asphalt. Sand asphalt is a mixture of sand, with or without added mineral filler, and asphalt cement. If mineral filler is added to the mix, a higher asphalt content and a denser, tougher, and more stable mix are possible.

Sand asphalt has been used alone for linings, as base courses for other linings, for revetments, and for groins, although not in the coastal zone. The largest use of sand asphalt for hydraulics purposes in the United States probably has been for bank paving along the Mississippi River by the U.S. Army Corps of Engineers. The Netherlands has made extensive use of sand asphalt in the construction of seawall revetments. Typically, the base thicknesses range up to 0.20 m (8 in) and are usually capped with a layer of asphalt concrete.

Local sand deposits generally can be used, because gradation is not particularly critical. The asphalt cement should be AC-20 (or equivalent AR or penetration grade) or a higher-viscosity grade. A typical mix has an asphalt content of about 6 percent. If about 5 percent mineral filler is added, the asphalt content will probably be around 8 percent. Sand asphalt mixes for linings are not as watertight as specially designed hydraulic asphalt concrete.

Asphalt primer treatment. The soil surface of a hydraulic structure is often primed with asphalt to seal it temporarily or to reduce seepage until such time as waterborne sediments in the impounded water settle and plug the soil pores. Asphalt primers have also been applied to sloped embankments before a sprayed-asphalt membrane is placed. In this case, the purpose is to anchor the membrane to the slope. Primes have also been used for much the same purpose as prime treatments of roadway surfaces prior to paving operations: to plug up voids and to provide a more stable surface on which to place asphalt construction. Prime treatments are neither watertight nor permanent. They are most applicable to silty sand soils that are quite permeable.

Asphalt injection. Asphalt injection is the subsurface application of asphalt pumped under pressure through pipes. The method is used to reduce leakage of a hydraulic structure through underground cracks, fissures, and cavities. Asphalt has been injected into the subsurface to prevent leaching of soils through rock-reveted embankments at commercially developed sites to prevent surface subsidence behind the embankment.

The hot, fluid asphalt is usually pumped through heated perforated pipes dropped into drilled holes at the leakage strata levels. Once in the leakage channel, the asphalt spreads out and hardens into a tight plug or water stop. If pumping pressure is sufficient, the asphalt will do that even in fissures filled with water. The asphalt plugs can adapt to slight movements in the formation and to changes in water pressure.

Asphalt mattresses. Asphalt mattresses are precast sections or blankets of asphalt mastic reinforced with wire mesh and steel cables or fiber netting and lines. Generally, they vary in thickness from 25 to 50 mm (1 to 2 in). Their length and width are limited only by the size of the molding platform and the capabilities of the equipment used to manipulate and place them.

The reinforced asphalt mattress was developed by the U.S. Army Corps of Engineers in 1932 to 1934 for use on underwater revetments on the banks of the lower Mississippi River. Continuous asphalt mattresses were cast on a special barge pulled into the water. Mattresses have since been adapted for use in European hydraulic structures and in Japan. Their principal function is to protect the surfaces on which they rest from erosion or scour by waves and currents. They are often used at the toes of revetments or linings. After short periods, the edges of the mattresses settle into the scour zone, thus stabilizing the erosive process. Asphalt mattresses are also used as linings and as protective blankets for hydraulic structures.

5.2 Preplaced-Aggregate Concrete

Preplaced-aggregate (PA) concrete derives its name from the unique placement methods by which it is made. Intrusion and grouted concrete are other common names for this type of concrete. In this method of construction, forms are first filled with clean, well-graded coarse aggregate. Structural quality grout is then injected into the voids of the aggregate mass to produce concrete. This method of placing concrete is especially adaptable to underwater construction, to concrete and masonry repairs, and in general to new structures when placement by conventional means is unusually difficult or when concrete of low volume change is required (U.S. Bureau of Reclamation, 1963). It has been used in the construction of bridge piers and atomic reactor shielding, plugs for outlet works in dams, tunnels, and mine workings, and for embedment of penstocks and turbine scroll cases, as well as a great variety of repair work. Recently the process has been used for exposed aggregate and other architectural treatments. Inasmuch as preplaced-aggregate concrete construction is relatively specialized, it is essential that the work be undertaken by well-qualified personnel who are experienced in this method of concrete construction.

Preplaced-aggregate concrete differs from conventional concrete in that the finished product contains a higher percentage of coarse aggregate. Because of point-to-point contact of the coarse aggregate, as placed, drying shrinkage is about one-half that which normally occurs in conventional concrete (U.S. Army Engineer, Waterways Experiment Station, 1954; Shideler and Litvin, 1964).

The higher percentage of coarse aggregate in the concrete has an influence on the modulus of elasticity, which is slightly higher than that of conventional concrete. The other physical properties also appear to be more affected by the properties of the coarse aggregate than are those of conventional concrete. In summary, the physical properties of preplaced-aggregate concrete are similar to those of conventional concrete except that overall drying shrinkage of the former is considerably less. Accordingly, with a properly proportioned and tested grout mix and with good construction practices, allowable working stresses used for conventional concrete structural design may be used (U.S. Army Engineer, WES, 1954).

The economics of use of preplaced-aggregate concrete are a function of site conditions and job requirements. Structural forms for PA concrete are usually more expensive than are those required for conventionally placed concrete because greater care is needed to prevent grout leaks and also because placements usually require additional lateral support. However, in underwater construction, higher placing rates have been achieved by the PA method than by conventional placing methods.

5.2.1 Types of grouts

Slurries of portland cement and water, with or without sand, have long been used in the construction industry for filling rock fissures. Unless sufficient pressure is applied to squeeze out excess water, settlement of solids may result in incomplete filling of voids. Clean sand cement or soil cement slurries may be used for low-pressure backfill grouting of rubble or rockfill when strength is not an important consideration.

As concrete technology has changed, basic grouts composed of portland cement, sand, and water have been modified to more effectively produce structural preplaced-aggregate concrete. Such grouts may be modified chemically by the inclusion of admixtures such as pozzolan, fluidifiers, expansion agents, air-entraining agents, and coloring additives, or the grout may be modified mechanically by use of specially designed high-speed mixers.

5.2.2 Grout and aggregate materials

Cement. Grout can be made with any one of the types of cement that comply with ASTM, Standard C150, Corps of Engineers specification CRD-C 201, which would be suitable for use in conventional concrete and produce the required conditions for preplaced aggregate concrete. The type of cement should be selected in accordance with controlling factors, job conditions, and service exposures, which would also influence the selection of cement for conventional concrete.

Coarse aggregate. Coarse aggregate must be clean, free of surface dust and fines, sound, and durable, and it should conform to ASTM Standard C33, Corps of Engineers specification CRD-C133, for aggregate acceptance, except as to grading. Most important, the coarse aggregate should not be susceptible to excessive breakage and attrition during handling and placing in the forms. The void content of the coarse aggregate after placement in the form will customarily range from 38 to 48 percent. For economy, it is desirable to keep the void content as low as possible to minimize the required volume of the intruded grout. Not only does a low void content result in a saving in cementing materials but, concomitantly, there is less volume change.

The maximum size of aggregate depends on availability, type of construction involved, and the usual limitations established for thickness of section and spacing of reinforcement bars (King, 1959). The minimum recommended size is dependent, essentially, on sand grading. Typical aggregate gradations are shown in Table 5.2. When grout is prepared with sand graded for use in conventional concrete, the minimum coarse aggregate size should be 38 mm (1.5 in). When mason or plaster sand grading is used, the minimum coarse aggregate size may be as low as 13 mm (0.5 in). No limit is placed on maximum size of the coarse aggregate. The coarse aggregate should be well graded up to

TABLE 5.2 Typical Aggregate Gradations for Preplaced-Aggregate Concrete Prepared with Fine Sand Grout Containing Pozzolan and Fluidifier

Typical fine-aggregate grading, cumulative percent passing given sieve						
No. 8 (2.36 mm)	No. 16 (1.18 mm)	No. 30 (600 μm)	No. 50 (300 μm)	No. 100 (150 μm)	Pan	FM
100	97	67	31	10	0	1.95
100	98	72	34	11	0	1.85
100	96	56	36	20	0	1.92

Typical coarse-aggregate grading, cumulative percent passing given sieve								
6 in (150 mm)	4½ in (114 mm)	3 in (75 mm)	1¾ in (45 mm)	1½ in (38.1 mm)	⅞ in (22.4 mm)	¾ in (19.0 mm)	⅝ in (16.0 mm)	½ in (12.5 mm)
			100	97	45	9	2	1
		100		62		4	2	1
	100	78		40		10	2	1
100		67		40		6	2	1

and including the largest size which can be placed economically in the forms without excessive segregation. Gap grading, using a ratio of minimum nominal size coarse aggregate to the maximum nominal size fine aggregate of 10:1 without intermediate sizes, has occasionally been used to achieve exceptionally low void cements. However, this grading is uneconomical for most work. Coarse aggregates as large as the largest stones capable of being carried by a man have been used with good results.

Fine aggregate.   Either crushed or natural sand may be used as fine aggregate. However, well-rounded sand grains from a natural source are preferable because such sands require less water to achieve acceptable grout fluidity. The sand should be hard, dense, durable, uncoated rock particles, and it should have a uniform, stable moisture content. It should conform to current ASTM Standard C33, except with respect to grading.

Pozzolan.   Pozzolan is used to reduce bleeding, to improve fluid properties of the mixture, and to reduce segregation of solid particles. The pozzolan combines with lime liberated during hydration of the cement to form strength-producing compounds at later ages. In massive structures such as dams and seawalls, pozzolans can often minimize the heat of hydration and reduce the buildup of internal temperature. Both natural and manufactured pozzolans have been used, but the pozzolan most generally used and preferred is fly ash conforming to ASTM Standard C618, Corps of Engineers Specification CRD C255. Some pozzolans have caused excessive abrasion of pumping equipment and increased water requirements, so preliminary tests should be made with the selected pozzolan.

Grout admixtures.   A water-reducing, set-retarding agent known as a grout fluidifier is commonly incorporated in the grout mixture to make the mixture more fluid, to reduce the amount of water required for a given fluidity, to delay setting time for ease in handling with pumping equipment, and to promote better penetration of the voids in the coarse aggregate. The agent is customarily a preblended material obtained commercially. It normally consists of a water-reducing agent, a suspending agent, aluminum powder, and a chemical buffer to assure properly timed reaction of the aluminum powder with alkalies in the cement. Reaction of the aluminum powder with alkalies during hydration of the cement generates hydrogen gas which causes expansion of the grout while it is fluid and provides small air bubbles within the grout. Normal dosage of the water-reducing agent in commercially available grout fluidifiers ranges from 0.20 to 0.30 percent by weight of cement plus pozzolan. Aluminum powder is normally employed in the range of 0.01 to 0.02 percent by weight of cement plus pozzolan. The fluidifier should be so proportioned that most of the expansion occurs within 3 h after initial mixing. Preblended grout fluidifiers should conform to Corps of Engineers Specification CRD C619.

5.2.3 Grout mix proportioning

Grout material proportions, as in conventional concrete practice, are influenced by structural design requirements. In addition, the grout must be so designed as to flow freely through the voids of the preplaced aggregate without appreciable segregation or water gain so that honeycombing is avoided and an intimate bond between grout and coarse aggregate particles is ensured. The importance of selecting maximum sand size, compatible with void size as determined by coarse aggregate grading, is reflected in Table 5.2.

Cement-to-sand ratios employed are commonly in the range of 1:1 to 1:2; although ratios as lean as 1:3 cement to sand have been used. Compressive strength and pumpability requirements limit the amount of sand which can be used in any grout (U.S. Army Engineer, WES, 1954). The grout must be sufficiently fluid that it will penetrate and fill all the voids in the aggregate mass yet be of such consistency that the suspended sand and cementing materials do not settle out. For normal structural work, the ratio of cementitious materials (cement plus pozzolan) to sand should be approximately 1:1. Usually, the proportions of cement to pozzolan are 2:1, although ratios as low as 1:1 and up to 9:1 have been used on various jobs. Occasionally, the pozzolan may be omitted entirely. For a structural grout, it is usually not desirable to exceed a cement-to-sand ratio of 1:2 by weight because higher ratios produce lower strengths and excessive segregation of sand in the grout mixture may occur. Mix proportions may be determined by Corps of Engineers Specifications CRD C615.

5.2.4 Physical properties

For structural preplaced-aggregate concrete, when strength and other physical properties are a consideration, the grout should be proportioned and test specimens, using the contemplated coarse aggregate grading, should be made to determine the grout mix proportions which will produce preplaced-aggregate concrete of the required physical properties. Such tests will also provide information as to the quantity of materials needed for the work. When necessary, the information on physical properties of the structural preplaced-aggregate concrete should include strength, resistance to freezing and thawing exposure, modulus of elasticity, drying shrinkage, volume change, or other structural criteria. Physical properties of preplaced-aggregate concrete made with a grout containing pozzolan and a fluidifier have been determined and compared with conventional concrete in a number of laboratory tests. This data can be found in published reports (U.S. Army Engineer, WES, 1954; U.S. Bureau of Reclamation, 1949).

Compressive strength of concrete with a given maximum size aggregate, grout fluidifier, and pozzolan is slightly lower at 28 days age than that of conventional concrete containing entrained air and an equal amount of cementing materials. At 90 days age and later, its strength is equivalent to that of conventional air-entrained concrete (U.S. Army Engineer, WES, 1954). Concrete containing aluminum powder or a grout fluidifier and pozzolan develops a higher bond strength with old concrete than does new, conventional concrete. The reason may be the greater fluidity of the grout as compared with mortar and the expansion of the grout, which develops a slight pressure during the formation of the hydrogen gas.

5.2.5 Placement

Foundation preparation is important in underwater placement. For example, if extremely fine material is left on the foundation or in heavy suspension just above the foundation, it will be displaced upward into the aggregate. The dispersed fine material then coats the aggregate or settles and becomes concentrated in void spaces in the aggregate, thus precluding proper intrusion and consolidation. Therefore, all loose fine material must be removed to the extent possible before placement of aggregate. Alternatively, if structural conditions permit, a layer of sand and gravel may first be deposited to serve as a filter bed to prevent contamination of preplaced coarse aggregate.

Aggregate placement. Coarse aggregate should be washed and screened immediately before it is placed in the forms so that it will be surface-moist at the time of grout injection. Dry aggregate will absorb water from the grout, which will thicken the grout within the aggregate mass and may result in ungrouted or honeycombed areas. If more than one size of coarse aggregate is used, the aggregate should be weighed, batched, and mixed in the proper proportions or discharged at proportional rates onto the wash screen. The wash screen can be either a vibrating deck or a revolving type. The latter is effective as a blender as well as a washer.

For structural concrete work, aggregate is commonly conveyed to the forms in concrete buckets. A flexible rubber elephant trunk is often used to limit the height of free fall (and thus prevent segregation) and for placement in constricted areas. The total fall distance and method of handling should be such that segregation and aggregate breakage are reduced to a minimum. Permissible fall distance depends on aggregate size and soundness. When coarse aggregate is being placed through water in mass concrete work, as in bridge piers, it can be discharged directly into the forms from bottom-dump barges or self-unloading ships.

Coarse aggregate has been placed successfully to depths of well over 30.5 m (100 ft) in water where the possibility of breakage is eliminated (Davis and Haltenhoff, 1956). Some segregation may occur, but segregation itself is not usually objectionable. It tends to result only in a somewhat greater void content and a nonuniform distribution of the void system throughout the aggregate mass. The result is to slightly increase the grout requirements; the effect on the strength of the mass concrete is insignificant. However, an accumulation of smaller sizes might reduce the void size sufficiently to preclude consolidation within the area by the grout.

Grout quality control. The pumpability of grout is controlled by the consistency test in which a standard flow cone is used in accordance with Corps of Engineers CRD C611. To maintain uniformity, time of outflow should be limited to between 18 and 22 seconds (s). However, grouts having an outflow of up to 30 s can be successfully pumped, depending on void content of the aggregate. Test cylinders should be made in accordance with Corps of Engineers Specification CRD C84 and tested in accordance with appropriate ASTM standards.

5.2.6 Curing

Curing should be in accordance with accepted conventional concrete practice. As with any concrete, extended periods of wet curing beyond the usual 7 days will be beneficial in improving the quality of the concrete.

5.3 Portland cement grout

Portland cement grout has a variety of uses in coastal structures; it is similarly varied in its makeup. A simple combination of cement and water, in a flowable consistency, is sometimes used to fill joints and voids in concrete, masonry, or rock. More often, other materials are added to improve various properties or reduce cost. They include sand and clay used as inert fillers and colloidal clays such as bentonite to stabilize fresh grout placed under water. Also included are special-purpose admixtures to increase strength, retard or accelerate set and strength gain, cause expansion of the grout, prevent shrinkage, or improve bond, penetration, impermeability, plasticity, or resistance to freeze-thaw damage or chemical attack. These are discussed more fully in Chap. 4, "Portland Cement Concrete," and in the previous discussion of preplaced-aggregate concrete. In some cases the desired properties can be satisfactorily obtained by using one of the previously described special types of portland cement.

5.3.1 Mixes

When grout can be pumped or poured into relatively open joints or voids, a mixture of one part cement to typically 3 or 4 parts sand is common, with just enough water for satisfactory placement. For very large voids, gravel can also be added. When the material will be pumped into the ground to raise a settled slab or behind a bulkhead to plug a hole or break, a mixture of clay or silt with about 10 percent cement can sometimes be used. For filling or repairing narrow joints or cracks, a neat cement grout consisting of 1 part cement mixed with 1 to 10 parts water to obtain proper consistency may be appropriate. In some cases the use of an admixture may be justified to enhance certain properties. In such a case care should be taken to ensure the suitability of the admixture for the conditions and materials involved and also to ensure that dosage and mixing are correct. Pertinent Corps of Engineers Specifications include CRD C615, CRD C619, CRD C611, and CRD C612.

5.3.2 Placement methods and effects

Portland cement grout is usually placed by one of the following methods:

- Dumping or pouring into large voids or onto flat or sloping surfaces
- Free discharge from a hose or tremie trunk, above or under water, into a form or into voids or cavities in rock, masonry, or concrete
- Free discharge from a hose or pipe
- Into soil to correct settlement of slab or light foundation by displacement or mud jacking or behind a bulkhead for sealing or patching
- Into cracks or joints in concrete, masonry, or rock
- Hand placement by pouring or dry packing

When grout is used to fill joints, cracks, or cavities in structural members, it is important to avoid or minimize shrinkage. That requires the best practicable combination of compaction, low water-cement ratio, and possibly the use of an expanding or plasticizing admixture in the grout. If a significant surface area of freshly placed grout is exposed to the air, especially when the grout has a high cement factor, the surface will need to be kept continuously moist for several days if shrinkage cracks are to be avoided.

If the grout is injected into a confined space, such as immediately behind a bulkhead or into cracks or cavities, caution must be used to avoid building up excessive pressure which could displace or rupture the confining structure. Excessive pressure can be avoided by carefully limiting the injection pressure or by grouting in properly sized increments or lifts and allowing adequate setting time between them. Tests for the setting time of grout are described in Corps of Engineers Specification CRD-C614.

5.3.3 Effects of the environment

Portland cement grout will be affected to varying degrees by the environmental conditions and forces acting on the concrete, masonry, or rock with which it is associated. When grout is placed in thin joints or cracks, it will have some protection from wave action, abrasion, periodic wetting and drying, and fire, but it may be vulnerable to water penetration, freeze-thaw cycles, chemical attack, and seismic forces. When it is used for surface repair, topping, or void filling, it may be exposed to all those environmental forces. Most of them will be adequately resisted by a grout having an optimum combination of strength, impermeability, entrained air content, and freedom from excessive shrinkage. When a certain type of exposure is likely to be severe, the added cost of a beneficial proprietary admixture may be warranted.

5.3.4 Functions in coastal structures

The many uses of portland cement grout can be broadly classified as follows:

- Filling voids in rock revetments (This may be for improved slope stability, erosion resistance against waves, currents, or floating debris, or rat-proofing.)

- Filling joints in precast block revetments

- Sealing voids in stone breakwaters to improve wave attenuation (This should be undertaken very judiciously lest it cause excessive pressure buildup in the breakwater structure under heavy wave action or trap and amplify resonant wave energy within the protected water area.)

- Grouting cyclopean or preplaced-aggregate concrete
- Setting steel piling or tieback anchors in rock
- Filling voids in hollow masonry walls
- Repairing spalled, broken, or cracked concrete
- Plugging breaks or holes in steel or concrete bulkheads
- Correcting foundation settlement

5.4   Soil Cement

Soil cement is a mixture of pulverized soil and measured amounts of portland cement and water compacted to a high density. As the cement hydrates, the mixture becomes hard and increases the stability of the soil.

The term "soil" includes native soils, gravels, sands, crushed materials, and miscellaneous materials such as cinders, slag, caliche, and chert.

5.4.1   Types of soil cement

There are three general types of soil cement mixtures depending on the quantity of cement and water added to the soil.

Compacted soil cement.   The compacted soil cement mixture contains sufficient cement and moisture for maximum compaction tests (ASTM D-558, D-559, and D-560) (CRD C-592, C-593, and C-594). It will withstand laboratory freeze-thaw and wet-dry tests and will meet weight loss criteria.

Cement-modified soil.   Cement-modified soil is an unhardened or semihardened mixture of soil and cement. When relatively small quantities of cement and moisture are added to a soil, the chemical and physical properties of that soil are changed. The soil's plasticity and volume change capacity are reduced and its bearing value is increased. In cement-modified soil, only enough cement is used to change the physical properties of the soil to the desired degree—less cement than is required to produce a hard soil cement. The use of cement to produce a cement-modified soil can be applied to both silt-clay and granular soils to increase the bearing values and reduce plasticity of soil materials.

Plastic soil cement.   Plastic soil cement is a hardened mixture of soil and cement that contains sufficient water, at the time of placing, to produce a consistency similar to that of plastering mortar. Plastic soil cement is used to line or pave steep or irregular slopes for erosion control of banks and ditches.

5.4.2 Mixing soil cement

Since soil cement obtains its stability primarily by the hydration of cement and not by cohesion and internal friction of the materials, practically all soils and soil combinations can be hardened with portland cement. The general suitability of soils for soil cement can be judged, before they are tested, on the basis of their gradation. On that basis, soils for soil cement can be divided into three broad groups.

Sandy and gravelly soils. Sandy and gravelly soils with about 10 to 35 percent silt and clay have the most favorable characteristics and generally require the least amount of cement for hardening if they contain 55 percent or more passing No. 4 sieve. These soils are readily pulverized, easily mixed, and can be built under a wide range of weather conditions.

Sandy soils deficient in fines. Sandy soils deficient in fines, such as some beach, glacial, and windblown sands, make good soil cement, although the amount of cement needed for adequate hardening may be higher than for the first group.

Silty and clayey soils. Silty and clayey soils make satisfactory soil cement, but those containing high clay contents are harder to pulverize. Generally, the more clayey the soil the higher the cement content required to harden it gradually.

Excessively wet soil is difficult to mix and pulverize. Experience has shown that cement can be mixed with sandy soils when the moisture content is as high as 2 percent above optimum. The moisture content of clayey soils should be below optimum for efficient mixing.

5.4.3 Curing

Compacted and finished soil cement contains sufficient moisture for adequate cement hydration. A moisture-retaining cover must be placed over the soil cement soon after completion to retain this moisture and permit the cement to hydrate. Materials such as waterproof paper or plastic sheets, wet straw or sand, wet burlap, or cotton mats are entirely satisfactory.

5.4.4 Engineering properties of soil cement

During construction, the soil cement is compacted to a high density. As the cement hydrates, the mixture hardens in this dense state to produce a structural slablike material and thus possesses engineering properties. The magnitude of these properties depends primarily on the type of soil, age, and curing conditions.

Depending on soil type, 7-day compressive strength of saturated specimens of the minimum cement content meeting soil cement criteria is generally higher than 2.1 MPa (300 lb/in^2). The 28-day flexural strength is approximately 20 percent of the compressive strength, and the modulus of elasticity is about 6900 MPa (10^6 lb/in^2). Soil cement tends to be brittle and to crack under impact and temperature stresses.

5.4.5 Functions of soil cement

Soil cement is used primarily as a base course for stabilizing and compacting soils for foundations, bank protection, and subbase construction. It has been used for earth dam cores, reservoir linings, and slope protection.

5.5 Sulfur Cement Concrete and Grouts

The years since the middle and late 1960s have seen a rapid increase in research and development work on sulfur, and two factors have been responsible. In the early 1960s, large quantities of sulfur were beginning to be recovered from sour natural gas and petroleum. Sulfur producers realized the need to create new end-use markets to absorb the sulfur, and they sponsored research to that end. As a result of the research, sulfur was discovered, or in some cases rediscovered, to have a number of interesting mechanical properties. Research workers who originally had envisioned sulfur as a substitute material now discovered that it had properties superior to those of some conventionally used materials and that it could outperform such materials both technologically and economically. The initial discoveries stimulated additional research on many aspects of sulfur. Many interesting new uses for sulfur in construction have been discovered. Some of the more promising are:

- Sulfur-asphalt paving materials
- Sulfur concretes
- Sulfur coatings
- Sulfur impregnation of materials
- Foamed sulfur

The acceptance of sulfur-asphalt technology by highway departments and contractors has been due not only to the desire to replace asphalt by readily available sulfur but also to the fact that the sulfur-asphalt materials have properties superior to those of asphalt and to the fact that sulfur asphalt permits the use of aggregates which would be unsuitable for use with asphalt.

It has been known for many years that mixing molten sulfur with sand or aggregate produces a sulfur concrete with excellent strength. However, the durability of simple sulfur concretes of this type has not been impressive, particularly under conditions of high humidity and wide temperature fluctuation. Research has centered on developing additives to sulfur to improve the durability. Work carried out by the U.S. Bureau of Mines at Boulder City, Nevada, and by Sulfur Innovations, Ltd., Calgary, Alberta, has resulted in sulfur concretes with greatly improved properties.

Porous materials can be impregnated with molten sulfur that, on solidifying, imparts additional strength to the materials. Resistance to freeze-thaw cycles and corrosion is also frequently improved by sulfur impregnation. Recent research indicates that with suitable additives, sulfur could be made into a rigid foam having excellent mechanical and insulating properties. The properties of sulfur-based construction materials generally equal or surpass those of conventional cementing materials.

5.5.1 Sulfur-asphalt (SA) materials

Basically, all the SA technologies involve combining molten sulfur and hot asphalt to produce a sulfur-asphalt binder, which is then mixed with mineral aggregate to give an SA hot-mix paving material. The individual technologies differ in the method and equipment used to produce the SA binder.

Mixes. Depending on the technology and type of aggregate used, from one-third to one-half (by weight) of the asphalt can be replaced by sulfur. Because sulfur is about twice as heavy as asphalt, sulfur-asphalt binders have densities higher than those of normal asphalt. Because a certain volume of binder is needed to obtain an acceptable void content of the compacted paving material, optimum stability of SA paving materials generally occurs at a somewhat higher binder content (by weight) than when straight asphalt is used. In practice, between 635 and 900 grams (g) (1.4 and 2 lb) of sulfur is needed to replace 454 g (1 lb) of asphalt. Because SA paving materials are three-component systems (sulfur, asphalt, and aggregate) they permit more flexibility of design than with regular two-component (asphalt-aggregate) paving materials.

When hot asphalt and sulfur are mixed, some of the sulfur, about 20 percent by weight of the asphalt, will dissolve in the asphalt. The remainder of the sulfur forms a dispersion of sulfur in asphalt. Both the dissolved and the dispersed sulfur modify the properties of the asphalt. Some of the dissolved sulfur reacts chemically with the asphalt to form polysulfides, which make the asphalt softer and more ductile. [At higher temperatures, above 160°C (320°F), dehydrogenation occurs with the formation of hydrogen sulfide (H_2S) and harder, more viscous asphalts. This type of reaction is undesirable, and it is one of the reasons why temperature control is important in SA technology.]

The dispersed sulfur is present as droplets, most of which are below 5 μm (0.2 mil) in size. These solidify as the paving material cools below the melting point of sulfur (120°C; 248°F), and the resulting solid sulfur phase imparts increased structural strength to the SA paving material.

Properties. Listed below are some of the improved properties of SA paving materials:

- The strength of SA, as measured by the Marshall method, can be designed to be considerably higher than that of regular asphalt.
- The increased strength obtainable with SA may permit the use of lower-quality aggregates or a reduction in paving thickness.
- SA increases the high-temperature stiffness of paving materials without a corresponding increase at low temperatures; therefore, softer asphalts can be used to minimize winter cracking with less danger of deformation at summer temperatures.
- SA has a lower viscosity than regular asphalt and can be mixed at lower temperatures; the result is reduced energy consumption at the hot-mix plant.
- SA improves resistance to water stripping (the breaking of the bond between asphalt and aggregate by a layer of water that forms on the surface of some types of aggregates); therefore, aggregates which would otherwise be unsuitable can be used, and the use of antistripping agents can be reduced or eliminated.
- Resistance to gasoline, diesel fuel, and other solvents is improved.
- Stress fatigue characteristics are improved.

Not only are many of the properties of SA superior to the properties of other concrete binder mixes but also some of these properties may be enhanced by combinations of SA with other materials. The combination of sulfur asphalt and a nonwoven fabric results in a high-performance product that has high tensile strength, is capable of withstanding considerable deformation without breaking, and remains waterproof even when placed on sharp aggregate. The lower viscosity of SA improves the impregnation of the fabric, and the ductility of SA improves the fabric's low-temperature performance. On solidification, the finely dispersed sulfur particles impart additional strength to the impregnated fabric.

Placement. The SA hot mix is handled like regular asphalt hot-mix paving material. Equipment and technology for transporting, placing, and compacting the SA material is identical to that used for regular asphalt hot mix.

5.5.2 Sulfur concrete

Interest in sulfur concrete (SC) is growing rapidly. Research and development on SC is currently being carried out by at least 50 companies and agencies, some of which are actively marketing SC products and materials.

Sulfur concretes are basically simple materials made by mixing sulfur plus certain additives with heated mineral aggregates. On cooling, SC sets to produce a high-strength material with superb corrosion resistance. Early attempts to make and use SC date back more than 100 years. However, current SC technology is a product of the 1960s and 1970s following the discovery and development of suitable additives or plasticizers for the sulfur which impart durability to SC. A considerable number of compounds have been screened as additives. The most popular ones are dicyclopentadiene (DCPD), dipentene (DP), certain proprietary polymeric unsaturated hydrocarbons, and combinations of these materials. Much of the current research work is concentrated on finding additives or combinations of additives which will further improve the durability and performance of SC.

Sulfur concretes can be designed to have compressive and tensile strengths twice or more those of comparable portland cement concretes (PCC), and full strength is reached in hours rather than weeks. Sulfur concretes are extremely corrosion-resistant to many industrial chemicals, including most acids and salts. They are highly resistant to salt water, and marine applications can be attractive.

Additives. Although satisfactory SC had been obtained by the addition of 5 percent DCPD as a modifier to the sulfur, U.S. Bureau of Mines work indicates that a superior product can be obtained by the use of mixed modifiers. These consist of a mixture of DCPD fractions from DCPD manufacture containing three, four, or more units of cyclopentadiene or methylcyclopentadiene per molecule. Several companies market materials of this type. The relative amounts of DCPD and oligomer used in the modifier will vary to some extent with the type of aggregate, but a mixture of 65 percent DCPD and 35 percent oligomer is about optimum in most cases. Five percent by weight of this mixed modifier is added with stirring to the molten sulfur at 130°C (266°F) and allowed to react with it for several hours. To ensure complete reaction of the sulfur with the modifier, the U.S. Bureau of Mines allows the mixture to react for 24 h, but there are indications that this time can be shortened considerably.

The use of mixed modifier rather than straight DCPD results in improved durability and corrosion resistance. It is also easier to prepare the modified sulfur by using the mixed modifier. The reaction of DCPD with sulfur is exothermic, and care must be exercised when adding straight DCPD to molten sulfur to prevent overheating. With mixed modifiers, the reaction is easier to control.

Aggregates. A metallurgist at the U.S. Bureau of Mines (W.C. McBee, Boulder City Laboratory, personal communication, 1981) stressed that each aggregate system must be analyzed and evaluated for its suitability for SC. Generally, limestone aggregates tend to give SC higher strength and freeze-thaw resistance, whereas quartz aggregates give higher corrosion resistance. The salts in chloride- and sulfate-containing aggregates have no effect on bonding, but some aggregates are unacceptable for SC because they react chemically with the binder. Aggregates that contain swelling clays also are undesirable.

Mix proportions. Design is normally for maximum compressive strength, and it is based on the VMA (voids in mineral aggregate).

Procedure. Optimum strength generally coincides with maximum workability and minimum voids level. Coarse and fine aggregates blended to give about 25 percent VMA have been found to be optimal. In the finished SC, 4 to 5 percent voids remain. These voids are not interconnected, and moisture absorption by SC is very low, 0.05 percent or less, whereas PCC often absorbs 3 percent or more. This is an important factor in the resistance of SC to corrosion.

Aggregate grading according to ASTM specification is unsatisfactory for SC. From 6 to 10 percent of fine (200 mesh) material should be included to provide good workability. To prevent dusting, the fine material can be mixed with the modified sulfur before it is added to the heat aggregate in the mixer.

Properties and uses. The quick-curing characteristics of SC make the material attractive in many situations. Eighty percent or more of final strength is reached within a few hours of pouring, compared to several weeks for PCC. Moreover, SC will cure equally well under freezing conditions, which are highly detrimental to PCC curing. SC can tolerate chloride- and sulphate-containing aggregates found in desert areas, because the bonding properties of sulfur are not affected by salts. The good heat insulation characteristics of sulfur and the elimination of water in the manufacturing process are two additional advantages of SC in desert areas (The Sulfur Institute, 1979).

Fire effects. The inherent flammability and low melting point of sulfur impose some limitations on SC use. Flammability can be controlled to some extent by the use of additives, and it is fortunate that the DCPD types of additives used to improve the durability of SC also impart a degree of fire resistance. Sulfur concretes are in any case considerably less of a fire hazard than wood. Because of the low thermal conductivity, heat penetration is slow, and SC can survive short exposures to fire without serious damage. Sulfur concretes do not support combustion, and flame spread is essentially zero.

Structural use limitations. The low melting point of sulfur limits the use of SC in applications such that loss of structural strength in event of a fire could be catastrophic. Thus, SC for load-bearing structures will probably not be used in high-rise apartment buildings. However, its properties appear to make SC fully acceptable for single-story dwellings, as well as for utility buildings and a wide range of prefabricated structures. The SC materials are well suited to specific uses in the coastal zone environment; and when used in a restricted manner, they may resist coastal environments for many years.

5.6 Environmental Considerations

5.6.1 Corrosive and pollutant attacks

Corrosive and pollutant attacks on the types of concrete and grout discussed in this chapter vary with the type of material and the exposure.

Bituminous concrete. Bituminous asphalt is stable in the presence of nearly all chemically laden substances; however, because asphalt is refined from petroleum-based products, exposure to petroleum solvents will cause its deterioration and disintegration. It is normally unaffected by usual concentrations (less than 1 normal) of acids, salts, and other waste materials.

Preplaced-aggregate concrete. Preplaced-aggregate concrete, like portland cement concrete, is rarely attacked by solid, dry chemicals. Corrosive chemicals must be in some minimum concentration and usually are sulfates or acids. Also, organic acids such as acetic, formic, and lactic can be quite destructive to preplaced-aggregate concrete.

Portland cement grout. Portland cement grout is attacked by the same chemicals as portland cement concrete; however, if the grout is not of equal density as concrete, then the chemical attack may be more aggressive.

Soil cement. Soil cement in its normal use is not attacked by most chemicals except for naturally occurring sulfates such as sodium, potassium, and calcium.

Sulfur cement and grout. Sulfur cement and grout are not generally affected by pollutants in the coastal or marine environment, and they can tolerate chloride- and sulfate-containing aggregates. They are highly resistant to the corrosive effects of salt water (The Sulfur Institute, 1979).

5.6.2 Sunlight exposure effects

Bituminous concrete. Bituminous concrete weathers only at the surface when exposed to sunlight and air. The weathering results in the very slow evaporation of solvents near the exposed surface; but because of the impermeability of asphalt, the solvent loss is very slow.

Preplaced-aggregate concrete. Preplaced-aggregate concrete is not affected by sunlight when it is properly protected from evaporation of mixing water during the curing period.

Portland cement grout. Portland cement grout is not affected by sunlight.

Soil cement. Soil cement is not affected by sunlight.

Sulfur cement concrete and grout. Sulfur cement concrete and grout do not appear to be affected by sunlight; but because their development is relatively recent, experience is limited.

5.6.3 Water penetration effects

Bituminous concrete. Bituminous concrete is highly resistant to water penetration because of the impermeability of asphalt. Formulations can be used in porous structures, but the asphalt does not deteriorate in the presence of water unless the water is accompanied by petroleum solvents.

Preplaced-aggregate concrete. Preplaced-aggregate concrete is not normally penetrated by water, but seawater with a high sulfate and chloride content may be moderately aggressive as on portland cement concrete.

Portland cement grout. Portland cement grout will be affected by the same environment as preplaced-aggregate concrete.

Soil cement. Soil cement will not be affected by water penetration.

Sulfur cement concrete and grout. Sulfur cement concrete and grout will not be affected by water penetration.

5.6.4 Wave and current effects

Bituminous concrete. Bituminous concrete is not affected by normal waves and currents unless they have sufficient force to carry suspended matter to cause severe abrasion. The abraded asphalt particles are a stable compound, and they are sufficiently diluted as to not cause any measurable impact on the adjacent environment.

Preplaced-aggregate concrete. Preplaced-aggregate concrete exhibits no effects from waves and currents.

Portland cement grout. Portland cement grout exhibits no effects from waves and currents.

Soil cement. Soil cement should not be used in a wave and current environment because it is brittle and subject to damage by wave impact.

Sulfur cement concrete and grout. Sulfur cement concrete and grout exhibit no direct effect due to waves and currents. Wear by abrasion and cavitation can result from severe wave action.

5.6.5 Effect of severe temperature and ice

Bituminous concrete. Bituminous concrete may be either a hot-mix or cold-mix design. The mixing temperature should be only sufficiently high to ensure dry aggregate; in no case should it exceed 163°C (325°F). Exposure to higher temperatures will cause solvents to dissipate and result in deterioration of the asphalt concrete. Ice does not affect bituminous concrete, but severe cold will cause the concrete to become brittle.

Preplaced-aggregate concrete. Preplaced-aggregate cement is generally not affected by severe temperatures. Air-entraining admixtures will further improve resistance to freezing and thawing as in the case of portland cement concrete.

Portland cement grout. Portland cement grout also is highly resistant to the effects of severe temperatures. Like preplaced-aggregate cement, its resistance can be further enhanced by addition of air-entraining admixtures.

Soil cement. Soil cement is seldom exposed to severe temperatures and ice because those conditions require an increase in cement content. The result is a substantial increase in cost that may eliminate the consideration of soil cement as a construction element in the environment. For severe exposures it is important to note that an excess of cement is not harmful but a deficiency of cement will result in inferior soil cement that will crack and spall.

Sulfur cement concrete and grout. Sulfur cement concrete and grout, because of the low melting temperature of sulfur (130°C; 266°F), will deteriorate at elevated temperatures. Low temperatures have little effect on the material.

5.6.6 Marine organisms

Bituminous concrete. Bituminous concrete is subject to deterioration from crustaceous organisms because it is comparatively soft as a coastal environment construction material. When it is used as a binder with other materials such as large stone, however, that type of deterioration has little effect on the structure.

Preplaced-aggregate concrete. Preplaced-aggregate concrete is not affected by marine organisms.

Portland cement grout. Portland cement grout is not affected by marine organisms if properly mixed and placed.

Soil cement. Soil cement is not usually exposed to an environment containing marine organisms. When it is, many organisms will become attached to the rough surface of an exposed soil cement surface and cause spalling. Boring animals can penetrate the surface but apparently do not find it a desirable environment; therefore, the condition is not common.

Sulfur cement concrete and grout. Sulfur cement concrete and grout have not shown any deterioration from marine organisms. Because of the limited experience with this material, however, long-range performance experience is not available.

5.6.7 Periodic wetting and drying

Bituminous concrete. Bituminous concrete surfaces may tend to develop fine cracks or alligator patterns on exposed surfaces when subjected to alternate wetting and drying. Such cracks seldom develop any significant depth in asphalt concrete and are not serious in coastal structures.

Preplaced-aggregate concrete. Preplaced-aggregate concrete is subject only to the same action as portland cement concrete, i.e., the formation on the surface of D cracks usually in random patterns. The cracks are normally restricted to the surface and do not contribute to serious deterioration.

Portland cement grout. Portland cement grout will react similarly to preplaced-aggregate concrete.

Soil cement. Soil cement may be exposed to alternate cycles of wetting and drying; but because of its nature, this material tends to harden until all of the cement content is fully hydrated. Sufficient experience is not available to predict the long-term effects; but because of the hardened condition, improved resistance to wetting and drying is a reasonable expectation.

Sulfur cement concrete and grout. Sulfur cement concrete and grout are unaffected by periodic wetting and drying.

5.6.8 Wind erosion

None of the five materials discussed in this section are subject to change or deterioration by wind action. Strong winds may pick up sand particles that cause some etching of exposed surfaces of the materials over extended periods of time.

5.6.9 Effects of burrowing animals

Bituminous concrete. Bituminous concrete, being a petroleum-base product, is an inhibitor of animal activity. The concrete has no food value and is therefore unaffected by animals.

Preplaced-aggregate concrete and portland cement grout. Preplaced-aggregate concrete and portland cement grout (like portland cement concrete) are very hard materials and are not disturbed by burrowing animals.

Soil cement.   Soil cement used primarily to stabilize soil is not as hard a material as portland cement concrete. It can be attacked by burrowing animals; but because of the shape and mass of these structural elements (foundations), any damage is usually minor and insignificant.

Sulfur cement concrete and grout.   Sulfur cement concrete and grout are hard and are resistant to activities of burrowing animals. The sulfur content also is a deterrent.

5.6.10   Effects of flora

There are no reports of flora growth having any effect on any of the five materials discussed in this section.

5.6.11   Effects of fire

Bituminous concrete.   Bituminous concrete, because of its petroleum-base content, is subject to serious damage by fire. Because the asphalt binder in it contains very little solvent, this concrete does not normally sustain burning itself.

Preplaced-aggregate concrete and portland cement grout.   Preplaced-aggregate concrete and portland cement grout are resistant to fire and extreme high temperatures.

Soil cement.   Soil cement is fire-resistant.

Sulfur cement concrete and grout .   The inherent flammability and the low melting point of sulfur (130°C; 266°F) result in the loss of structural strength and cause the immediate deterioration of a structure.

5.6.12   Abrasion

Bituminous concrete.   Bituminous concrete has a substantial resistance to wearing away by rubbing and friction. Since it is as hard a material as portland cement concrete or steel, it is not highly resistant to severe impact by large particles; but it has a high degree of resistance to normal sand or wind abrasion. Precise limits for abrasion resistance are not possible to determine, and it is usually best to rely on an analysis of specific environmental conditions to evaluate the physical properties of the designed bituminous concrete to be used.

Preplaced-aggregate concrete and portland cement grout.   Preplaced-aggregate concrete and portland cement grout are subject to abrading or etching by wind or waterborne particles; but the slow rate of the abrasion process in the coastal zone is usually unnoticeable. Abrasion of structures above the water surface may result in some minor dusting problem.

Soil cement. Soil cement, when used as an integral part of a foundation, is not subject to abrasion. When used as a surface stabilizer, it will wear quite readily if subjected to surface traffic, and the result will be increased dusting.

Sulfur cement concrete and grout. Sulfur cement concrete and grout have a relatively high resistance to abrasion. It is believed that admixtures can enhance those properties. Because of the short field experience with this relatively new material, research and testing to develop admixtures that can improve abrasion resistance properties of the material are still underway.

5.6.13 Seismic effect

Bituminous concrete. Bituminous concrete, not being a structurally rigid material, does not resist seismic forces or movement of the earth. Instead, its properties of plasticity and the ability of asphalt to flex allow it to move with a seismic event and thus reduce the possible damage to structures.

Preplaced-aggregate concrete and portland cement grout. The resistance of preplaced-aggregate concrete and portland cement grout to seismic effects is primarily a design problem. Severe seismic forces can cause structural failure that is due to direct damage or to alteration of the foundation resulting in structure settlement and failure or deterioration.

Soil cement. Soil cement used in coastal structures is usually unaffected structurally by seismic activity, but it may shift with earth movement resulting from seismic-induced stress.

Sulfur cement concrete and grout. The resistance of sulfur cement concrete and grout to seismic effects is similar to that of portland cement concrete. Seismic forces can cause structural failure by producing excessive stress in the material, resulting in structural deterioration.

5.6.14 Human activity

Human activity has very little impact on any of the structures discussed above except for the visual impact of graffiti or other defacing action.

5.7   Uses In Coastal Structures

Bituminous concrete is used to perform three basic functions in coastal construction: as a binder or filler to stabilize quarrystone work or soils, as a sealant to prevent the migration or flow of liquids, and, in its asphaltic cement form, as a wearing surface that can be easily repaired or replaced as it is eroded. Bituminous materials are also used in the preservative treatment of wood as discussed in Chap. 8.

Preplaced-aggregate concrete is usually used in large-dimension mass concrete structures when aggregate larger than can be conveniently handled by ordinary mixing methods is desired. Portland cement grout is used generally as a filler and binder for quarrystone work and as a stabilizer for soils. Soil cement is used to strengthen foundation soils and to resist erosion of selected layers of soil.

Sulfur cement and grout are resistant to many environmental attacks in the coastal zone, and they may become economical to use because of their increasing abundance. However, sulfur cement and sulfur asphalt have had limited general use and they must therefore be considered unproven materials. The property of reaching full strength on cooling could be especially useful in making repairs to structures and embankments when the cost of delay is high. In busy cargo terminals, on heavily used roads, or in coastal structures subject to imminent assault by storms, quick repairs to structures not immersed in water could be made. No practical techniques for placing sulfur concrete under water have been developed to date; fast cooling is the problem.

5.7.1   Application to structures

Bituminous concrete is used for reinforcement or grout to fill and plug the voids in stone or rubble mound breakwaters. The binding action of the mastic tends to produce a large firm mass that is flexible enough to conform to some differential settlement of the structure.

An impermeable breakwater could be made by placing uniformly graded stones in layers along the contours of a rubble mound breakwater. Each layer would be bound together with tremie-placed portland cement concrete grout. The resulting mass concrete structure would be accomplished by the preplaced-concrete method except that no forms would be used.

Breakwaters, jetties, and groins could be made in the same manner as offshore structures by using bituminous concrete or preplaced-aggregate methods. In addition to the types of uses described for offshore structures, bituminous concrete products and sealers may be required to make impermeable membranes when required.

Preplaced-aggregate concrete techniques can be used for mass concrete seawalls. Forms for the vertical or specially contoured faces would be required. Otherwise, the placement methods would be as described for offshore structures except that a tremie might not be necessary for layers placed above the waterline. For good void filling, a vibrator would be used to ensure that the concrete grout flowed into all the voids and filled the form.

Portland cement grout could be used to fill synthetic mesh bags and tubes used to form seawall units. Bituminous concrete is used to bind stone blankets together to form a stable mass. It can be used by itself as slope protection. Portland cement grout can be used to bind stones together in a blanket to form a stable but brittle mass. It can also be pumped into weak soils to firm them for foundations or be used to fill voids in earth layers to obtain foundation continuity.

Soil cement techniques can be used where the slope is composed of the right type of soil and the exposure is not subject to severe wave action. It has the advantage of not requiring aggregate materials to be hauled to the site.

Bituminous concrete wearing surfaces are sometimes used to provide economical surfaces to be sacrificed to the wear of ordinary use. Replacement or refurbishment is easily and cheaply done, and the underlying structure remains undisturbed.

Preplaced-aggregate concrete has been used along the perimeter of landfills to act as a combination seawall, retaining wall, and wharf structure. Such structures are usually topped with large reinforced concrete walls that support the fender system and form the "face of wharf."

6

Structural and
Sheet Metals

6.1 Steel

Steel has been an important construction material for marine service since the late 1800s. It obtained this dominance at the expense of wood and iron because of its greater strength and availability. Although other materials may have advantages such as corrosion resistance, steel is relatively inexpensive and strong and is available in various shapes and sizes for marine application.

To ensure the quality of the material used in construction, materials are purchased to specifications. ASTM standard specifications define the requirements to be satisfied by the particular material and indicate the procedure by which it can be determined that these requirements are met. ASTM Standard A6, Standard Specification for General Requirements for Rolled Steel Plates, Shapes, Sheet Piling, and Bars for Structural Use, lists a number of specifications for materials that are suitable for marine application. Other ASTM specifications cover pipe, mechanical tubing, fittings, forgings, and other materials. ASTM specifications suitable for marine application, from both ASTM Standard A6 and other appropriate ASTM specifications for steel materials, are included in Appendix A.

6.1.1 Metallurgy

Metallurgists define carbon steel as an alloy of iron and carbon with the carbon content under 2 percent. Structural steel specifications limit carbon to 0.35 percent or less for weldability considerations.

This chapter was written by Dr. Albert L. Roebuck.

Manganese is added to improve strength and toughness, but mostly to aid in the deoxidation of steel during refining and to modify the detrimental effects of sulfur. Sulfur causes steel to be "hot short," i.e., to be brittle at high temperatures, which can lead to cracking during hot rolling and forging. Manganese combines with sulfur in the molten steel to form insoluble manganese sulfide, some of which is removed as slag and the balance of which remains as well-distributed inclusions throughout the steel. The shape of the inclusions can be controlled by special processing during the steelmaking process.

Phosphorus also is present as an impurity. Most steel specifications require phosphorus to be less than 0.05 percent because larger amounts decrease ductility and toughness and cause the steel to be "cold short." Silicon is often added as a deoxidizing agent during the melting and refining of steel. Copper also may be present up to 0.25 percent.

6.1.2 Carbon steel alloys

Depending on the alloy content, carbon steel can be classified as low, medium, or high alloy. Carbon steel contains only residual elements and elements, such as manganese and silicon, added during the melting and refining stage to obtain a workable product. Low-alloy steel contains up to 1.5 percent of elements added to obtain increased strength or heat treatment capability. Medium-alloy steel contains 1.5 to 11 percent added alloy elements. Above 11 percent alloy element content, the steel is classified as high alloy. The high-alloy steels include both the ferritic and austenitic stainless steels.

6.1.3 Processing

Steels are classified on the basis of the amount of gases evolved during solidification. In the manufacture of steel from pig iron, excess carbon is removed by the exposure of the molten metal to oxygen or air until the desired carbon content is reached. Oxygen dissolves in the molten metal and reacts with carbon to form bubbles of carbon monoxide, which rise to the surface. Other sources of oxygen include oxygen contained in materials added to the slag or molten metal and that which is present as a product of chemical reactions that occur during the steel-making process. Because carbon and oxygen may continue to react during solidification or the steel may have an unacceptably high oxygen content, deoxidation in the ladle of the molten steel may be required. The deoxidation practice, which may or may not be part of the steel specification, is often an important consideration in choosing a steel for a particular application. Steels with decreasing degrees of gas evolution or deoxidation are termed "rimmed," "capped," "semikilled," and "killed." Figure 6.1 shows sketches of typical ingot cross sections corresponding to those degrees of deoxidation.

Rimmed steels do not receive any silicon or aluminum deoxidation before being poured into ingots. Carbon and oxygen dissolved in the molten steel continue to combine and form small bubbles which effervesce during solidification. Also, some effervescence is due to the much lower solubility of oxygen in solid steel compared to that in molten steel. Chemical composition varies widely throughout rimmed steel ingots. The area near the surface is much lower in carbon, sulfur, and phosphorus than the remainder of the ingot. This low-carbon skin persists in the finished mill product and contributes to the superior deep-drawing properties of rimmed steel.

Capped steel is similar to rimmed steel except that the effervescence of gas from the molten steel in the ingot mold is allowed to occur for only a minute or so before a cast-iron cap is placed on the mold. Capped steel has a thinner low-carbon rim than rimmed steel, but it is of more uniform composition. Capped steel is used for plate, strip, pipe, trim plates, wire, and bars.

Semikilled steel is produced by silicon and other deoxidation elements added during the manufacture. The additions are carefully made in order to balance evolution of gases with solidification shrinkage. Semikilled steel is used for structural shapes, plate, pipe, forging billet and bars.

Killed steel is produced by deoxidizers in excess of the amounts required to fully remove all oxygen from the ladle. As the molten killed steel solidifies, a large shrinkage cavity forms. It is called a pipe, and it is shown in Fig. 6.1. It is necessary to cut off the top of a killed steel

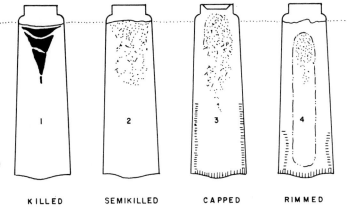

KILLED   SEMIKILLED   CAPPED   RIMMED

Figure 6.1 Typical ingot structures.

ingot at the bottom of the pipe to avoid producing a defect, known as a seam, in the rolled product. Killed steel is more uniform than semikilled steel in both composition and mechanical properties. However, because of the low yield of product per ingot, it is more costly than semikilled steel.

The choice of deoxidizer used is often dictated by the specification chosen by the user after considering the end use. If high-temperature use is contemplated (e.g., for use in a steam boiler) silicon coarse-grain killed steel will be specified because of its improved resistance to deformation at high temperatures. When improved resistance to brittle failure is desired, particularly for service at temperatures below 20°C (68°F) a silicon-aluminum dioxidation practice will be specified to produce a steel having a fine-grain structure. Vacuum degassing also is used when premium quality is required.

Heat treatment. Mechanical properties of steel can be altered by heat treatment. When steel is heated above a critical temperature (the specific temperature depends on the composition), transformation of the microstructure into a single-phase solid solution occurs. This solid solution is called austenite. The temperature at which transformation takes place is called the austenitizing temperature. Steel heated to the austenitizing temperature and allowed to cool in the furnace to a temperature low enough for the steel to be handled is said to be annealed. Annealing is performed to reduce hardness, improve machinability, and facilitate cold working.

Normalizing is the process of heating the steel above the austenitizing temperature, allowing sufficient time for transformation to occur, and then removing the steel from the furnace and cooling in air. Normalizing is performed to refine grain size and homogenize microstructure, improve machinability, and/or provide the desired mechanical properties. Normalizing is usually performed on steels that require additional heat treatment for hardening and on hot-formed pressure vessel heads and when specified by the applicable material specification.

Steels having sufficient carbon can be hardened by heating above the austenitizing temperature, holding at that temperature long enough for transformation to occur, removing the steel from the furnace, and immediately quenching it in water or oil. The resulting surface hardness depends on the carbon content, section size, and quenching medium. The depth of hardening also depends on alloy content and grain size. After quenching, steels are tempered by heating them to specified temperatures (below the transformation temperatures) and holding them at those temperatures for specified times, usually an hour per 2.54 cm (1 in) of thickness. The process restores ductility. The particular tempering temperature used depends on the alloy content, mechanical strength requirements, and end use.

Low-carbon steels are often stress-relieved by heating them between 593°C (1100°F) and the austenite transformation temperature to remove residual stresses resulting from prior forming or welding operations. Stress relieving restores ductility and toughness. It may also improve fatigue life. Welds areas are often postweld heat-treated locally, i.e., stress-relieved, by using proprietary portable heating equipment.

Alloy additions. Alloying elements are added during the steel-making process to improve mechanical properties or to improve corrosion resistance. Small additions, singly, of copper, nickel, chromium, silicon, and phosphorus have been shown to be effective in improving the corrosion resistance of steel. The greatest improvements in corrosion resistance are obtained by adding specific combinations of the alloying elements such as those specified by ASTM Standard A690 for H-piles and sheet piles intended for service in the splash zone. Other steels that are suitable for marine applications and have improved atmospheric corrosion resistance are ASTM Standards A242, A441, and A588.

Chromium and molybdenum improve the high-temperature oxidation resistance as well as the high-temperature strength of steel. High-pressure steam tubes and piping are often 1.25 percent chromium-alloy steel. Stainless steels that meet ASTM 400 series specifications include type 410, with 12 percent chromium, and type 430 with 18 percent chromium. However, because of a tendency to pit, the 400 series stainless steels are not recommended for marine service.

6.2 Aluminum

Aluminum, in high-purity form, is soft and ductile but does not possess enough strength for most commercial applications. Alloying elements, added either singly or in combination, impart strength to the metal. Aluminum alloys can be put into two categories: non-heat-treatable and heat-treatable. The non-heat-treatable wrought alloys can be strengthened by cold working only and are usually designated as the 1000, 3000, 4000, or 5000 series. The degree of cold working is termed the aluminum strain hardening or temper, which is denoted by an H followed by a number.

Certain alloying elements, such as copper, magnesium, zinc, and silicon, show increasing solid solubility in aluminum with increasing temperatures. Many aluminum alloys that contain these elements can be heat-treated to enhance initial strengths. The alloy is heat-treated by first raising it to an elevated temperature below the melting point, called the solutioning temperature, which puts the soluble element or elements into solid solution. This is followed by quickly cooling the material, usually by quenching in water, to retain the elements in solid solution at room temperature. At this stage the freshly quenched alloy structure is very workable.

By storing such material at below-freezing temperatures, the workable alloy structure can be retained until the fabrication is ready to form the alloy into the desired final shape. Such an alloy is not stable at room or elevated temperatures after quenching because precipitation of the constituents from the supersaturated solution takes place. After a period of several days at room temperature or hours at an elevated temperature, the alloy is considerably stronger. This process is called age hardening or precipitation hardening. The degree of hardening or temper produced by heat treatment is denoted by a T followed by a number.

6.2.1   Identification of aluminum alloys

Aluminum alloys are identified by specific numbers. Alloys belong to certain series depending on the particular alloying elements. The 1000 series consists of the high-purity aluminums containing at least 99 percent aluminum. These alloys are characterized by having high thermal and electrical conductivity, excellent corrosion resistance, excellent workability, but low strength. They can be hardened only by cold working. Major impurities are iron and silicon.

Copper is the major alloying element of the 2000 series. The alloys of the series are solution-heat-treated to obtain optimum properties. Some alloys of the series are aged at slightly elevated temperatures, a process called artificial aging, to obtain increased yield strengths. The corrosion resistance of the alloys in the 2000 series is less than that of most of the other aluminum alloys. Sheet forms of these alloys are often clad with a high-purity alloy or a magnesium-silicon alloy of the 6000 series, which provides galvanic protection to the core material and therefore increases resistance to corrosion. Manganese is the principal alloying element of the 3000 series alloys. Alloys of this group generally cannot be heat-treated, but they can be hardened by cold working.

Silicon is the major alloying element of the 4000 series, which, when present in sufficient quantities, lowers the melting point without producing brittleness. Aluminum-silicon alloys are used in welding and brazing wire when a lower melting point is beneficial in joining other aluminum alloys. Although most alloys of this group are non-heat-treatable during welding of heat-treatable alloys, some elements from the parent material may be picked up by the weld metal and provide joints that can be strengthened by heat treatment.

The alloys of the 5000 series contain magnesium. Although they are non-heat-treatable, the magnesium makes them have moderate to high strength, good welding characteristics, and good corrosion resistance to marine atmospheres. They are subject to stress corrosion cracking if employed in the cold-worked condition in services such that the temperature exceeds about 65°C (150°F).

The aluminum alloys of the 6000 series contain both silicon and magnesium, in approximately equal proportions, which combine during melting to form magnesium silicide. The alloys are heat-treatable, and they possess good formability and corrosion resistance with medium strength. One of the most versatile heat-treatable alloys is the major alloy of this series, 6061.

Zinc is the major alloying element of the 7000 series; and when it is coupled with a smaller percentage of magnesium, it produces heat-treatable alloys of very high strength. Small amounts of other elements such as chromium and copper also may be added. Alloys in this series are used in airframe structures and for high-stressed parts. Among the high-strength aluminum alloys, 7075 can be heat-treated to 565 MPa (82,000 lb/in^2) tensile strength and 496 MPa (72,000 lb/in^2) yield strength.

The complete designation of aluminum alloys includes the temper designation, which is separated from the alloy designation by a hyphen as in 7061-T6. The basic temper designations are as follows:

F   As fabricated. No special control is exercised over thermal conditions or strain hardening.

O   Annealed. Heat-treated to obtain lowest-strength temper and improved ductility.

H   Strain-hardened (wrought products only).

W   Solution heat-treated (applies only to alloys hardenable by thermal heat treatment) is an unstable temper. Subfreezing is sometimes used to preserve this temper against natural aging.

T   Thermally treated to produce stable tempers other than F, O, or H.

Numbers following the basic temper designations further describe the specific combination of operations affecting the temper and in turn the mechanical properties. Specifications, such as those of the American Society for Testing and Materials (ASTM), fully define the alloy composition, mechanical properties, and other requirements for applicable aluminum materials. Alloys 5083, 5086, 5052, and 6061 are the most popular aluminum alloys for applications in marine atmospheres. The 5000 series of alloys are the most corrosion-resistant, but the 1000, 3000, and 6000 alloys have been used in marine atmospheres. These aluminums can also be employed in the splash zone, but they are not recommended for continuous immersion in seawater.

6.2.2 Copper and copper alloys

Copper has several unique properties. High thermal and electrical conductivity, excellent corrosion resistance in normal atmospheric conditions, good workability, and availability at reasonable cost make it a first choice for conductors in electrical equipment. Its alloys can have improved strength, corrosion resistance, creep resistance, and machinability.

The U.S. copper industry, through the Copper Development Association, formerly designated alloys by using a three-digit identification system, but now it uses a five-digit number with a prefixed letter C to conform to the Unified Numbering System for Metals and Alloys (UNC) developed and managed jointly by ASTM and the Society of Automotive Engineers (SAE). In the UNS system, numbers C10000 through C79999 denote wrought alloys. Cast alloys are numbered from C80000 through C99990. Within those two categories, the alloy compositions are grouped into families of coppers and copper alloys as presented in Table 6.1.

Copper, like other metals that have recrystallization temperatures, or softening temperatures, above room temperature can be hardened by cold working. If the cold-worked metal is exposed to temperatures above a certain critical temperature determined by the amount of cold work received and the composition of the metal or alloy, the microstructure changes from marked distortion to a recrystallized structure. Yield strength, tensile strength, and hardness are reduced to the same levels the alloy had before cold working. The recrystallization temperature, or softening temperature, of copper can be raised by adding sufficient quantities of silver, phosphorus, cadmium, tin, arsenic, or antimony. Copper is often alloyed to raise the softening temperature to above the temperature at which soldering is to be performed so that the benefits of increased strength due to cold working can be retained in the final product.

Copper alloys that are precipitation-hardenable contain beryllium, chromium, zirconium, or nickel in combination with silicon or phosphorus. Alpha aluminum bronze containing cobalt or nickel also is precipitation-hardenable. During hardening, the alloy is heated to an elevated temperature, held a sufficient time for solid solutioning to occur, and then rapidly cooled to room temperature. That is followed by aging at an intermediate temperature. Beryllium copper (C17200) in

TABLE 6.1   Classification of Coppers and Copper Alloys

Copper No.	Classification
C10000–C15999	Coppers (Cu 99.3%)
C16000–C19999	High-copper alloys (96% <Cu < 99.3%)
C20000–C29999	Copper-zinc alloys (brasses)
C30000–C39999	Copper-zinc-lead alloys (leaded brasses)
C40000–C49999	Copper-zinc-tin alloys (tin brasses)
C50000–C52999	Copper-tin alloys (phosphor bronzes)
C53000–C59999	Copper-tin-lead alloys (leaded phosphor bronzes)
C60000–C64699	Copper-aluminum alloys (aluminum bronzes)
C64700–C66399	Copper-silicon alloys (silicon bronzes)
C66400–C69999	Miscellaneous copper-zinc alloys
C70000–C72999	Copper-nickel alloys
C73000–C79999	Copper-nickel-zinc alloys (nickel-silver)
C80000–C81199	Coppers (Cu 99.3%), cast
C81200–C82999	High-copper alloys (96% < Cu < 99.3%), cast
C83000–C83999	Copper-tin-zinc and copper-tin-zinc-lead alloys (red brasses and leaded red brasses), cast
C84000–C84999	Semired brasses and leaded semired brasses, cast
C85000–C85999	Yellow brasses and leaded yellow brasses, cast
C86000–C86999	Manganese and leaded manganese bronze alloys, cast
C87000–C87999	Copper-zinc-silicon alloys (silicon bronzes and silicon brasses), cast
C90000–C91999	Copper-tin alloys (tin bronzes), cast
C92000–C94699	Copper-tin-lead alloys (leaded tin bronzes and high-leaded tin bronzes), cast
C94700–C94999	Copper-tin-nickel alloys (nickel-tin bronzes)
C95000–C95999	Copper-aluminum-iron and copper-aluminum-iron-nickel alloys (aluminum bronzes), cast
C96000–C96999	Copper-nickel-iron alloys (copper-nickel), cast
C97000–C97999	Copper-nickel-zinc alloys (nickel-silver), cast
C98000–C98999	Copper-lead alloys (leaded copper), cast
C99000–C99990	Special alloys, cast

the solution-annealed and aged condition usually has a tensile strength of 1210 MPa (175,000 lb/in^2).

Copper and copper alloys have useful corrosion resistance for marine applications. Most corrosion-resistant to seawater are aluminum brass, classified as a miscellaneous copper-zinc alloy, inhibited admiralty, a tin brass containing elements which inhibit the loss of zinc, and the copper-nickel alloys. These alloys form films of corrosion products that provide protection even in flowing seawater. The limiting velocity at which these films are lost depends on the alloy. Copper and copper alloys are attacked by ammonium hydroxide because a soluble component is formed. Copper alloys containing more than 15 percent zinc are made susceptible to stress corrosion cracking by ammonium ion and also dezincification, i.e. the loss of zinc due to selective corrosion. Stress corrosion cracking occurs at areas of high stress that can become more anodic than the surrounding metal. Corrosion occurs at the interfaces of the metal crystals that are perpendicular to the stress, and the bonding between crystals is weakened until cracking occurs. Dezincification occurs in waters having a high oxygen and carbon dioxide content.

6.2.3 Other alloys

Nickel aluminum bronzes and two-phase aluminum bronzes are transformation-hardenable. Such an alloy is heat-treated at an elevated temperature to form a single-phase solid solution and held a sufficient time for solution to occur. It is then cooled rapidly to produce a metastable, ordered, close-packed-hexagonal beta phase structure much like the transformation structure that is formed during the quenching of high-carbon steel from a temperature above the austenitizing temperature. This structure is very hard, but it is too brittle for most engineering purposes. It must be tempered by heating it to an intermediate temperature, typically 595 to 650°C (1100 to 1200°F), and holding it for a sufficient time to reprecipitate fine acicular alpha phase particles in the tempered beta phase structure. Tempering stabilizes the structure and restores ductility and toughness.

6.2.4 Galvanic coupling

When two dissimilar metals are in electrical contact with each other and are immersed in an electrolyte, a potential is established and electric current may flow. This potential is related to the relative tendency of each of the metals to go into solution. The more active metal acts as the anode and corrodes at a faster rate than it would by itself. The more noble (stable) metal acts as the cathode and is protected. This phenomenon is known as galvanic corrosion. The two dissimilar metals electrically connected are called a galvanic couple. Table 6.2

TABLE 6.2 Galvanic Series in Flowing Seawater [2.4 to 4.0 m/s (7.9 to 13 ft/s)] at Ambient Temperature

Magnesium

Zinc

Aluminum alloys

Calcium

Carbon steel

Cast iron

Austenitic nickel cast iron

Copper-nickel alloys

Ferritic and mortensitic stainless steel (passive)

Nickel-copper alloys, 400, K-500

Austenitic stainless steels (passive)

Alloy 20

Nickel-chromium-molybdenum alloy C

Titanium

Graphite

Platinum

presents a galvanic series for flowing seawater at ambient temperature for several metals and alloys. The series is based on practical measurements of corrosion potentials at equilibrium in seawater. Galvanic corrosion is most likely to occur if the two metals are widely separated in the series. The rate of corrosion is dependent on current density.

If two dissimilar metals must be joined, several steps may be taken to minimize galvanic corrosion:

- Choose metals close together in the galvanic series to reduce the potential.

- Avoid unfavorable area effects by keeping the cathodic area small in relation to the anode and thereby reducing current density.

- Insulate the two metals from each other while making sure that contact is not restored in service by grounding or by corrosion products bridging the insulator.

- Use coatings: The anodic material must be completely covered to prevent rapid attack at holidays in the coating. Sometimes it is also beneficial to coat the cathodic material to reduce current density.

- Place a more anodic third metal in contact with the other two metals so that the third metal provides sacrificial protection.

6.3   Joining, Cutting, and Repairing Metals

At one time, riveting was the primary means of joining metals together. Today the importance of riveting in construction has lessened because of the development of welding and high-strength bolting. The American Society of Mechanical Engineers (ASME) Boiler and Pressure Vessel Code no longer lists riveting as an acceptable method for pressure vessel fabrication, although repairs to riveted vessels can be made in accordance with the Code requirements that were used for the vessel construction. Riveted joints have one important advantage over bolted joints: when properly set, rivets do not loosen. In spite of its lack of favor in construction, riveting is an important joining method in manufacturing.

Rivets are made from bar stock by hot or cold forming the heat. Round button heads are most common, but flattened and countersunk types also are produced. For structural steel fabrication, steel rivets should be specified to ASTM Standard A502, Steel Structural Rivets. This standard lists three grades, all of which are intended to be hot-driven. Grade 1 is a carbon steel rivet for general purposes, usually for joining steel conforming to ASTM Standard A36. Grade 2 is a carbon-manganese steel rivet used for joining high-strength carbon and high-strength low-alloy structural steels. Grade 3 has about the same strength as grade 2 rivet steel; but because copper and chromium are required in the steel composition, grade 3 rivets have enhanced atmospheric corrosion resistance to approximately four times that of carbon steel without copper. Grade 3 rivets correspond to steels conforming to ASTM Standard 588, High Strength Low-alloy Structural Steel with 340 MPa (50,000 lb/in^2) Minimum Yield Point to 10.2 cm (4 in) Thick. Steel rivets also are listed in ASTM Standard A31, Boiler Rivet Steel and Rivets, for repair of riveted boilers and pressure vessels, and in ASTM A131, Structural Steel for Ships. Rivets meeting the requirements of ASTM A31 or ASTM A131 are not suitable for structural construction unless they have also met requirements of ASTM A502. One important difference between these standards is that ASTM A502 requires hardness tests, whereas the other standards specify tensile tests on the rivet steel.

Holes for rivets may be punched or drilled. If they are punched, it is recommended that they also be reamed to remove distorted metal, particularly if the structure is to be subjected to vibration. For steel construction, the diameter of the holes is usually 1.6 mm (1/16 in) larger than the nominal diameter of the undriven rivets. Flame cutting of holes is not recommended because of the microstructural changes that occur in steel.

Temporary bolts are often inserted in a few holes to serve as an alignment aid and to help draw the structural steel members together. Steel rivets are usually driven hot by heating to 982°C (1800°F). During driving, a second head is formed and the rivet shank may be expanded to fill the hole. As the rivet cools, it shrinks and squeezes the connected pieces together. The magnitude of this clamping force depends on the driving and finishing temperature of the riveting operation, the overall grip length, and the driving pressure. Because these are variables and are difficult to control, no credit may be claimed for clamping force in design calculations.

Riveting is used to advantage in joining aluminum structural alloys that have been heat-treated for greater strength. The high heat encountered in welding reduces the strength of heat-treated aluminum. Less skill is required for riveting than for welding. The specification covering wire and rod to be used in aluminum rivet manufacture is ASTM Standard B316-75, Aluminum-alloy Rivet and Cold-Heating Wire and Rods. Of the alloys listed, 1100, 3003, 5005, 5052, 5056, and 6061 are most suitable for joining aluminum alloys in coated structures. Alloy 6061 is the only alloy that can be heat-treated to obtain higher strength levels.

Bolts are made from bar stock. High-strength steel bolts are made by the open-hearth, basic oxygen, or electric furnace process. The bolts are fine-grained and must meet ASTM Standard A588-75. The atmospheric corrosion resistance of the steel is approximately twice that of carbon structural steel with copper. High-strength bolts are made by using various types of quenching and tempering processes and are used in structural connections such that high stress and corrosion resistance are required.

Bolts are used to advantage in structural installation where welding is not practical and where working connections such as tongue-and-groove pile connections, bulkhead wales, and tiebacks are necessary. It is common practice to oversize bolt requirements in marine exposures as an allowance for excessive metal loss from corrosion.

6.3.1 Welding

Welding processes most likely to be used during coastal structure construction include gas welding, arc welding, and thermite welding. Other processes, such as resistance welding, friction welding, and induction welding, are used during fabrication of mechanical and electrical equipment. Each welding process has areas of application where its use is the most economical for the desired level of quality.

Gas welding is classified by the gases used; i.e., air-acetylene, oxyacetylene, oxystabilized methylacetylene-propadine, and oxyhydrogen. The oxyacetylene flame has the highest temperature, about 3371°C (6100°F), that can be obtained with commercially available gases. Because the temperature produced by the oxyacetylene flame is far above the melting points of most metals, the rapid localized melting necessary for welding is produced.

Oxyacetylene is suitable for welding carbon and alloy steel, cast iron, copper, nickel, aluminum, and zinc alloys. Lower-melting-temperature alloys, such as those of aluminum, magnesium, zinc, lead, and some precious metals, can be gas-welded by using hydrogen, methane, or propane fuel gases. Gas welding is not suitable for joining the refractioning metals such as columbium, tantalum, molybdenum, and tungsten nor a reactive metal such as titanium or zirconium.

By varying the relative amounts of fuel gas and oxygen in the gases flowing to the tip of the welding torch, the characteristics of the flame can be altered. When fuel gas and oxygen are supplied in the stoichiometric ratio for complete combustion, a neutral flame is produced. As more oxygen is introduced, an oxidizing flame is produced. Slightly less oxygen than is required for a neutral flame results in a reducing flame. Still less oxygen results in a carburizing or carbonimpregnating flame. In any flame, the highest temperature is reached at the tip of the inner cone.

In oxyacetylene welding, an oxidizing flame is never used to weld steel but is sometimes used to weld copper and copper base alloys. The copper oxide slag that forms on top of the weld provides shielding from the weld puddle. Temperatures exceeding 315°C (600°F) can be obtained in oxidizing oxyacetylene flames. The reducing flame is frequently used for welding with low-alloy steel rods. Flame temperature at the tip of the inner cone is usually 2930° to 3040°C (5300 to 5500°F). The carburizing flame has a tendency to soot the cold work, but it is useful when lower temperatures are required, as for silver brazing, soldering, and the melting of lead. For most oxyacetylene welding, a neutral flame is used. When steel is welded, the outer envelope provides protection to the molten weld puddle and no flux is required.

Fluxes are required when stainless steel, cast iron, and most nonferrous metals are oxyacetylene welded. There is no flux suitable for all metals. The functions of the flux are to clean the metal surfaces to be joined and to provide protection to the weld puddle by lowering the melting point of the metal oxides or dissolving the oxides so they rise to the top of the weld pool and form a protective slag covering.

Fluxes are not required for welding lead, zinc, and some precious metals.

6.3.2   Underwater arc welding

Although many experts consider underwater welding to be suitable only for emergency ship repairs of a semipermanent nature, satisfactory permanent welds can be accomplished by using special techniques. Sometimes underwater welding is the only practical method of making attachments to or repairs of such underwater structures as drilling platforms or bulkheads. Three different techniques have been used for performing welding below the waterline: wet welding, dry welding, and welding using either a caisson open to the surface attached to the area to be welded or a special habitat constructed around the area to be welded.

Underwater wet arc welding requires the use of divers in full deep-diving suits. The helmets are fitted with supplementary hinged faceplates with appropriate welding glass. It is advisable for the diver's head to be insulated from the helmet by a cap, and the metal should be covered with insulating tape. Scuba diving is suitable only at shallow depths because it is necessary for the diver to be in voice communication with a topside assistant. The assistant operates the power source at the command of the diver-welder.

Electrodes for underwater welding must be waterproofed by using proprietary products or coatings of cellulose acetate, although some brands of electrodes are satisfactory without additional coatings. Because the coatings can be considered to provide only temporary protection, divers should carry only a few electrodes at a time. The Navy recommends 4.75-mm (3/16-in) electrodes of type E6013 for all positions except when the section is too thin for that size electrode. Other electrodes, such as waterproofed iron powder electrodes, may be satisfactory. Qualification testing should be performed. Electrodes for underwater welding are designed for straight polarity; that is, the electrode is negative. If reverse polarity is used underwater, the electrode holder is consumed by electrolytic action. It is important that the electrode holder be insulated and be designed to permit easy changing of electrodes by the diver.

Power sources for underwater arc welding should be capable of delivering at least 300 A of rectified or direct current. Because welding is usually done at a considerable distance from the power source, the welding cables should be at least size 2/0. To facilitate maneuverability, the last 3 m (10 ft) of cable at the electrode holder is usually size 1/0. A safety switch is installed in the circuit, and it is closed only while the welder is actually welding. For good electrical countinuity, the ground cable must be securely attached to the work after the contact area is cleaned. Provided the pieces to be welded fit together properly, 4.75-mm (0.19-in) fillet welds can develop 44 kn (10,000 lb) tensile strength per 25.4 mm (1 in). By using 4.75-mm (0.19-in) electrodes and the drag technique whereby the electrode is allowed to consume itself as it is pressed against the work, 4.75-mm (0.19-in) fillet welds are produced in a single pass. The stringer bead technique should be used if additional weld reinforcement is required. Because visibility under water is poor, multipass welds are difficult to finish after the first bead is laid because the guiding groove is filled. Fillet welds can be made in the horizontal, vertical, and overhead positions.

Bubbles generated during welding interfere with visibility. Welders minimize the problem by welding toward themselves when making horizontal welds and from the top down when making vertical welds.

Underwater welds in mild steel plate develop 80 percent or more of the tensile strength but only 50 percent of the ductility of similar welds made in air. This substantial decrease in ductility is explained by the hardening resulting from the drastic quenching of the surrounding water. It is not possible to preheat weld areas wet by water. Therefore, to avoid cracking, underwater welding should not be attempted on base materials having carbon contents above 0.25 percent or carbon equivalents (percent carbon plus 0.17 percent manganese) above 0.40 percent. To assure sound welds, the area to be welded must be free from marine growth, paint, mill scale, and rust. Electric shock is a hazard that must be taken into account by equipment and safety procedures. Another hazard that could be overlooked is a possible explosion resulting from the accumulation of hydrogen and oxygen gas in closed or inadequately ventilated compartments or spaces. Bubbles generated during arc welding are about 70 percent hydrogen and are produced by electrolysis of the water.

Underwater work including welding has been accomplished dry in air by using a caisson open to the surface. Such a structure must be strongly constructed to stand the pressure of the water, approximately 9.8 kPa per meter of depth (62.4 lb/ft^2 per foot of depth), depending on salt content and temperature. A caisson has been used to repair a tear, 13.7 m (45 ft) below the surface, in the stainless steel liner of a waterfilled storage pool. Habitats have been used to make underwater modifications to a drilling platform in the Gulf of Mexico. Habitats can be constructed to surround the areas to be welded and filled with air so that preheating of the weld areas is possible. A weld produced under these conditions will basically have the same strength and ductility as a weld produced under the same conditions topside. A habitat used in welding is usually open at the bottom; and because of buoyancy, it must be securely attached and weighted. A constant flow of air through the habitat is necessary to remove the fumes produced by welding, but hydrogen formation by electrolysis is avoided because water is kept away from the arc.

6.3.3 Underwater cutting

Underwater cutting is used in salvage work and whenever cutting below the waterline is required on steel structures such as docks, piers, drilling platforms, and ships. The two most widely used methods are flame cutting and oxygen arc cutting. The technique used for underwater flame cutting is not too different from that used for flame cutting steel in air. In each case, a fuel gas in a torch is mixed with oxygen and burned to produce a flame that preheats the steel and a cutting jet is provided to supply oxygen to cut the steel. The underwater cutting torch, however, has one important difference in construction. The underwater torch supplies its own ambient gas atmosphere, an air bubble around the flame, by means of compressed air that is ejected through a special nozzle surrounding the tip. An adjustable shield on the top of the torch is also usually supplied to help control the formation of the air bubble and to allow the torch to be held at the optimum distance from the work, even under conditions of poor visibility and constraint due to the cumbersome diving suits that must be worn. Slots are cut in the shield to allow gases to escape. The underwater torch is furnished with three hoses: for compressed air, oxygen, and fuel gas.

Underwater flame cutting is most effective in severing a steel plate in the thickness range of 12.7 to 152 mm ($\frac{1}{2}$ to 6 in). Below 12.7 mm ($\frac{1}{2}$ in), the quenching effect of the water retards the cutting action greatly.

It is important that the air hose never be used for oxygen. Compressed air may contain some oil which can coat the hose and cause an explosion when oxygen is introduced. The fuel gas is usually hydrogen or natural gas because each gas can be used at any depth without liquefying. Acetylene is almost never used in underwater cutting because, at pressures more than 207 kPa (30 lb/in^2) acetylene becomes unstable and may decompose violently even if no oxygen is present.

A standard welding power source capable of supplying 300 A of direct current, straight polarity, is satisfactory for oxygen arc cutting under water. Electrode holders are fully insulated and are of a special design so that both oxygen and current can be supplied to the electrode. To reduce resistance losses, cables should be size 2/0 except for the last 3 m (10 ft) at the torch, which may be 1/0 for added flexibility. If the power source is more than 120 m (400 ft) from the work, parallel cables of size 1/0 or 2/0 are required. All underwater cable connections should be wrapped with rubber tape. A safety switch must be provided so that the torch is energized only while cutting is in progress.

Electrodes for underwater oxygen arc cutting are either tubular carborundum or steel. Steel electrodes are available in 4.75- and 7.93-mm (3/16- and 5/16-in) diameters with a 1.6-mm- (1/16-in) diameter bore. These electrodes are provided with a waterproof coating which serves as an insulator during cutting.

Cutting underwater requires that positive pressure be maintained by the electrode against the metal being cut; whereas in air the electrode is dragged along the intended line of cut. Particular attention must be paid to safety. The power sources must be grounded to the tender, and ground cables must be securely connected to the work. All parts of the power cables and torches must be fully insulated and periodically inspected. An operating disconnecting switch must be part of the cutting electric circuit. To prevent possible explosions, enclosed spaces must be vented so that gases generated during cutting cannot accumulate.

6.4   Environmental Considerations

In contrast to organic materials, sunlight does not cause deterioration of metals. Under some conditions, however, sunlight can be a contributing factor in the stress corrosion cracking of some stainless steels. Stress corrosion cracking occurs when high stress accelerates corrosion along intercrystalline boundaries and leads to weakening of intercrystalline bonds and eventual cracking. Austenitic stainless steels, such as types 304, 316, 321, 347, and even 216, are susceptible to stress corrosion cracking when they are exposed to heat, stress, chloride ion, and oxygen simultaneously. Cold-worked materials are most susceptible, but even annealed austenitic stainless steel contains some residual stresses from fabrication and can crack. Stress corrosion cracking is believed to be time-dependent, but the exact threshold conditions for this phenomenon to occur have not been established. Process equipment constructed of austenitic stainless steel and hydrotested with seawater but not properly drained and flushed has been ruined by stress corrosion cracking when the only heat applied was the heat of the sun.

6.4.1   Effect of severe temperatures

At one time, harbor facilities were located only in the torrid or temperate zones. Many ships were built by riveting, and welded ships were constructed by fitting each plate individually in turn. World War II created a demand for cargo ships that only mass-production techniques could meet. These techniques involved constructing large hull sections offsite and then moving them into position for welding together to complete the hull. Because alignment of the sections was not perfect, force was applied to obtain sufficient alignment for welding. Many of these ships broke in two in the North Atlantic, and investigations revealed that brittle failure was the cause of these losses. Today, knowledge of the relation between notch toughness and brittle failure makes possible the design of marine structures that will survive the most severe temperatures.

Carbon and most alloy steels suffer a decrease in toughness as temperatures are reduced. When slow rates of loading are applied, these materials exhibit increased tensile and yield strength and only slight losses of elongation and reduction of area at reduced temperatures. When the load is rapidly applied, as in the Charpy impact test, the amount of energy absorbed during fracture decreases gradually as testing is performed at progressively lower temperatures until, at some temperature, the absorbed energy drops dramatically. This temperature is known as the nil ductility transformation (NDT) temperature, the temperature at which the specimen exhibits little ductility before fracture. The NDT temperature can be defined by Charpy V-notch testing as (1) the temperature at which a certain absorbed energy is attained, (2) the temperature at which 50 percent shear fracture on the broken specimen is attained, or (3) the temperature at which a certain lateral expansion on the specimen opposite the notch is attained. A common value for minimum absorbed energy at the NDT temperature for ordinary constructional steels is 20 newton meters (N • m) [15 foot-pounds (ft • lb)], but acceptable impact values are often stated in ASTM or other material specifications. Complete procedures for conducting many mechanical tests on metals, including impact tests, are given in ASTM A370. Figure 6.2 presents a repre-

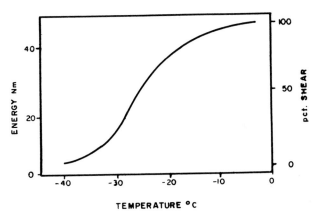

Figure 6.2   Plot of Charpy V-notch test on a low-carbon steel.

sentative plot of absorbed energy vs. temperature for Charpy V-notch tests on a typical low-carbon steel.

Values obtained from Charpy V-notch testing cannot be used directly in engineering calculations for design. Notch toughness values become significant only when they are correlated with a particular type of structure in a particular service; they are useful to compare different materials. The NDT temperature determined by Charpy V-notch testing has correlated rather well with temperatures at which service failures of components of the same steel have occurred.

Factors affecting the notch toughness of a metal are:

- Chemical composition

- Gas content

- Microstructure (e.g., size, shape and orientation of grains, and grain boundaries of a structure)

- Grain size

- Section size (physical cross section dimensions)

- Hot and cold working temperatures

- Method of fabrication

- Specimen orientation in relation to working direction

Figure 6.3 presents representative plots of Charpy V-notch absorbed energy curves for several materials used in construction. Fully killed carbon steel made by using a fine-grain melt practice in the normalized heat treatment has the best notch toughness at lower temperatures of the carbon steels. If sulfide shape control is used during processing, improved notch toughness can be obtained in the transverse (across grain) and through (with grain) section directions. Austenitic stainless steel, type 304, and 9 percent nickel are candidate materials for handling liquefied gases. Note that gray cast iron exhibits little notch toughness at any temperature shown.

6.4.2 Wind erosion

Wind erosion does not have a severe effect on metals. Wind-driven sand, however, can destroy paint and therefore increase the formation of rust on steel structures. Appearance is the property most often affected.

6.4.3 Exposure to flora

The major effect of flora on metals is a slight increase in corrosion rates where the plants, by their root systems, may transport addi-

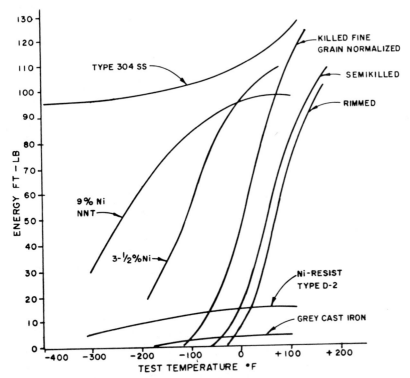

Figure 6.3   Representative Charpy V-notch absorbed energy curves for several materials. *(Data supplied by International Nickel Company and American Society for Metals)*

tional moisture to the metal surfaces. In some soils, there are aerobic bacteria that oxidize sulfur which either is present in the soil or is obtained from decaying organic matter. By oxidation of sulfur, a strong solution of sulfuric acid is formed, and it reacts with any basic material that is present. Either anaerobic or aerobic bacteria can cause soils to be corrosive to metals even though usual mineral analyses of the soil and water do not reveal that a corrosive condition exists.

6.4.4   Exposure to burrowing animals

Most metals have high hardness, which prevents burrowing animals from penetrating. Animals have damaged the insulation covering of some buried electric cables, which resulted in failure of the cables.

6.4.5   Exposure to fresh water

Exposed surfaces of metals are subject to some degree of corrosion from water. The corrosiveness of water is basically dependent on three factors: acidity, oxygen content, and electrical conductivity. Many rivers are polluted by industrial wastes or runoff from mines, which causes the water to become acid. Such water may be more corrosive to carbon steel than seawater would be. Because it becomes saturated with carbon dioxide as it falls to earth, rainwater becomes slightly acid. As the rain contacts the earth, it becomes altered by reaction with minerals and soil. Depending on the acidity and mineral composition, stream water may be less corrosive to carbon steel than rainwater.

Water-penetration effects.   Metals are impervious to water penetration.

Freezing and thawing.   The volume expansion of ice in fine-grained soils, such as very fine sand, silt, or clay, may produce lateral thrusts to sheet pile structures. Placement of free-draining coarse granular soil above the frostline behind sheet pile walls should eliminate the possibility of lateral thrust from ice or frozen ground. Steel sheet piling can yield laterally to relieve any thrust load due to ice. Plugged and broken waterlines caused by ice are inconveniences that require cooperation between design and construction personnel and between operations and maintenance personnel to eliminate the problem. Heat tracing and insulation are possible solutions, but other methods, such as pumping out fire hydrants and closing doors to heated loading docks, may be more practical.

6.4.6 Exposure to seawater

Corrosion rates of metals in seawater are higher than in pure water because ions of halogen compounds, such as sodium chloride, have the power to cause localized breakdown of oxide films that are responsible for passivity and corrosion resistance. Halogen ions can form soluble acidic corrosion products, such as ferric chloride, which interfere with the restoration of passivity to steel and lead to localized corrosion in the form of pitting. Tests have shown that rates of corrosion of carbon steel in the atmosphere at the shoreline are 10 times the rates of plates 460 m (500 yd) from the shoreline.

It has been shown that the rates of corrosion of steel in seawater and fresh water are governed to a large extent by the oxygen content of the waters. Carbon steel in contact with fresh water saturated with oxygen at ambient temperature usually exhibits a corrosion rate of 220 μm (9 mils) per year general corrosion plus an additional 220 μm per year of pitting. When fresh water is oxygen-free, the corrosion rate for carbon steel is usually only 25 micrometers (μm) (1 mil) per year or less, provided no corrosive pollutants are present.

Variable oxygen content. The patterns of corrosion found on steel pilings in the atmosphere, splash zone, and tidal zone, submerged in clean seawater, and in the mud zone vary considerably. A principal variable related to position is the oxygen content. The high corrosion rate in the splash zone is attributed to the constant wetting of the steel by highly aerated seawater. In the tidal zone, differential aeration produces a protective cell effect that results in a considerably lower corrosion rate. At deeper positions, less oxygen is present and the rate of corrosion of steel usually drops to the range of 76 to 152 μm (3 to 6 mils) per year. Carbon steel in seawater that has been treated to remove dissolved oxygen and marine bacteria exhibits an even lower corrosion rate under low velocities.

Austenitic stainless steel and aluminum alloys exhibit satisfactory corrosion resistance in the splash zone because the high oxygen content helps keep passivating films intact. Aluminum has better corrosion resistance in the splash zone than at greater depths, at which less oxygen is present. The high corrosion rates on carbon steel piling in the splash zone may also be attributed to the severe electrochemical corrosion cells set up in the pile. Piles made from high-copper-bearing, high-strength, low-alloy steel conforming to ASTM Standard A690 have 2 to 3 times the resistance to seawater corrosion in the splash zone of ordinary carbon steel, although such steels exhibit no better corrosion resistance at greater depths.

Effects of polluted seawater. Polluted waters often contain hydrogen sulfide, which has severe effects on metals sensitive to sulfides. Hardened steel and welds not stress-relieved in medium-carbon steel may crack because of stress corrosion. Hydrogen sulfide can lead to corrosion of the vapor side of copper alloy heat exchangers. Small amounts of ammonia may also be present in polluted seawater; the ammonia causes aggressive attack and stress corrosion cracking of copper-zinc alloys. The copper-nickel alloys are preferred when ammonia pollution is expected, and the 90-10 copper nickel alloy (UNS No. C70600) has demonstrated satisfactory performance in many applications where sulfide pollution has been present.

6.4.7 Effects of marine organisms

"Biofouling" and "biological fouling" are common terms that refer to the settlement and growth of living organisms on materials exposed to the marine environment. Some metals, such as titanium and the nickel–chromium–high molybdenum alloys, are completely corrosion-resistant under fouling. Copper-base alloys exhibit varying degrees of resistance to biofouling. Other materials, such as aluminum, carbon steel, and stainless steel, both foul and suffer increased corrosion because of biofouling. On structures such as wharves and breakwaters, biofouling may not be as important but it does cause increased wave action loadings. Increased flow blockage and decreased heat transfer efficiency are other problems encountered as a result of marine biofouling of metal structures in marine service.

Copper and the 90-10 copper-nickel alloy have the highest resistance to biofouling. Brass and bronze have good resistance, but 70-30 copper-nickel alloy, aluminum bronze, zinc (galvanizing), and Monel alloy 400 have only fair resistance. The high resistances to biofouling of many of the copper-base materials have been attributed by some researchers to the inhospitable nature of the green cupric hydroxychloride corrosion product that forms on those materials. The film is itself loosely attached, so that any marine organisms that do attach to it are soon removed. Monel, carbon steel, aluminum, and stainless steel exhibit poor corrosion resistance to biofouling. Carbon steel suffers general corrosion, whereas Monel, aluminum, and stainless steel exhibit pitting and crevice corrosion. Crevice corrosion is caused by differential oxygen cells produced when oxygen is prevented from reaching the metal surface under barnacles.

Corrosion rates on carbon steel may be reduced a little because of the reduced velocity of water at the metal surface when biofouling is present, but they remain relatively high. Biofouling will periodically slough off when the corrosion product breaks off. The high general corrosion rate of carbon steel in seawater is attributed to marine organisms known as anaerobic bacteria. The principal groups are the sulfate-reducing and the iron-consuming bacteria. Sulfate-reducing bacteria require oxygen, which is derived from the reduction of compounds such as sulfates, sulfites, thiosulfates, or organic substances rather than dissolved oxygen. These bacteria liberate hydrogen sulfite which attacks iron severely by removing hydrogen from the cathodic areas of the steel with the formation of iron sulfide. The iron-consuming bacteria do not actually consume iron as food, but they do require iron in solution for growth.

Biofouling has been controlled by using copper-base alloys, antifouling coatings, mechanical cleaning of the surfaces, and environmental controls. The latter include increased flow velocity, elevated temperature, and chlorination. When such control measures are employed, the corrosion performance of the base metals must be considered.

In addition to biofouling, there are several organisms (such as limnoria, teredo, and termites) commonly found in the marine environment that cause deterioration of structures through boring. Because metal structures are impenetrable, these organisms do not cause the deterioration usually undergone by marine pilings constructed of wood.

6.4.8   Wave and current effects

Fouling diminishes as water velocities in contact with a structure reach the 1 to 2 meter per minute (m/min) [3 to 6 foot per minute (ft/min)] range. Pitting of the more noble materials slows down and may even cease. As velocities continue to increase, stainless steel and nickel-base materials remain passive and inert but corrosion barriers are stripped away from carbon steel and copper alloys. Although wave or current velocities are seldom too high to allow the use of carbon steel, velocity is a factor to consider in the design of such equipment as piping and pumps.

6.4.9   Abrasion

Abrasion of metal structures is caused by the movement of the elements in the coastal zone and their ability to transport particles with force against all structures. Significant particle transport is caused by wind and water. Structures on wind-swept beaches and shores are subject to severe abrasion from sand that is wind-driven with substantial force. In the case of steel or other metal structures, significant metal wear can result. Abrasion of steel pile structures at the mud line is far in excess of the metal loss from corrosion. Additional abrasion resistance of some structures, usually on land, or of structural elements located above the waterline in the ocean can be provided by protection in the form of concrete, wood, or hard-surfaced alloys. Most hard-surfaced metals require special heat treating and the addition of small amounts of other elements such as manganese.

The latest improvement in steel piling was the development of mariner steel for seawater exposure. Mariner steel was developed primarily to improve the corrosion resistance of steel in seawater by alloying about 0.5 percent each of copper and nickel in addition to about 0.1 percent of phosphorus (ASTM Standard Specification A 690-77). Mariner steel has a somewhat improved strength, but its hardness is little different from that of normal steel piling and its abrasion resistance also is little different. Although cathodically protected steel structures in seawater usually take on a calcareous coating (chemicals deposited from the seawater), the coating is too soft to offer any resistance to abrasion.

6.4.10   Seismic effects

Metals are well suited for marine construction in areas of seismic activity. They possess high tensile strength, good ductility, and, when properly specified, good toughness. In addition, they can consistently meet specified minimum seismic requirements. Steel, the most economical of the metals for construction of harbor facilities, is available in several shapes. The inherent high-bending strength of steel H-piles permits the development of required resistance to lateral forces when the piles are used in foundation designs for service where resistance to seismic forces is required. The ability of metals to be loaded in shear, compression, or tension within calculated limits facilitates the design of earthquake-proof structures.

6.4.11 Effects of fire

Metals in the shapes and sizes used in construction will not burn in a fire, but they do have reduced strength as temperatures rise. Carbon steel is affected above 340°C (650°F); at 480°C (900°F), it has only about half the strength that it has at room temperature. During fires, structures of all-steel construction have collapsed. Structures are protected from fire loss by sprinkler systems, which spray water on the building and roof supports to keep them cool, and by concrete coverings on steel beams and supports to provide insulation. Metals that have experienced a fire should be tested for suitability before being reused, because strength and toughness may be reduced. Heat treatment may be required to restore properties.

6.4.12 Human activity

Accidents and theft are two elements of human activity that the designer must consider. Piers and wharves are constructed with fender systems to minimize damage from impact by ships. When accidents do cause damage, metal structures may be repairable by welding.

Legs on drilling platforms are often sheathed with copper-nickel alloys in the splash zone areas to minimize corrosion. Attempts to use similar methods in harbors have not been completely successful because the sheathings have been stolen for their value as scrap metal.

6.5 Uses of Structural and Sheet Metal

The uses described below are those generally or frequently encountered. Other uses are possible, but metals must be protected if durability is desired.

6.5.1 Steel

Various parts of coastal structures are made of steel. Steel H-piles or pipe piles are used to support foundations. The former are used in preference to the latter because they are more easily driven into soils containing hard strata or obstructions such as boulders. Steel H-piles are easily spliced by welding, which makes possible driving to deep rock strata if necessary. They are also frequently used to support fender systems immediately in front of wharves. Steel bolts are used to attach rubber bumpers to the fender system; the bumpers prevent damage to the structure by absorbing the loads of ship impacts. Cast-steel bollards and mooring posts are used to take up ships' lines.

Steel is an ideal material for breasting and mooring dolphins because it can be easily jointed and has high tensile strength, good ductility, and good toughness. Structural steel shapes are used for framing structures. Even the fences around coastal installations are most often made of chain-link steel.

Large quantities of steel are used in components that appear to be all concrete; concrete piles, beams, structure foundations, walls, roadways, and pads contain steel reinforcing bar, wire, or wire fabric. Materials for concrete reinforcement are covered in Chap. 5.

6.5.2 Aluminum

Many alloys of aluminum possess high corrosion resistance to marine atmospheres as well as good strength-to-weight ratios. Those properties make aluminum an economic choice for many applications in coastal structures, particularly where freedom from maintenance is desired. Door and window frames are usually made of 6063 aluminum and roofing and siding of alclad 3004 aluminum. Tread plate, such as that used for decking and catwalks, is 6061 aluminum heat-treated to T4, T42, or T6 temper. Aluminum alloys are also used for architectural trim, hardware, and gutters and downspouts. Insulation is often faced with aluminum foil to reflect heat, which makes the insulation more effective.

Electrical wire and bus bars are either copper or 1350 aluminum. Because the conductivity of 1350 aluminum is approximately one-half that of copper, the cross section of the aluminum conductor must be approximately twice that of the copper for equivalent current capacity. The specific gravity of aluminum is so much less than that of copper that conductors of 1350 aluminum weigh only half as much as equivalent conductors of copper, making the choice one of economics. Lamp poles and standards are made of 6063 aluminum. Even the lamp bases may be 3004 or 5050 aluminum.

Tanks and equipment for liquid natural gas facilities must be constructed of materials having high notch toughness. Aluminum alloys 5083 and 5456 have been specified for liquid natural gas storage tanks and vaporizers because they have good corrosion resistance in marine atmospheres and high notch toughness. In accordance with ASTM

A370, aluminum alloys do not require Charpy impact testing because aluminum alloys do not become brittle at cryogenic temperatures.

6.5.3 Copper

Electrical conductors such as wire and bus bars are the largest applications of copper; other important applications are pipe and sheathing. Copper has many hidden applications in supporting equipment at coastal structures, including radiators in air conditioners and powered equipment, springs and contacts in communication and control systems, and even tools of beryllium copper for use in areas where sparking must be prevented to avoid fires and explosions. Copper alloys are used in such equipment as heat exchangers, pumps, valves, and hardware for sluice gates and traveling water screens.

6.5.4 Nickel

Nickel-base alloys have good corrosion resistance to seawater and generally high resistance to cavitation damage. Most resistant are nickel-chromium-molybdenum-columbium alloy 625 (Inconel 625, produced by Huntington Alloys, Huntington, West Virginia) and nickel-molybdenum-chromium alloy C (Hastelloy C, produced by Stellite Division, Cabot Corporation, Kokomo, Indiana). These two alloys are used for springs, cable connectors, bellows-type expansion joints, rupture disks, and pump seal rings in coastal facilities. Nickel-copper alloy 400 (Monel 400, produced by Huntington Alloys) is the lowest-cost nickel-base alloy for marine service. It is used for valve and pump trim, fasteners, heat exchangers, and piping.

6.5.5 Titanium

The major uses of titanium in coastal structures are in steam condensers employing seawater cooling, ball valves, and desalination equipment. Titanium will tolerate polluted seawater under conditions in which other materials fail. As a result, many coastal power plants are installing steam condensers that have titanium tubes.

6.6 Sheet Pile Structures

Bulkheads in waterfront facilities are subjected to lateral pressure resulting from earth movement and from unbalanced hydrostatic and seepage forces acting on opposite sides of the walls. A higher water level may exist in the backfill behind a wall than in front of it as a result of a receding tide, receding high water, or a heavy rainstorm. Other lateral loads that may be encountered are ice thrust, wave forces, ship impact, mooring pull, and earthquake forces. Because of its material strength, sheet pile is often used in marine construction for bulkheads.

The designer, after evaluating the lateral pressure and forces, must determine the required depth of piling penetration, the maximum bending moments in the piling, and the maximum bending stresses in the wall. An appropriate sheet pile section must be selected by taking into account yield strength and moment of inertia of the selected section. Some typical steel sheet pile profiles are shown in Fig. 6.4. A choice may be made between a cantilevered or anchored wall.

Anchored sheet pile walls can be designed for greater height than is possible with the cantilever-type design of a similar sheet pile section. For heights to about 11 m (35 ft) (depending on soil conditions), sufficient support can be obtained from anchor tie-rods near the top of the wall and lateral support of the embedded part of the wall. For greater heights, higher-yield-strength steel or multiple tie-rods at lower levels are required. Anchorage systems in use include deadman, H-pile, and sheet pile anchors; sketches are shown in Fig. 6.5. Regardless of which anchorage system is used, the anchor must be located outside the potential active fracture zone behind the sheet pile wall. Passive resistance of the anchor is not possible if the ground is unstable.

A complete sheet pile wall system may consist of the wall, wale, tie-rods, and anchor. The wale is a flexible member attached to the wall; it distributes the horizontal reactive force from the anchor tie-rods to the wall section. Locating the wale on the outside of the wall where the piling will bear against it in compression is preferred for engineer-

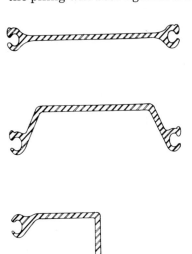

Figure 6.4 Typical steel sheet pile profiles. Top profile, straight; middle profile, arch; bottom profile, angle.

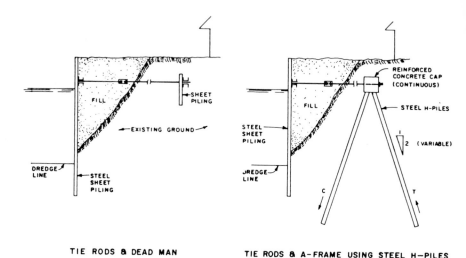

TIE RODS & DEAD MAN   TIE RODS & A-FRAME USING STEEL H—PILES

Figure 6.5   Typical steel pile anchorage systems.

ing purposes. However, the wale is sometimes bolted onto the inside face to provide a clear outside face.

Wales are often constructed of steel structural channels conforming to ASTM Standard A36 and mounted with their webs back to back and separated by enough space to clear the end of tie-rod between them. When the wale is located on the inside face, each sheet pile section is bolted to it. Standard designs for wales located on both outside and inside faces are shown in Fig. 6.6.

Tie-rods are usually round steel bars, conforming to ASTM Standard A36, that have been upset and threaded at each end so as to maintain cross-sectional area in the threaded part. Usually a turnbuckle is used between two tie-rod sections to allow removal of slack. Sagging of the tie-rods may occur because soil settlement around them drags them downward and causes increased tension in them. There are two ways to avoid this condition: (1) Use light vertical piles at 6- to

DOUBLE CHANNEL INSIDE WALL   DOUBLE CHANNEL OUTSIDE WALL

Figure 6.6   Standard wale designs.

9-m (20- to 30-ft) intervals to support the rods. (2) Encase the rods in large conduits. Tie-rods are subject to corrosion and must, therefore, be adequately coated and wrapped.

6.6.1 Construction

Steam hammers are commonly employed for pile driving in the United States. During driving, the steam hammer, consisting of a housing and the moving part called the ram, rests on top of the pile. A single-acting steam hammer is a free-falling ram with steam pressure acting on a piston to raise the ram prior to fall. In a double-acting hammer, steam is used not only to lift the ram but also to help drive the ram downward. Double-acting hammers are able to deliver blows faster than single-acting hammers of the same energy output because they use a shorter stroke and higher ram acceleration. Both drop hammers and diesel hammers are available. A drop hammer consists of a heavy weight or ram that is allowed to fall by gravity on top of the pile. Fall height must be controlled to avoid damage to pile heads by excessive impact from rams moving with high velocity.

Excessive impact or improper cushioning during pile driving may result in mashed pile heads. Vertical misalignment of the pile as a result of obstructions encountered below the ground surface or of poor pile-driving conditions may cause failure of pile interlocks. If excessive misalignment occurs, sheet piles can become overstressed and result in bulkhead failure.

The method used for the construction of steel breakwaters depends on the soil conditions and the height of the waves. If the waves are below 10 ft and the bottom is soft to a great depth, steel sheet pile topped with concrete and supported with batter piles may be used.

Bulkheads for small-boat harbors have been constructed by using sheet piling made of aluminum alloy 5052-H141. Coping was of 6063-T6; tie-rods and stiffener bar beams were of 6061-T6. Deadman anchors were constructed of 5052-H141 alloy. Aluminum sheet pile is available in 3.6-m (12-ft) lengths, which limit application to shallow facilities.

6.7 Gabions

Gabions, compartmented rectangular containers made of galvanized steel hexagonal wire mesh and filled with small stones, have been used to reinforce the shoulders of seawalls constructed of rock. They have also been used to construct jetties, as well as revetments and seawalls to control shoreline erosion. Gabion mattresses can also be used as foundations or filter layers under rubble mound structures and caisson structures. For seawater use, gabions of galvanized wire should be coated with plastic to reduce corrosion. The Alaska District limits the use of gabions in the wave zone where ice occurs because the gabions are burst by the ice. Also, if the gabions are not rigidly filled, the rock filling can move and abrade the wire.

7

Wood

7.1 General

Wood is widely used in the coastal zone because it is strong and resilient and is easily installed with common tools and equipment. Also, it is a common material available nearly everywhere at a reasonable cost. When properly treated, it is very durable. Its ability to absorb energy (resiliency) is a feature that makes it especially desirable for such uses as fender piles.

The main problem when using wood in the coastal zone is that it is an organic material that is the natural food supply and habitat for fungi, bacteria, insects, and marine organisms. The first three occur on land and are more active in the high-moisture conditions at the coast. Wood treatments to prevent attack by natural enemies are very effective in combating such damage.

7.2 Physical Properties

Wood is a cellular organic material made up principally of cellulose, which comprises the structural units, and lignin, which cements the structural units together. It also contains certain extractive and ash-forming minerals. Wood cells are hollow and vary from about 1000 to 8000 μm (40 to 330 mils) in length and from 10 to 80 μm (0.4 to 3.3 mils) in diameter. Most cells are elongated and are oriented vertically in the growing tree, but some, called rays, are oriented horizontally and extend from the bark toward the center, or pith, of the tree.

7.2.1 Hardwoods and softwoods

Species of trees are divided into two classes: hardwoods, which have broad leaves, and softwoods or conifers, which have needlelike or

---

This chapter was written by Lawrence L. Whiteneck and Lester A. Hockney.

scalelike leaves. Hardwoods shed their leaves at the end of each growing season, but most softwoods are evergreens. The terms "hardwoods" and "softwoods" are often misleading because they do not directly indicate the hardness or softness of wood. In fact, there are hardwoods which are softer than certain softwoods.

7.2.2 Heartwood and sapwood

Several distinct zones are distinguishable in the cross section of a log: the bark, a light-colored zone called sapwood, an inner zone, generally of darker color, called heartwood, and, at the center, the pith (Fig. 7.1). A tree increases in diameter by adding new layers of cells from the pith outward. For a time, this new layer contains living cells which produce sap and store food, but eventually, as the tree increases in diameter, cells toward the center become inactive and serve only as support for the tree. The inactive inner layer is the heartwood; the outer layer containing living cells is the sapwood. There is no consistent difference between the weight and strength properties of heartwood and sapwood. Heartwood, however, is more resistant to decay fungi than is sapwood, although there is a great range in the durability of heartwood from various species.

Annual rings. In climates where temperature limits the growing season of a tree, each annual increment of growth usually is readily distinguishable. Such an increment is known as an annual growth ring or annual ring; it consists of an earlywood and a latewood band.

Earlywood (springwood) and latewood (summerwood). In many woods, large thin-walled cells are formed in the spring when growth is greatest, whereas smaller, thicker-walled cells are formed later in the year.

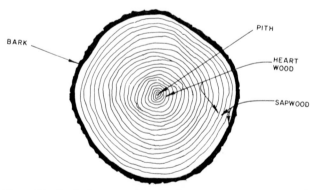

Figure 7.1 Typical cross section of a log.

The areas of fast growth are called earlywood, and the areas of slower growth, latewood. In annual rings, the inner, lighter-colored area is the earlywood and the outer, darker layer is the latewood. Latewood contains more solid wood substance than does earlywood and is therefore denser and stronger. The proportion of width of latewood to width of annual ring is sometimes used as one of the visual measures of the quality and strength of wood.

Grain and texture. The terms "grain" and "texture" are used in many ways to describe the characteristics of wood; in fact, however, they do not have definite meanings. "Grain" often refers to the width of the annual rings, as in "close-grained" or "coarse-grained." Sometimes it indicates whether the fibers are parallel to or at an angle with the sides of the pieces, as in "straight-grained" or "cross-grained." "Texture" usually refers to the fineness of wood structure rather than to the annual rings. When these terms are used in connection with wood, the meanings intended should be defined.

7.2.3 Moisture content

Wood may contain moisture as "free water" in the cell cavities and as "absorbed water" in the capillaries of the cell walls. When green wood begins to lose moisture in the seasoning process, the cell walls remain saturated until the free water has been evaporated. The point at which evaporation of free water is complete and the cell walls begin to lose their moisture is called the fiber saturation point (fsp). In most species, this point occurs at between 25 and 30 percent moisture.

Moisture in wood is expressed as a percent of the ovendry weight and is determined most accurately by weighing a representative sample, drying it at slightly more than 100°C (212°F), until no further loss of weight takes place, reweighing, and then dividing the difference between the original and final weights by the final (ovendry) weight. Electric moisture meters offer a simpler though less exact method of determining moisture content. With slight seasonal variations, wood in use over a period of time attains an equilibrium moisture content (emc) corresponding to the humidity and temperature of the surrounding atmosphere. When exposed to similar atmospheric conditions, different woods will have the same moisture content regardless of their density.

Moisture content has an important effect upon susceptibility to decay. To develop, most decay fungi require a moisture content above the fiber saturation point. In addition, a favorable temperature, an adequate supply of air, and a source of food are essential. Wood that is continuously water-soaked (as when submerged) or continuously dry (with a moisture content of 20 percent or less) will not decay. Moisture content variations above the fiber saturation point have no effect upon the volume or strength of wood. As wood dries below the fiber saturation point and begins to lose moisture from the cell walls, shrinkage begins and strength increases.

7.2.4 Directional properties

Because of the orientation of its cells and the manner in which it increases in diameter, wood is not isotropic. It has different mechanical properties with respect to its three principal axes of symmetry: longitudinal, parallel to the grain; radial, perpendicular to the grain; and tangential, perpendicular to the grain (Fig. 7.2.). Strength and elastic properties corresponding to those three axes may be used in design. The difference between properties in the radial and tangential directions is seldom of practical importance in most structural designs; for structural purposes it is sufficient to differentiate only between properties parallel and perpendicular to the grain.

7.2.5 Specific gravity

Solid wood substance is heavier than water; its specific gravity is about 1.5 regardless of species. Despite this fact, dry wood of most species floats in water because a part of its volume is occupied by air-filled cell cavities. Variation among species in the size of cells and in the thickness of cell walls affects the amount of solid wood substance present and hence the specific gravity. Thus, specific gravity of wood is a measure of solid wood substance and an index of strength properties. Specific gravity values may, however, be somewhat affected by gums, resins, and extractives which contribute little to strength. The relation between specific gravity and wood strength is implied by the

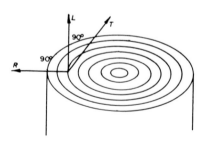

Figure 7.2 The principal axes of wood. *L*, longitudinal; *R*, radial; *T*, tangential.

practice of assigning higher basic stress values to lumber designated as dense.

7.2.6 Dimensional stability

Wood, like most other solids, expands on heating and contracts on cooling. In most structural designs, the increase of wood in length because of a rise in temperature is negligible; as a result, the secondary stresses that are due to temperature changes may in most cases be neglected. This increase in length is important only in certain structures that are subject to considerable temperature changes or in members with very long spans.

The increase in length per unit of length for a rise in temperature of $1°$ is designated the coefficient of linear thermal expansion. It differs in the three structural directions of wood. Radially and tangentially (perpendicular to grain), the coefficient of linear thermal expansion varies directly with the specific gravity of the species. It is in the range of 45×10^6 meters per meter per degree Celcius [m/(m • °C)] $\{25 \times 10^6$ feet per foot per degree Fahrenheit [ft/(ft • °F)]} (times specific gravity) for a dense hardwood such as sugar maple to 81×10^6 m/(M • °C) [45×10^6 ft/(ft • °F)] (times specific gravity) for softwoods such as Douglas fir, Sitka spruce, redwood, and white fir.

Radial or tangential dimensional changes for common sizes of wood structural members are relatively small. Longitudinally (parallel to grain), the coefficient is independent of specific gravity and varies from 3.08×10^6 m/(m • °C) [1.7×10^6 ft/(ft • °F)] to 4.5×10^6 m/(m • °C) [2.5×10^6 ft/(ft • °F)] for different species. This is from one-tenth to one-third of the values for other common structural materials and glass. For that reason, consideration must be given to the difference in thermal expansion coefficients of various materials used in conjunction with wood. The average coefficient of linear thermal expansion for plywood is 6.12×10^6 m/(m • °C) [3.4×10^6 ft/(ft • °F)]. The coefficient of thermal expansion of plywood thickness is essentially the same as that of solid lumber.

7.2.7 Moisture content

Between zero moisture content and the fiber saturation point, wood shrinks as it loses moisture and swells as it absorbs moisture. Above the fiber saturation point there is no dimensional change with variation in moisture content. The amount of shrinkage and swelling differs in the tangential, radial, and longitudinal dimensions of the piece. Engineering design should consider shrinkage and swelling in the detailing and use of lumber.

Shrinkage occurs when the moisture content is reduced to a value below the fiber saturation point (for purposes of dimensional change, commonly assumed to be 30 percent of the moisture content at the fiber saturation point) and is proportional to the amount of moisture lost below that point. Swelling occurs when the moisture content is increased until the fiber saturation point is reached, and then the increase ceases. For each 1 percent decrease in moisture content below the fiber saturation point, wood shrinks about one-thirtieth of the total possible shrinkage; and for each 1 percent increase in moisture content, the piece swells about one-thirtieth of the total possible swelling. The total swelling is equal numerically to the total shrinkage. Shrinking and swelling are expressed as percents based on the green wood dimensions. Wood shrinks most in a direction tangent to the annual growth rings and somewhat less in the radial direction, or across the rings. In general, shrinkage is greater in heavier pieces than in lighter pieces of the same species and greater in hardwoods than in softwoods.

As a piece of green or wet wood dries, the outer parts are reduced to a moisture content below the fiber saturation point much sooner than are the inner parts. Thus the whole piece may show some shrinkage before the average moisture content reaches the fiber saturation point. Neither the initial nor the final moisture content (M_i nor M_f) can be greater than 30 percent when shrinkage is calculated because that is the moisture content at which, when drying, wood starts to shrink or at which, when absorbing moisture, it reaches its maximum dimension. Values for longitudinal shrinkage with a change in moisture content are ordinarily negligible. The total longitudinal shrinkage of normal species from fiber saturation to ovendry condition usually ranges from 0.1 to 0.3 percent of the green wood dimension. Abnormal longitudinal shrinkage may occur in compression wood, wood with steep slope of grain, and exceptionally lightweight wood of any species.

Cross-laminated construction gives plywood its relatively good dimensional stability in its plane. The average coefficient of hydroscopic expansion (or contraction) is about 0.002 m per meter (0.0002 ft per foot) of length or width for each 10 percent change in relative humidity, or 0.2 percent ovendry to complete saturation.

7.3 Mechanical Properties

Wood is not an isotropic material, because strength properties differ along its different axes. It is strongest when loaded to induce stress, in either tension or compression, parallel to grain. However, this condition is not always possible, and loading perpendicular to grain may be accomplished in a satisfactory manner.

The anisotropic nature of wood may be confusing to designers during their first experiences with wood use; but as they get to know the material, they find that engineering design with wood can be interesting as well as productive in the way of lower construction costs. The discussion which follows provides a brief description of the various mechanical properties of structural wood as they affect engineering design.

7.3.1 Tension parallel to grain

A force-generating tension parallel to grain, as shown in Fig. 7.3, creates a tendency to elongate the wood fibers and to cause them to slip by each other. Resistance to tension applied strictly parallel to grain is the highest strength of wood. It is, however, substantially reduced when the force is applied at an angle to the grain or when the cross section of the wood is reduced by knots or holes.

7.3.2 Tension perpendicular to grain

A force-generating tension perpendicular to grain tends to separate the wood fibers along the grain. This is the direction in which wood has the least strength. Because it is not good practice to apply loading to induce tension across grain, design values are not provided for this strength property except for special applications.

7.3.3 Compression parallel to grain

A force-generating compression parallel to grain, as shown in Fig. 7.4, has a tendency to compress the wood fibers in the lengthwise position. As with tension, resistance to compression parallel to grain is affected by the angle of load to grain and by the presence of knots or holes.

7.3.4 Compression perpendicular to grain

A force applied perpendicular to grain, such as the bearing under the ends of the beam shown in Fig. 7.5, tends to compress the wood at its surface. Although the wood becomes more dense as it is compressed, the action of the force causes slight displacement of the supported member. Thus, limits are placed on loading in bearing perpendicular to grain.

7.3.5 Shear parallel to grain

A force applied in the manner illustrated in Fig. 7.6 causes one section of the piece to shear or slide along the other section in a direction par-

Figure 7.3 Tension parallel to the grain.

Figure 7.4   Compression parallel to the grain.

allel to grain. In a loaded beam in which the induced stress on one side is compression and on the other side is tension, as illustrated in Fig. 7.6, shearing stress parallel to grain is created. The largest shear stress parallel to grain usually occurs along the neutral axis on the plane at which the induced stress changes from compression to tension and generally increases to its maximum at the supports or ends of the beam. Shakes, checks, and splits, which may occur during the drying of lumber, have the effect of reducing the area in the plane of shear resistance. Consequently, laboratory test values for shear strength parallel to grain are substantially reduced for design purposes in order to accommodate the probability of the occurrence of shakes, checks, and splits after drying.

7.3.6   Shear perpendicular to grain

Shear perpendicular to grain is not a factor in solid wood design because effective control is applied through limits on design stresses in shear parallel to grain and compression or bearing perpendicular to grain.

7.3.7   Fiber stress in bending

A force or set of forces applied perpendicular to a beam, as shown in Fig. 7.7, creates compression in the fibers on the side to which the force is applied and it also creates tension in the fibers on the opposite side. Thus there is a tendency to compress the fibers on the compression side and to elongate those on the tension side. As the stress is distributed from the extreme fibers or outside faces toward the center or neutral axis of the piece, it is reduced in intensity. Thus deviations in slope of grain and the presence of knots or holes in the outside faces

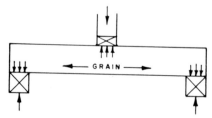

Figure 7.5   Compression perpendicular to the grain.

Figure 7.6 Shear parallel to the grain.

tend to reduce the resistance in the extreme fibers and the bending strength of the beam.

7.3.8 Proportional limit, static bending

The proportional limit occurs at the point where the induced strain or deformation ceases to be proportional to the stress or applied load as determined by the standard test method. Stress at the proportional limit is computed by the standard method. All conventional methods of structural design for wood are within the proportional or elastic limit.

7.3.9 Modulus of rupture, static bending

The modulus of rupture is computed from the ultimate load or the point at which the piece breaks under the standard bending test. Loading by test beyond the proportional limit shows an increasing rate of deformation, without a specific yield point, until ultimate load is reached.

7.3.10 Modulus of elasticity, static bending

The modulus of elasticity is a measure of stiffness; it is computed on the basis of the load and deformation within the proportional limit.

7.4 Design Values

Design values are assigned to lumber in a scientific manner to provide material with predictable strength properties to meet the requirements of engineering design. Because of the varying nature of the different species of trees, the designer has a wide range of stress values from which to make a selection. However, to avoid delay during con-

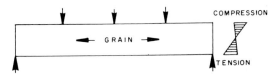

Figure 7.7 Fiber stress in bending.

struction, it is advisable to determine which species and grades are available locally before design values are selected.

7.4.1 Classification of structural lumber

Because the effects of knots, slope of grain, checks, and shakes on the strength of lumber vary with the loading to which the piece is subjected, structural lumber is often classified by size and use. The three major classifications are discussed in the following subsections.

Dimension lumber. Pieces of rectangular or square cross section, 2 to 4 in thick and 2 in or more wide (nominal dimensions) are graded primarily for strength in bending edgewise or flatwise but also used where tensile or compressive strength is important. Dimension lumber may be further classified as joists and planks, for material 5 in or more in nominal width, and as light framing or structural light framing for material 2 to 4 in wide.

Beams and stringers. Pieces of rectangular cross section, 5 to 8 in (nominal dimensions) and larger, are graded for strength in bending when loaded on the narrow face.

Posts and timbers. Pieces of square or nearly square cross section at least 5 by 5 in (nominal dimensions) are graded primarily for use as posts or columns but are adaptable to miscellaneous uses in which bending strength is not especially important.

7.4.2 Characteristics affecting strength

Aside from the natural properties of the species, the major characteristics affecting the strength of a piece of lumber are the sizes and locations of knots or holes, the sizes and locations of checks or shakes and splits, the amount of wane (absence of wood), slope of grain, degree of density (rings per inch), and the condition of seasoning. All these characteristics are taken into consideration in the stress grading of a piece of lumber; they are illustrated in Fig. 7.8.

7.4.3 ASTM standards

There are two ASTM standards which serve as principal references in the assignment of working stresses of lumber. One standard is ASTM

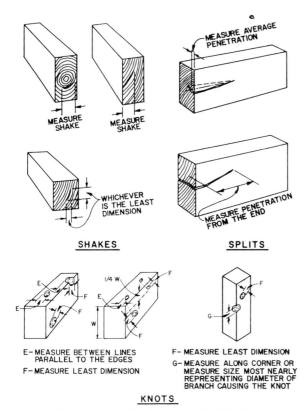

E—MEASURE BETWEEN LINES
PARALLEL TO THE EDGES

F—MEASURE LEAST DIMENSION

F— MEASURE LEAST DIMENSION

G— MEASURE ALONG CORNER OR
MEASURE SIZE MOST NEARLY
REPRESENTING DIAMETER OF
BRANCH CAUSING THE KNOT

KNOTS

Figure 7.8   Defects that affect the strength of lumber.

D2555, "Methods for Establishing Clear Wood Strength Values," which sets forth procedures for establishing strength values for clear wood of different species in the unseasoned condition and unadjusted for end use. Such procedures can be applied to a single species or to a group of species when growth and marketing conditions justify the grouping. The other standard is ASTM D245, "Methods for Establishing Structural Grades for Visually Graded Lumber," which sets forth reduction factors to be applied to the clear wood values and provides procedures for determining strength ratios based on knots and other characteristics. When applied to the adjusted clear wood values, the ratios are measures of the working stresses of the various commercial grades of any species. This standard also provides adjustments for degree of density and for condition of seasoning.

7.4.4   Lumber-grading rules

Lumber-grading rules are, in effect, specifications of quality. In the rules, the maximum knots, slope of grain, and other strength-reducing characteristics are described in sufficient detail that the procedures of ASTM D245 can be applied and working stresses can be assigned to the specified quality. It is common practice to give each grade a commercial designation such as No. 1 for best, No. 2 for next best. This means that the purchaser orders the commercial grade which qualifies for the values used in design.

7.4.5   Machine-graded lumber

Although most structural lumber is assigned design values on the basis of visual grading to meet minimum quality specifications, there is a growing trend toward the nondestructive testing of lumber by machine. In this method a piece of lumber is passed flatwise through a series of loading rollers and the stiffness, or modulus of elasticity, is automatically recorded. Through correlation with previously established test data, bending strength and other strength properties are assigned to each piece tested. At present, machine grading is supplemented by visual grading, particularly in the assignment of horizontal or longitudinal shear values.

7.4.6   National design specification

The principal reference for working stresses of commercial grades of structural lumber is the National Design Specification for Wood Construction; it is available from the National Forest Products Association, Washington, D.C. The design value information in this specification is taken from the published rules written by the American Lumber Standards Committee (ALSC) and other grading rules-writing agencies. When the values are used, each piece of lumber is required to be identified by the grade mark of a lumber grading or inspection agency recognized as being competent.

The National Design Specification provides for design of single-member uses of lumber and other structural timbers and also for repetitive member uses of lumber where load sharing is known to exist between repetitive framing members which are spaced not more than 0.6 m (24 in), are not less than three in number, and are joined by floor, roof, or other load-distributing elements adequate to support the design loads. For repetitive member uses, the design values in bending are higher than those for single-member uses, as provided in the National Design Specification.

7.5   Timber Piles

Recommendations for the use of timber piles in foundations can be found in material published by the American Wood Preservers Institute (1967). ASTM D25-73, Standard Specifications for Round Timber Piles, classifies round timber piles according to the manner in which their load-carrying capacity is developed. There are two classes: friction piles and end-bearing piles.

7.5.1 Friction piles

Friction piles are used when pile capacity is determined by the friction developed in contact with the surrounding soil, along with the compressive strength of the timber piles used. Table 7.1, from ASTM Standard D25-73, lists size requirements for friction piles.

7.5.2 End-bearing piles

End-bearing piles are used when pile capacity is determined primarily by the end-bearing capacity of the soil at the pile tip, along with the compressive strength of the timber piles used. Table 7.2, from ASTM Standard D25-73 (75), lists size requirements for end-bearing piles.

7.5.3 Wood sheet piles

Wood sheet piles are sometimes used for groins, bulkheads, and subterranean cutoff walls in a salt water environment. Wood used as sheet piling is subject to environmental attack and must therefore be treated with preservatives if it is to have a useful life of more than a few months. Wood sheet pile should be beveled at the bottom on one

TABLE 7.1 Friction Piles: Specified Butt Circumferences with Minimum Tip Circumference (ASTM)

Length, m	Required minimum circumference 914 mm from butt, mm										
	559	635	711	787	889	965	1041	1118	1194	1270	1448
	Minimum tip circumferences, mm										
6.1	406	406	406	457	559	635	711				
9.1	406	406	406	406	483	559	635	711			
12.2				406	432	508	584	660	737		
15.2					406	432	483	559	635	711	
18.3						406	406	472	549	625	803
21.3						406	406	406	411	488	665
24.4							406	406	406	406	554
27.4							406	406	406	406	495
30.5							406	406	406	406	457
33.5										406	406
36.6											406

Note: Where the taper applied to the butt circumferences calculates to a circumference at the tip of less than 406 mm, the individual values have been increased to 406 mm to ensure a minimum of 127 mm tip for purposes of driving.

TABLE 7.2   End-Bearing Piles: Specified Tip Circumferences with Minimum Butt Circumferences (ASTM)

Length, m	Minimum circumferences 0.9 m from butt, mm, for following minimum tip circumferences, mm							
	406	483	559	653	711	787	889	965
6.1	559	610	686	762	838	914	1016	1092
9.1	597	673	749	826	902	978	1080	1156
12.2	660	737	813	889	965	1041	1143	1219
15.2	724	800	876	953	1029	1105	1206	1283
18.3	787	864	940	1016	1092	1168	1270	1346
21.3	851	927	1003	1080	1156	1232	1334	1410
24.4	914	991	1067	1143	1219	1295	1197	1473
27.4	980	1057	1133	1209	1285	1361	1463	1537
30.5	1041	1118	1194	1270	1346	1422	1524	
33.5	1107	1184	1260	1336	1412	1549		
36.3	1168	1245	1321	1397	1473			

side and one edge to facilitate driving and to cause each succeeding pile to wedge firmly against the adjacent pile. Sheet pile should not be driven more than a meter. If deeper penetration is needed, the area along the line of piles should be excavated before driving so that the piles need be driven only a meter to final tip elevation. There are two types of wood sheet pile in general use. Members are sized according to the loads and conditions to be resisted by the sheeting.

Tongue and groove.   Tongue-and-groove sheeting consists of planks so milled that on one edge there is a projecting tongue and on the opposite edge a groove into which the tongue of the adjoining plank is fitted when driven.

Wakefield sheet piling.   Wakefield sheet piling is made up of three layers of planks spiked or bolted together to form a sheet pile such that the middle plank projects beyond the edges of the planks on each side and thus forms a tongue on one edge and a groove on the other (Fig. 7.9).

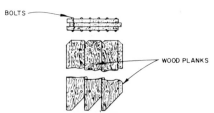

BOLTS

WOOD PLANKS

Figure 7.9   Wakefield sheet piling.

7.6   Characteristics of Common Construction Species

Woods normally used in the coastal zone are the domestic softwoods generally available in the United States: Douglas fir, southern pine, spruce, hemlock, redwood, cedar, and a number of species of pine, including lodgepole, ponderosa, and white. Hardwoods are less commonly used not because of inferior quality but because of cost or availability. Hardwoods in general are more difficult to treat with preservatives, but special situations may call for them. For instance, an imported hardwood called greenheart has gained some acceptance for use as fender piles because it appears to be fairly resistant to marine borers in its untreated state. Tables 7.3 and 7.4 list significant characteristics of domestic softwoods and hardwoods, respectively.

7.7   Destructive Biota

Although there are many life-forms that may eat, live in, or make use of wood in a way that we call destructive, many are so rare or do so little damage during the useful life of wood structures that they can be ignored relative to the use of wood in the coastal zone. Those that most seriously affect the useful life of wood are the shipworms (teredos) of the family Teredinidae and small (2-mm; 80-mil) crustaceans of the genus Limnoria. These marine biota are generally more active in clean water with high dissolved oxygen. On land, the most destructive insects are termites. Also on land but more in air, and very destructive in the presence of moisture or intermittent wetting, are the fungi and bacteria. Preservative treatment can reduce the destructive effects of the various biota and extend the useful life of wood but cannot completely prevent the attacks. Cracks or holes in the wood or leaching of the preservatives will eventually allow access for some marine borer or nest of termites.

TABLE 7.3   Domestic Softwoods

General characteristics	Douglas fir	Redwood	Cedar	Spruce	Southern pine
Shrinkage in volume from green to oven dry, %	10.3	11.5	11.2	10.4	12.4
Modulus of rupture, MPa	43.7 (green)	51.4 (green)	41.6 (green)	31.4 (green)	44.1 (green)
Modulus of elasticity at 6.895 MPa	1131 (green)	1101 (green)	754 (green)	866 (green)	1140 (green)

TABLE 7.4   Domestic and Imported Hardwoods

General characteristics	Domestic				Imported greenheart
	Oak	Maple	Ash	Birch	
Shrinkage in volume from green to ovendry, %	12.7 to 17.7	12.0 to 14.5	11.7	15.0 to 16.8	3
Modulus of rupture, MPa	49.5 to 73.8 (green) 54.2 to 77.3 (dry)	40.1 to 62.5 (green) 105.2 (dry)	41.4 to 68.9 (green) 108.0 (dry)	61.7 133.8 (dry)	12.3 to 206.8 (dry)
Modulus of elasticity at 6.895 MPa	877 to 1687 (green) 1321 to 2083 (dry)	943 to 1474 (green) 1761 to 1878 (dry)	1030 to 1480 (green) 1395 to 1778 (dry)	1490 to 1543 (green) 2396 (dry)	2900 (green)

7.7.1 Teredinidae

The Teredinidae are marine bivalve mollusks that have evolved into long wormlike shapes. The shell parts have become a set of grinders at one end, and it is with the grinders that the teredo bores holes in wood. An adult can be 50 to 100 mm (2 to 4 in) long and 5 to 10 mm (0.2 to 0.4 in) in diameter. The individual enters the wood as a larva by making a small hole that is never enlarged at the surface. As it grows, the teredo bores a larger hole into the wood at the rate of 20 to 300 mm (0.8 to 11.8 in) per month to accommodate its whole body and apparently to feed itself. An infestation of teredos can destroy an untreated pile at the mud line in 5 to 6 months (Kofoid and Miller, 1927). Species found in abundance in U.S. waters are *Teredo diegensis* and *T. navalis*. Teredos are sensitive to coal-tar creosote treatment.

7.7.2 Limnoria

The Limnoria are small marine crustaceans about 2 mm (80 mils) long and less than 1 mm (40 mils) wide (Fig. 7.10) that either enter the wood in the adult stage or are hatched in and remain in the same piece of wood. They use the wood as habitat and apparently as food supply, because they continue to bore holes after they are securely entrenched in the wood (Ray, 1959). They bore at the rate of about 0.5 mm (20 mils) per day (Kofoid and Miller, 1927). At that rate, a heavy infestation of limnoria could eat through a 30-cm (12-in) untreated pile in about a year. One species, *Limnoria tripunctata,* is present off most of the U.S. coastline (Fig. 7.11). A subspecies, *Tripunctata*

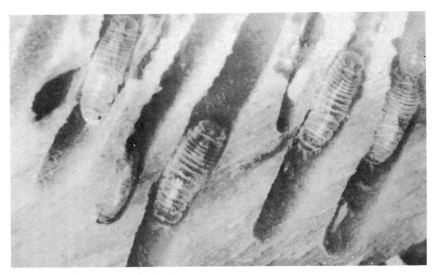

Figure 7.10   Live limnoria in their burrows.

mengies, is found all along the Atlantic seaboard and in the Pacific ocean from the southern end of South Island, New Zealand, to several hundred miles north of Vancouver, Canada. *L. tripunctata* is particularly troublesome because it apparently is not repelled by coal-tar creosote preservative (American Society for Testing Materials,

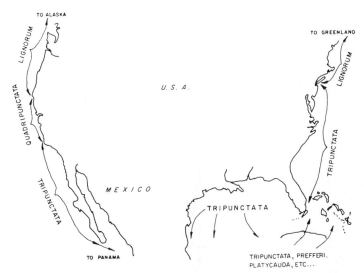

Figure 7.11   Distribution of limnoria in North America. *(Courtesy of ASTM)*

1957). Where *L. tripunctata* is present, the dual treatment, described in USDA, *Preservative Treatment of Wood*, paragraph 6, is required. Figure 7.12 (Ray, 1959) shows the damage that can occur from limnoria attack.

7.7.3 Termites

The principal termite species attacking wooden structures in the United States is a subterranean type named *Reticulitermes hesperus*. The typical life cycle of this species starts with winged reproductive adults that fly from the nest for the purpose of establishing new colonies. When a pair finds a suitable environment, it starts a colony. In 5 or 6 years, a colony may contain several thousand individuals (Palermo, 1951).

Termites are antlike insects about 5 mm (0.2 in) long that spend their lives inside an earth nest or gnawing tubes through available wood (except for the winged adults). Termite damage is not evident on casual observation because the termites leave the outer layer of wood untouched for their own protection. The usual evidence of their presence consists of the piles of fecal pellets that are pushed out of the way through small vent holes in the wood about 1 mm (40 mils) in diameter. A structure attacked by termites will eventually fail unless the infestation is discovered early and the termites are destroyed.

Figure 7.12 Collapse of wood wharf under railroad gondolas. *(Courtesy of Los Angeles Harbor Department)*

Termite control can be accomplished in one of several ways. Separation of structural wood from the ground and removal of all cellulose material from the ground in the vicinity of the structure are accomplished in the design and construction phases. Dry ground, good ventilation, and exposure to sunlight also discourage termites from nesting. If contact with the ground cannot be avoided as in the case of power poles, pressure treatment with preservatives will discourage termites. Poison can be injected into the wood and nesting areas where termites are established.

7.7.4 Fungi

Decay fungi, which are of primary concern, consist of microscopic threadlike strands known as hyphae; the hyphae aggregate into masses called mycelia. Under suitable conditions, the mycelia form fan sheets, especially when they are developing in a very moist locality. They may give rise to the fruiting body of the fungus which, in the case of the decay fungi, are relatively flat. These fruiting bodies bear enormous numbers of microscopic spores which are similar in function to the seeds of higher-order plants. The spores are readily distributed by water or air currents or by men and animals. Spores germinate and penetrate wood by means of hyphae. The fungus may also be spread from decayed material to sound material by the hyphae.

In the United States there are many species of fungi that cause wood decay. Two important species are the building poria, *Poria incrassata,* and the tear fungus, *Merulius lachrymans.* The tear fungus is most common in northern United States and Canada; the poria fungus prevails in the south and west. Timber-destroying fungi require both moisture and oxygen at a temperature of about 20 to 36°C (68 to 97°F) for optimum growth. Therefore, wood that is kept very dry will not decay, nor will wood that is submerged where the oxygen is excluded. Because wood must be kept moist, the term "dry rot" is a misnomer for the crumbly brown rot that results from the action of fungi. Figure 7.13 shows specimens of wharf timbers heavily damaged by fungi.

Control of fungi in wood structures can be accomplished by proper design and by chemical treatment. Design criteria should anticipate meteorological conditions such as fog, rain, and dew, which deposit moisture on wood surfaces. Wood structures should be designed to provide for drainage of wood surfaces and eliminate joints and pockets where moisture can collect. Where exposure to moisture is severe and cannot be eliminated by design, pressure treatment with a wood preservative is required. Coal-tar creosote, copper naphthenate, pentachlorophenol, and

Figure 7.13   Specimens of wharf timbers damaged by dry rot.

salt preservatives such as chromated zinc chloride are used separately or in combination for fungi control.

7.8   Preservative Treatment of Wood

To extend its life for both economical and practical use in the coastal zone, wood must be protected from its natural enemies: fungi, bacteria, insects, and marine organisms. Effective preservative treatments have been found to discourage the natural enemies and extend the useful life of wood to about four to five times that of untreated wood. Untreated wood can be used effectively for temporary structures and facilities.

7.8.1   Pressure processes

The most effective method of treating wood with preservatives is by means of pressure. There are a number of pressure processes that employ the same general principle but differ in the details of application. Treatment includes loading the timber on tramcars, which are run into a large steel cylinder. The cylinder door is bolted, and the preservative is applied under pressure until the required absorption has been obtained. Two principal types of pressure treatment, the full cell and empty cell, are in common use (U.S. Department of Agriculture, 1952).

Full cell processes. In pressure treatments with the so-called full cell or Bethell process, a preliminary vacuum is first applied to remove as much air as practicable from the wood cells. The preservative is then admitted into the treating cylinder without admitting air. After the cylinder is filled with preservative, pressure is applied until the required absorption is obtained. A final vacuum is commonly applied immediately after the cylinder has been emptied of preservative to free the timber (or charge) of dripping preservative. When the timber is given a preliminary steaming and vacuum treatment, the preservative is admitted at the end of the vacuum period following steaming.

It is impossible to remove all the air from the wood cells regardless of the method of treatment employed. For that reason, even under the most favorable conditions, there is some unfilled airspace in the cell cavities of the treated wood after impregnation by the full cell process.

Empty cell processes. Two empty cell treatments, the Lowry and the Rueping, are commonly used; both of them depend on compressed air in the wood to force part of the absorbed preservative out of the cell cavities after preservative pressure has been released. In the Lowry process, which is also designated as the empty cell process without initial air, the preservative is admitted to the treating cylinder at atmospheric pressure. When the cylinder is filled, pressure is applied and the preservative is forced into the wood against the air originally in the cell cavities. After the required absorption has been obtained, pressure is released, a vacuum is drawn, and the air under pressure in the wood forces out part of the preservative absorbed during the pressure period. This makes it possible, with a limited net retention, to inject a greater amount of preservative into the wood and to obtain deeper penetration than when the same net retention is obtained with the full cell process. The Lowry process is convenient to use in any pressure-treating plant, since no additional equipment is required.

The Rueping process is called the empty cell process with initial air; it differs from the Lowry empty cell process in that air is forced into the treating cylinder before the preservative is admitted. The air pressure is then maintained while the cylinder is filled with preservative; thus the wood cells are left more or less impregnated with air pressure.

7.8.2   Classification of preservatives

Wood preservatives can be grouped into two broad classes: preservative oils and waterborne preservatives. Each of these classes can be further subdivided in various ways. For example, preservative oils include petroleum-refining byproduct oils such as coal-tar and other creosotes, mixtures of coal-tar creosote with coal tar, petroleum, or other oils, solutions of toxic chemicals such as pentachlorophenol or copper naphthenate in selected petroleum oils or other solvents, and various mixtures of these solutions with the byproduct oils and mixtures. The waterborne preservatives include solutions of single chemicals such as chromated zinc chloride (CZC) or chromated copper arsenate (CCA), which are not resistant to leaching, and various formulations of two or more chemicals that react after impregnation and drying to form compounds with limited solubility and sometimes with high resistance to leaching.

Preservatives vary greatly in effectiveness and in suitability for different purposes and use conditions. The effectiveness of any preservative depends not only on the materials of which it is composed but also on the quantity injected into the wood, the depth of penetration, and the conditions to which the treated material is exposed in service.

Coal-tar creosote. Coal-tar creosote is defined by the American Wood Preservers Association as a preservative oil obtained by the distillation "of coal tar produced by high-temperature carbonization of bituminous coal; it consists principally of liquid and solid aromatic hydrocarbons and contains appreciable quantities of tar acids and tar bases; it is heavier than water; and has a continuous boiling range of at least 125°C, beginning at about 200°C." It is highly effective, and it is the most important and most extensively used wood preservative for general purposes. Its solutions vary, but they usually contain from 30 to 70 percent of coal tar by volume; the most prevalent mix contains 50 percent coal tar.

Chemicals dissolved in solvents other than water. Preservatives composed of toxic chemicals carried in nonaqueous solvents, such as petroleum distillates, are now being used to an increasing extent. They were originally devised for the purpose of providing a clean treatment without causing swelling of the wood and were originally applied by nonpressure methods.

A shortage of creosote that developed during World War II created an active interest in the use of other preservatives as a possible substitute for creosote, especially in the pressure treatment of poles. Particular attention was directed to the chlorinated phenols, which are known to have a high degree of toxicity. Pentachlorophenol is the best known and most widely used in this group. Other preservatives of this type, which in the past have been largely limited to use in surface treatments, are the metallic naphthenates, such as copper naphthenate. The latter has also been used to a limited extent for pressure-treating poles. Although some of these toxic chemicals, particularly pentachlorophenol, have given excellent results over a considerable period of time, service records are still inadequate to evaluate them completely against coal-tar creosotes.

Waterborne preservatives. A variety of chemicals in water solution are used as wood preservatives. They include zinc chloride, sodium fluoride, arsenic in various forms, copper sulfate, and similar toxic chemicals. Most of these salts are used in combination with one or more other chemicals, frequently including a chromium compound. Chromated zinc chloride (CZC), which is composed of a mixture of zinc chloride and sodium dichromate, has come into wide use in recent years. The preservative is now much more extensively used than straight zinc chloride, which was formerly the most widely used waterborne salt. A less widely used compound, fluorochromearsenate phenol (FCAP), is one of the preservatives listed in the current standards and government specifications.

Arsenic compounds have been used as preservatives for many years. They are important ingredients of a number of proprietary preservatives, some of which have demonstrated high effectiveness and are extensively used. Three effective compounds commonly used are chromated copper arsenate (CCA), acid copper chromate (ACC), and ammoniacal copper arsenite (ACA). Two types of CCA are specified in TT-W-571. The type is chosen according to availability and economics.

Copper sulfate, although extensively used in Europe for many years and demonstrated to be moderately effective in retarding decay, has found little use for wood preservation in the United States except in certain proprietary preservatives, in which it is combined with other chemicals. Several of these preservatives are of high effectiveness and are extensively used. Copper sulfate is corrosive to iron and steel; therefore, it cannot be used alone in ordinary treating equipment.

Proprietary preservatives. Various patented or proprietary preservatives for pressure treatment are sold under trade names. For the most part they are composed of various waterborne salts and are injected in water solutions. Others employ a volatile solvent to carry the toxic substance into the wood. Some of the waterborne preservatives contain chemicals that are intended to react after injection into the wood and to form substances that are of low solubility and resistant to leaching.

Wolman salts is one of several proprietary names for a waterborne salt, chromated copper arsenate (CCA), also known as green salt. Chemonite is another proprietary name for ammoniacal copper arsenate (ACA). Other proprietary names for preservatives can be found in the *American Wood Preservers Association Standard* (AWPA)

M9. It lists a number of proprietary names for each of the standard preservatives.

7.9 Applications for Treated Wood

Tables 7.5 and 7.6 indicate the amount of preservative to be retained in various wood forms by using approved practices for preservative treatment with creosote and solutions containing creosote, pentachlorophenol, and waterborne preservatives. The net retentions in the tables are minimum penetration requirements. Higher net retentions may be needed for severe use conditions and should be specified when applicable. Data in the tables are taken from Federal Specification TT-W-571.

Coal-tar creosote, creosote–coal tar solution, creosote-petroleum so-

TABLE 7.5   Preservative Retention for Timber Treatment

		Retention by species, lb/ft^3			
Application	Type of preservative	Southern pine Coast Douglas fir Western hemlock Larch	Softwoods: Interior Douglas fir Ponderosa pine Redwood Jack pine Lodgepole pine White pine Red pine Sugar pine	Oak	Gum
Submerged	Creosote^a	25.0	25.0	10.0	12.0
	CCA or ACA	2.5	2.5	NR	NR
Splash zone	Creosote^a	12.0	8.0	5.0^2	5.0^b
	Pentachlorophenol	0.6	0.4	0.35	0.4
	ACC	NR	0.25	0.5	0.5
	CCA or ACA	0.6	0.6	0.4	0.4
Above	Creosote^a		8.0	5.0	6.0
ground	Pentachlorophenol		0.4	0.3	0.3
	ACC, ACA, CCA		2.5	2.5	2.5
	CZC		0.45	0.45	0.45
	FCAP		0.25	0.25	0.25
Soil	Creosote^a	10.0	10.0	0.35	0.4
contact	Pentachlorophenol	0.5	0.5	0.5	0.5
	ACC	0.5	0.5	0.4	0.4
	CCA or ACA	0.4	0.4		
Dual	Creosote^a	20.0			
treatment	CCA or ACA	1.5			

^{a}Includes creosote–coal tar.
^{b}6 lb/ft^3 for members under 16 mm (5 in) thick.
FROM AWPA-C2.

TABLE 7.6   Preservative Retention for Treatment of Wood Piles

		Retention by species, lb/ft^3				
Application	Types of Preservative	Southern pine	Douglas fir Pacific coast	Oak	Red pine Ponderosa pine Jack pine	Lodgepole pine Western larch Interior Douglas fir
Foundation, land, and fresh water	Creosote^a	12.0	17.0	6.0	12.0	17.0
	Penta-	0.6	0.85	0.3	0.6	0.85
	chloro-phenol CCA or ACA	0.8	1.0	NR	0.8	1.0
Marine (submerged)	Single:					
	Creosote^a	20.0	20.0	10.0		
	CCA or ACA	2.5	2.5	NR		
	Dual:					
	Creosote^a	20.0	20.0			
	CCA or ACA	1.0	1.0			

^{a}Includes creosote–coal tar.
From AWPA-C3.

lution, and pentachlorophenol in heavy petroleum solvent and the three waterborne preservatives, ACA, CCA Type I, and CCA Type II, are ordinarily used for wood exposed to severe weathering conditions, such as contact with soil or water, and for important aboveground structures exposed to the weather. Because oil-type preservatives afford protection against weathering and checking as well as against decay, they are generally preferable to waterborne preservatives for the treatment of sawn wood that is to be used in contact with the ground. If cleanliness, freedom from odor, or paintability is essential, either of the waterborne preservatives mentioned above may be expected to give good protection to sawn wood that is selected for its receptiveness to treatment and is treated to meet the minimum penetration requirements. For moderate weathering conditions, retentions should be at least one-third higher than the recommended minimums.

Pentachlorophenol in a volatile petroleum solvent (Table 7.5) is ordinarily used in aboveground structures, particularly where cleanliness and paintability are required. All of the waterborne preservatives (Table 7.6) are suitable for such use. Pentachlorophenol in a light petroleum solvent also is generally limited to aboveground use, especially where moderate cleanliness is desired and freedom from residual solvent is not essential. If water repellency also is desired in order to avoid surface damage due to wetting during storage, it should be stipulated by the purchaser. In some harbors, conditions are highly favorable for limnoria, and the life of creosoted piling may be extended by mechanical barriers. AWPA Standard C3 includes a dual treatment that is recommended for trial in harbors where experience has shown that a high limnoria hazard exists.

Painting of treated wood involves special considerations. Wood treated with creosote, solutions containing creosote, and pentachlorophenol in heavy petroleum solvent cannot ordinarily be painted satisfactorily. When requested, it can be conditioned by the producer to improve its cleanliness. Difficulties may be encountered in painting wood treated with pentachlorophenol in a light petroleum solvent. Wood treated with waterborne preservatives should be properly seasoned after treatment, and it may require light brushing or sanding in order to provide a paintable product. Since "cleanliness" is a relative term, it is recommended that the purchaser make known the specific requirements and the end use of the material and that the supplier be required to furnish evidence that the material is suitable for that use. In the absence of accepted methods for determining cleanliness, paintability, and water repellency of pentachlorophenol-treated wood, the purchaser may elect to use arbitrary test methods which should be described to the supplier.

7.9.1 Timber and lumber

The treatment required for marine timbers and lumber depends on the environment in which the timbers and lumber function. Timbers subject to the marine environment but not submerged or intermittently submerged are treated differently from those that are submerged. The reason is that submerged timbers are subject to marine borer attack and must be treated in accordance with the anticipated attack. Unsubmerged timbers are highly subject to fungus attack, particularly where water spray or airborne moisture can frequently wet them, but cannot be attacked by marine borers.

Examples of submerged use are the framing and bracing members of a wood pier. Wales also are examples, particularly the lower wales of fender piles and the wales at the top of wood groins; they are frequently located at or below the water level. The planks of Wakefield piling used as a groin and wood cribbing below water are other examples.

When timber and lumber are used above the water but near enough to be frequently wetted by splash and spray, they are in the spray or splash zone. Pier decks and wood fittings such as handrails are frequently in this zone of use. Timber bulkheads and cribbing above water are also frequently in the splash zone.

When wood is used away from the immediate contact with salt water or its splash and spray, two different treatments are called for. They are pressure treatments that have different retention requirements depending on whether the wood is placed in contact with the soil or is above the soil in air. Retention requirements for those uses are shown in Table 7.5.

Examples of wood in contact with soil are bulkheads and retaining walls using Wakefield piles or horizontal planking supported by vertical piles. Sometimes boardwalks are incorporated into a bulkhead structure and are frequently in firm contact with the soil. Sand fences and cribs placed above the tide line are usually in direct contact with the soil.

Wood in air is probably best visualized in causeway decking (far enough removed from the water to be free of the direct influence of splash and mist), buildings, towers, navigation aids, and other such structures built on piles or concrete foundations. Whatever the foundations, wood in air must be clear of the ground by at least 200 mm (8 in) and well ventilated. In the southern United States or in especially warm and moist climates, additional clearance should be considered (ASCE, 1975).

The adequacy of preservative treatment may be determined by the quantity of preservative retention or by the penetration of preservative into the wood. The quantity of preservative required for adequate protection is given by the American Wood Preserving Association in pounds per cubic foot. The retained quantity is measured by assaying the contents of core samples.

Timber and lumber used in submerged locations should be pressure-treated by using the full cell process to achieve retention equal to or greater than the amounts shown in Table 7.5. In parts of the world where teredo and pholad attack are expected and where *Limnoria tripunctata* attack is not prevalent, creosote or creosote–coal tar treatment will provide adequate protection. Where *L. tripunctata* attack is expected, and where either teredo or pholad attack is expected, the dual treatment with creosote or creosote–coal tar and either CCA or ACA preservatives to the retentions shown in Table 7.6 will give the best protection known (AWPA C3).

Timber and lumber in the splash zone can be protected by using either of the oil-base preservatives, creosote or creosote–coal tar, or one of the waterborne preservatives (CCA or ACA) to the retentions shown in Table 7.5. The creosote or creosote–coal tar preservatives are usually preferred because the waterborne preservatives are subject to leaching.

Creosote and creosote–coal tar mixtures are commonly employed for sawed material (such as bridge timbers) used under relatively severe conditions. Retentions specified for such timbers vary from about 96 to 320 kg/m^3 (6 to 20 lb/ft^3), but about 160 to 192 kg/m^3 (10 to 12 lb/ft^3) is most common. Both empty cell and full cell methods are employed, depending on the amount of sapwood, retention required, size of timbers, and similar factors. The full cell process is commonly employed in the treatment of resistant heartwood timbers and timber for use in salt water. Waterborne salts are widely applied in the treatment of sawed lumber under conditions that make the employment of preservative oils impractical.

Specifications for retentions of both preservative oils and waterborne salts often fail to take into consideration the relation of the timber dimensions to penetration and retention. The specifications may require a net retention in large heartwood timbers that cannot be obtained because of the small ratio of surface area to volume, although the same retention might be obtained without difficulty in heartwood timbers of tie size or smaller or in large-size timbers containing a large proportion of sapwood. In timbers containing 50 percent or more sapwood, it is recommended that at least 1570 g/m^3 (98 lb/ft^3) preservative oil be specified.

The AWPA standards for adequate penetration of the preservative indicate the required penetration in inches or percent of the thickness of the sapwood, whichever is greater. The penetration requirement of preservatives in timbers and lumber generally varies with the species, but for some species it also varies by size. Timbers and lumber smaller than 5 in (127 mm) require less penetration when the species is coastal Douglas fir, hemlock, or a species of pine other than southern pine and ponderosa.

Requirements for penetration are found in tables of AWPA Standard C2. The following are representative examples of preservative penetration requirements found in the tables. For use aboveground or in fresh water, the penetration required for southern pine is 635 mm (2.5 in) and 85 percent of the sapwood for all sizes. Coastal Douglas fir requires a penetration of 12.7 mm (0.5 in) and 90 percent of the sapwood for sizes 127 mm (5 in) and larger. For sizes under 127 mm (5 in), the requirements are 10.16 mm (0.4 in) and 90 percent of the sapwood. Oak, for the same uses, has only the percent of sapwood requirement; white oak requires 90 percent; and red oak requires 65 percent. In the marine environment, the penetration requirements are similar for each preservative of the dual treatment.

Treatment of cuts and holes.   Insofar as practical, wood pieces should be trimmed, dapped, bored, and counterbored before pressure treatment because field treatment cannot match the thorough penetration and distribution of preservatives obtainable in the pressure retort. However, it is not always possible and practical to avoid all field cuts and bores. When pretreated wood is cut in the field, it is essential that the exposed wood be generously mopped with the same preservatives. The top faces of field cuts are particularly vulnerable to fungus attack and should be given extra careful field treatment. Wood submerged in salt water is vulnerable to marine borers that can enter the wood in very small cracks or exposed areas.

7.9.2 Piles

The principal woods used for piling are southern pine and coastal Douglas fir, although a few other woods, such as red pine, lodgepole pine, western larch, and oak, are used in some localities. No untreated wood commercially available for pilings, either domestic or imported, will resist borer attack for more than several years. However, one species of tropical tree known as greenheart (*Ocotea rodiaei* or *Nectandra rodiaei*), which is not treatable, may last 2 or 3 years longer than treated Douglas fir in the same water. Timber piles should conform to the requirements of ASTM Standard D25.

Untreated pine and fir piles usually last no longer than 2 years in the ocean and often less than 1 year where marine borers, such as *L. tripunctata,* are present in great numbers. Treated piles have a life expectancy averaging 8 to 10 years where *tripunctata* are present. *L. tripunctata* were selected as an example because they are the only known species of Limnoria which will attack and destroy heavily creosoted piling [Civil Engineering Laboratory (CEL), 1974].

The Forest Products Laboratory has tested a large number of preservatives to study their effectiveness in protecting wood against marine borers. Results obtained in these experiments, as well as experience in general, have shown that heavy retentions of coal tar creosote are essential if the best protection is to be obtained (USDA 1952). The heavy retentions ensure better penetrations and also furnish a reserve supply of creosote to provide against early depletion by leaching. Over much of the coastal region of the United States, marine timbers are exposed to severe borer attack, and it is poor economy to specify retentions that will not give the maximum protection under such conditions. Specifications for such timbers should require treatment to refusal by the full cell process, and the specified retention should be the minimum that will be accepted. No maximum should be specified.

The Civil Engineering Laboratory (CEL, 1974) reports that a compound that is toxic to *L. tripunctata* does not prevent *Toredo diegensis* attack, and vice versa. Experiments by CEL indicate that a dual treatment of wood piles should be used in moderate or warm waters to defend against marine borer attack effectively. The dual treatment consists of metallic salts, either ammoniacal copper arsenite (ACA) or chromated copper arsenate (CCA) and coal-tar creosote. A 26 N/m^3 (1.5 lb/yd^3) treatment of metallic salts is applied in water solution. After drying, the wood is pressure-treated with coal-tar creosote to a 3140 N/m^3 (20 lb/ft^3) retention.

The above treatment can be specified as conforming to American Wood Preservers Association Standard C2. It significantly increases the expected life of wood piles used in moderate or warm waters, but it also reduces the strength and toughness of the wood. Studies show that a 157 N/m^3 (1 lb/ft^3) metallic salt retention could reduce strength by 5 percent and toughness by 13 percent and cause increased brittleness. At 470 N/m^3 (3 lb/ft^3), the reductions will be 17 and 38 percent, respectively. Because of the decreased strength and toughness, dual treatment may not be recommended for uses such as fender piles when breakage from impact may limit the useful life before marine borer attack. Regardless, the Navy uses dual-treated fender piles.

In northern waters or where attack by *L. tripunctata* is not anticipated and Teredo is the only threat, pressure treatment of 3140 N/m^3 (20 lb/ft^3) of coal-tar creosote is sufficient. Wood members can be designed to full strength. Other chemicals, such as pentachlorophenol, should not be used in seawater because the chemical will hydrolyze. That is, water will split the chemical bonds and unite with the radical ions of the original compound to form acids and bases.

Table 7.6 shows preservative retention requirements as set forth in AWPA Standard C3 for single and dual treatment of wood species most likely to be used for piles in the United States. Preservative retentions, in pounds per cubic foot, are measured by assay of bore samples.

Requirements for adequate preservative treatment of piles include minimum penetrations. Penetration requirements for various species of wood piles and use conditions also are set forth in AWPA Standard C3. Penetration tests are made by gaging the penetration distance from the outside force of the pile. Representative preservative penetration requirements for the wood species most frequently used for piles are presented in Table 7.7.

TABLE 7.7   Representative Preservative Penetration Requirements

	Foundation or fresh water	Marine dual treatment
Southern pine	7.6 cm (19 in) and 90% of the sapwood	8.9 cm (23 in) and 90% of the sapwood
Coastal Douglas fir	1.9 cm (4.8 in) and 85% of the sapwood	2.5 to 4.4 cm (6.4 to 11 in) and 85% of the sapwood

7.9.3 Treatment of pile cutoffs, framing cuts, and holes

After driving treated wood piles in a wood wharf or other structure, excess wood in the piles is sawed off at the desired elevation. This exposes untreated wood at the cutoff, which necessitates some kind of preservative treatment in place. The usual method is to swab the cutoff with creosote, cover that with Irish flax, and add another coat of creosote before placing the pile cap. On inspection after only a few years service, cones of dry rot have been found in pile tops that were given the foregoing treatment.

A method by which cutoff areas have been adequately protected and which is inexpensive consists of boring five or six 19.1-mm (0.75-in) holes about 25.4 mm (1 in) apart in a circular pattern in the untreated area of the cutoff. This is shown in Fig. 7.14. The holes are then filled with a 50-50 mix of liquid coal tar and creosote. A layer of Irish flax is placed on top and is covered with a 3.8-mm (150-mil) layer of high-density polyethylene before the pile cap is placed. Side and end grain penetration of the preservative completely impregnates the entire pile top to a depth of more than 25.4 mm (1 in) in less than 2 years.

7.9.4 Poles

Prior to World War II, most of the pressure-treated poles used in the United States were treated with American Wood Preservers Associa-

Figure 7.14 Preparing cut-off piles for coal-tar creosote treatment.

tion specification grade 1 coal-tar creosote with a specified distillation residue above 355°C (670°F) of not more than 20 to 25 percent. Coal-tar creosote treatment may still be the preferred preservative under conditions such that waterborne preservatives could leach away or cost may be the controlling factor.

In recent years, solutions of pentachlorophenol have attracted attention as substitutes for creosote or for use in mixtures with creosote, and large quantities have been used. Thousands of poles have been treated with pentachlorophenol dissolved in the lighter petroleum oils or with solutions containing various proportions of coal-tar creosote and pentachlorophenol dissolved in a petroleum solvent. These poles have not been in service for sufficient time to determine how the results will ultimately compare with those obtained with creosoted poles. Experimental installations under observation, however, are giving excellent results, so that this preservative may find a wide field of use in the future.

Most of the poles that have been pressure-treated, and on which the best service records are available, are southern yellow pine and coastal Douglas fir. Preservative retention quantities for these and other species are listed in Table 7.8. The data are taken from Federal

TABLE 7.8   Preservative Retention for Treatment of Poles

	Retention by species,[a] lb/ft^3				
Preservative	Southern pine[a] Ponderosa pine[a]	Coastal Douglas fir[a]	Red pine[a]	Jack pine Lodgepole pine	Western larch Western red cedar Inland Douglas fir
Coal-tar creosote[b]	7.5 to 9[c] 12[d]	9 to 12[c]	10.5 to 13.5[c]	12 to 16	16
Pentachlorophenol in heavy petroleum	0.38 to 0.45[c] 0.60[d]	0.45 to 0.60[c]	0.53 to 0.68[c]	0.60 to 0.80	0.80
ACA	0.60[c]	0.60[c]	0.60[c]	0.60	0.60
CCA	0.60[c]	0.60[c]	0.60[c]	0.60	0.60

[a]Retentions are for use as utility poles except for southern, ponderosa, and red pines and coastal Douglas fir, which are used for building poles as noted in footnotes c and d.
[b]According to AWPA C4, creosote–coal tar also may be used for utility poles.
[c]According to AWPA C23, the highest retentions are used for building poles as well as utility poles.
[d]Fed. Spec. TT-W-571J requires these high retentions for building poles but not utility poles.
Federal Specification TT-W-571J and AWPA C4 and C23 combined

Specification TT-W-571, which gives a more complete specification on the treatment of wood poles.

7.10   Joining Materials

The various members and parts of wooden coastal structures are in nearly all cases joined together by metal. Most common are the bolts, spikes, and driftpins which fasten heavy timbers in structures such as groins, jetties, bulkheads, and piers (Fig. 7.15). Another category includes such items as spike grids and split-ring connectors for increasing the shear capacity of bolted joints (Fig. 7.16), and sheet-metal framing anchors for lighter structural framing and miscellaneous hardware such as bearing plates and straps. A third category of metal connection material includes tying items such as rods, wire rope, and

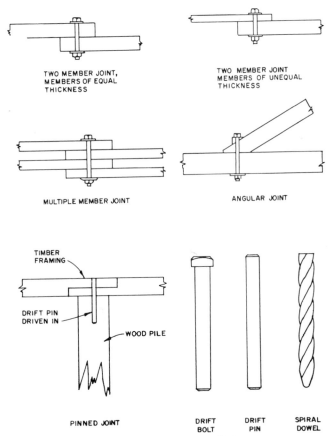

Figure 7.15   Typical bolted and pinned wood joints.

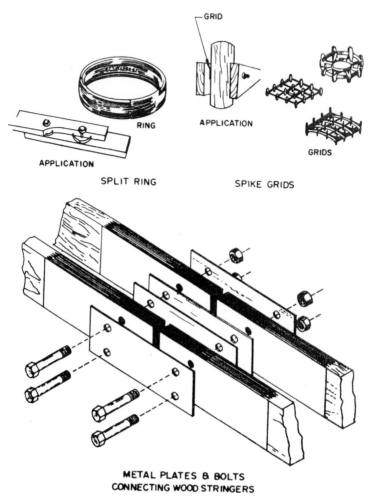

Figure 7.16   Split rings, spike grids, and metal plate connectors.

chain (Fig. 7.17). Metal connection material for a timber structure is subject to many of the same deteriorating factors in a coastal environment as are metal structures. These are predominantly corrosion and, in some cases, abrasion. They are discussed further in Chaps. 6 and 10.

In addition to resisting corrosion, the material may also have to resist chafing, or abrasion by drifting sand, floating debris, or moored vessels. This factor should be considered in selecting the anticorrosion coating or system, as discussed in Chap. 10. Because even the best protective system will have only limited life in a severe marine expo-

Figure 7.17 Pile dolphin tied with wire rope.

sure, a program of periodic inspection and preventive maintenance will probably be needed.

7.10.1 Adhesives

At present, the use of adhesives to form or assemble wood structural members is largely confined to factory production of building components. There the wood parts to be joined can be milled to close tolerances and the joining and curing processes can be closely controlled. For use in a coastal structure where they are exposed to the weather or subject to immersion, a wet use adhesive, phenol or resorcinol resin or a blend of the two, should be specified for shop-fabricated members. Such members have only limited use in coastal structures, primarily for such items as footbridge girders and trusses and small craft docks in marinas.

There is also some use of adhesives in the field assembly of wood structural members primarily for buildings. This use is at present largely in secondary connections in which a failure would not be hazardous to life or property. The necessary gluing pressure is often provided by nailing. Because in-field gluing and alignment of material may be much less precise than in the shop, it is necessary to use different adhesives, which, until recently, have not provided the joint strength and rigidity obtainable in factory gluing. The relatively recent development of fast-curing, gap-filling phenolic and phenol-resorcinol resin adhesives for construction may let onsite gluing further expand into the area of primary load-bearing connections [American Institute of Timber Construction (AITC), 1974].

Conditions of service determine the type of adhesive required. In general, dry use (water-resistant) adhesive should be used for interior locations and wet use (waterproof) adhesive for exterior locations. Under some conditions, however, a member glued with dry use adhesive may be used satisfactorily on an exterior member for certain uses. It is not practical to use both types of adhesives within the length of the same members. If any part of a member's length requires them, wet use adhesives must be used throughout the length. It should be kept in mind that using a wet use adhesive will generally increase the cost of a laminated member; therefore, the adhesive should not be specified unless actually needed.

Dry use adhesives. Casein adhesive with a suitable mold inhibitor is the standard dry use adhesive of the structural glued laminated timber industry. It has proved its dependability for over two generations in Europe and North America. It is used in large quantities by other wood products manufacturers as well as this industry. Casein adhesive with mold inhibitor is satisfactory in properly designed, constructed, and maintained buildings as long as the members are not subjected to repeated wettings or high humidity over a long period of time. Although casein adhesives can withstand some wetting during erection of the members, special attention should be given to the protection of the top face of beams, rafters, or arches during shipment and erection when end or beveled faces are exposed. Angular cuts that pass through one or more laminations and result in feathered ends on the individual laminations are often made. These surfaces have greater than average moisture absorption and should be properly end-sealed to prevent delamination of the feathered ends. Although such damage is not likely to be of structural concern, it may be unsightly. All end cuts should be well sealed (American Institute of Timber Construction, 1974).

Two major requirements must be met to ensure proper performance of casein adhesives in exterior locations. If the requirements cannot be met, wet use adhesives must be used. Complete protection from the direct effects of precipitation on members must be provided either by undercutting the ends to keep off wind-driven moisture or by using fascia boards or end caps to prevent water from collecting on vertical surfaces of the members. Ends of members should be coated with white lead paste or treated with water-repellent sealer before the cap is applied. Casein adhesives are not considered suitable for laminated members intended for exterior use where the moisture content of the wood exceeds 16 percent for repeated or prolonged periods of service.

Only one major condition must be satisfied to ensure proper performance of casein adhesives in interior locations: The moisture content of the wood must not exceed 16 percent for repeated or prolonged periods of service. Dry use adhesives should comply with the requirements of ASTM Standard D3024-72 for structural glued laminated timber.

Wet use adhesives. Phenol-, resorcinol-, and melamine-base adhesives will withstand the most severe conditions of exposure. They are more expensive than water-resistant adhesives. Phenol-resorcinol base or resorcinol-base adhesives are the wet use adhesives most widely used in structural glued laminated members.

Although the wet use adhesives can be employed for all conditions of use, they are generally used only when the equilibrium moisture content of the members in service exceeds 16 percent. The following are examples:

- Members which must be pressure-treated
- Marine vessels and structures such as barges, ships, piers, wharves, docks, slips, and dredge spuds and structures and members exposed to the weather, such as bridges and bridge girders (other than for temporary construction such as falsework and centering)

Wet use adhesives should comply with the requirements of ASTM Standard D2559-72 for structural glued laminated timber. Only adhesives meeting the requirements for wet use should be used with California redwood.

7.11 Repair Materials and Methods

Field repair methods for wood structures are varied and on the whole creative. New methods are proposed regularly, but those presented in this chapter are the methods now in general use. Adhesives are rarely used in field repairs.

7.11.1 Concrete encasements

Wood bearing piles which have received damage from marine borers (either partially or totally destroyed) can be restored to their design capacity, in place, by encasing them in reinforced concrete jackets (Fig. 7.18). Piles are enclosed with nylon jackets (with steel reinforcement added as necessary), and the jackets are filled with tremie concrete. A similar successful method is to enclose the damaged part of the pile in a fiberglass form, install the necessary reinforcing, and fill the space between the form and the pile with a hydrophilic epoxy (Fig. 7.19).

Figure 7.18   Sketch of typical damaged pile and fiberglass form. *(Courtesy of U.S. Navy Civil Engineering Laboratory)*

Figure 7.19   Piles repaired by using fiberglass forms and grout. *(Courtesy of Aquatic Marine Systems)*

Wood pilings that have been severed by marine borers have been successfully restored, in place, by literally dozens of methods. Most of these are based on external reinforcement, such as heavy-wall steel pipe, overlapping the upper and lower pile sections, followed by corrosion protection of the steel by a plastic wrap. Small-scale tests in the laboratory have proved the efficiency of these methods in repairing piles, both in the bearing and bending capacity, to achieve their full design loads.

7.11.2   Synthetic materials wrap

Wood piles can be protected, in place, from marine borer attack by wrapping them with flexible polyvinyl chloride or polyethylene sheeting (cigarette fashion) from 305 mm (12 in) below the mud line to 305 mm (12 in) above the highest tide line. In preparing a wood pile for a jacket, it is important that all sharp protrusions be removed. In the case of barnacles, sharp edges can be smoothed by various simple hand or mechanical devices (Fig. 7.20).

Modular kits that permit fast and positive application from above or below water are available. This system effectively removes wood piles from their environment. Marine borers attacking the piles while encapsulated under the wraps die from lack of oxygen within 48 h, and the synthetic sheathing prevents further intrusion. This system has

Figure 7.20   Wood bearing piles with PVC wrapping. *(Courtesy of the Port of Los Angeles)*

been successfully used on both coasts of the United States and in Germany, Australia, the Bahamas, and elsewhere for more than 20 years. The U.S. Navy also has used the system on numerous projects (see NAVFAC specification 75M-B10a).

Other methods of wrapping wood piles with synthetic film before driving have been tried, as shown in Fig. 7.21. Unless the pile-driving crew is very careful, the PVC jackets can be ripped during driving. Fortunately, permanent repairs can be made by nailing patches of synthetic film over torn areas with aluminum alloy 5056 roofing nails.

7.11.3 In-place treatment of timber cracks

If timber cracks make the pile structurally unsound and also expose it to internal marine borer action, the cracks can be bolted together. One or more bolts are used with form-fitting steel washers on each side of the pile. The entire area should then be jacketed with a synthetic film jacket.

Replacement. In wood structures, parts such as framing members are frequently repaired by replacing the damaged member. Replacement is relatively easy because the fasteners are usually in accessible places and the wood members are in discrete sizes of individual pieces.

Figure 7.21 PVC-wrapped pile ready for driving. *(Courtesy of the Port of Los Angeles)*

When treated wood is repaired, treatment of the new member must be to the same specification as the original. Cuts and bored holes are to be made prior to pressure treatment in the shop when possible. Cut faces of pieces to remain in place must be coated with material similar to that used originally. New fasteners should be used when there is any damage to or corrosion of the original.

Damaged wood piles are removed by pulling after deck and stringers, if present, have been removed. Before pulling, remnants of the damaged pile and fastenings must be cleared away to make space for the replacement pile. A replacement pile can then be set in the same hole as the original and driven to refusal. If it is to support a deck, the pile is driven alongside the pile cap, cut off below the pile cap, and pulled over into place. The cut is treated with preservatives, and shims are inserted to fill the space between the pile and pile cap. A driftpin is then hammered into place to secure the pile.

7.12   Environmental Considerations

Chemical actions of three general types may affect the strength of wood. The first causes swelling and the resultant weakening of wood. This action is almost completely reversible when the swelling solution is removed. The second type of action brings about permanent changes in the wood such as hydrolysis of the cellulose by acids or salts. The third action also brings about permanent change in wood and involves delignification of wood and dissolving of hemicelluloses by alkalies.

7.12.1   Salt water and fresh water

Both salt water and fresh water penetrate wood fibers. When the moisture content is between zero and the fiber saturation point (about 30 percent moisture content), wood will swell. The rate of swelling is proportional to the moisture content up to the fiber saturation point. As wet wood dries, the outer part of the wood loses moisture faster than the inner parts; therefore, the shrinkage rate is uneven, and that can result in the development of checks or cracks. When wood is immersed over extended periods, water can soften the fibers.

Water, and particularly saltwater, carries dissolved oxygen and marine biota that can severely affect wood or wood fastenings. As a bearer of oxygen, water enhances corrosion of iron and steel fastenings. When wood is periodically wetted and dried in the presence of oxygen, it becomes susceptible to the fungus which causes dry rot.

7.12.2   Strong acids

Strong acids (such as nitric and hydrochloric) and highly acidic salts (such as zinc chloride) tend to hydrolyze wood and cause serious strength loss if they are present in sufficiently high concentrations. When the pH of aqueous solutions of weak acids is above 2, the rate of hydrolysis of wood is small and is dependent on the temperature.

7.12.3 Wood oxidation

Wood oxidation by air in dry locations is slow; it attacks the spring wood first to produce a rough or weathered-looking surface. Very dry wood can resist hundreds of years of normal exposure to oxidation. Wood can be dissolved by strong acids, but basically wood is considered to be somewhat resistant to the action of acids and basic hydroxides. Wood is also resistant to most commercial solvents.

7.12.4 Pollutant attacks

Pollution in both the air and water environments can have the effect of prolonging the useful life of wood by reducing the oxygen supply that oxidizes the wood and supports the biota that attack wood.

7.12.5 Sunlight exposure

In sunlight, wood will expand because of the increase in temperature. In most structures, the increase in length of wood for a normal rise in temperature is negligible and secondary stresses that are due to the change can be neglected. Cut pieces of wood will warp toward the sun unless restrained or dried before use.

7.12.6 Wave and current effects

Because wood has less strength than some other commonly used structural materials, a larger member is needed to protect wood adequately against the force developed by water currents and waves even when a solid face is presented to the wave and current forces. The resilient characteristic of wood, however, allows wood members to absorb impact energy and rebound intact better than concrete and steel.

7.12.7 Effects of severe temperature and ice

The temperature effect upon wood strength is immediate, and its magnitude depends on the moisture content of the wood. If the exposure is above normal atmospheric conditions for a limited period and the temperature is not excessive, wood can be expected to recover essentially all its original strength. Indications are that air-dry wood can be exposed to temperatures of about 65.6°C (150°F) for a year or more without significant permanent loss of any of its strength properties. Ice or freezing conditions will cause fiber failure and thus loss of strength through a reduction of section.

7.12.8 Marine organisms

As discussed in Sec. 7.7, the principal marine organisms that cause wood destruction in the coastal zone are Teredo, Limnoria, Poria, and Merulius. Most of these animals attack wood as free-swimming organisms. They bore entrance holes in the wood, attach themselves, and grow in size as they bore tunnels into the wood. Wood structures are protected from them by proper treating with creosote or coal-tar solutions or by protective enclosures.

7.12.9 Periodic wetting and drying

Wood in a marine environment should always be protected from excessive moisture or water, and therefore it has little change in its structural or mechanical properties. If the preservatives eventually leach out of the wood cells, alternate expansion and contraction of the cells can result in gradual and slow deterioration. Wood structures have a history of long service life even when they are subjected to alternate wetting and drying.

7.12.10 Wind erosion

Because wood is a relatively soft construction material as compared to concrete or metal, it can be eroded by wind action. Wind does not erode wood directly, but strong winds picking up particles of sand or other materials will cause a wood surface to wear. Erosion of this kind will usually take place near the ground line.

7.12.11 Effects of burrowing animals

Marine animals will burrow into wood very rapidly unless the wood is protected by an appropriate preservative. Wood is sufficiently soft to offer little resistance to burrowing attack; in addition, it serves as a food source for the animals. On shore, termites are very destructive to wood. These attacks, if left unchecked, will eventually result in the loss of all the structural properties of wood.

7.12.12 Effects of flora

There are no reported effects of flora growth on wood.

7.12.13 Effects of fire

When exposed to fire, wood forms a self-insulating surface layer of char and thus provides a degree of its own fire protection. The undamaged wood below the char retains its strength. Heavy timber members will retain their structural integrity throughout long periods of fire exposure because of their size and the slow rate at which charring penetrates inward from the wood surface.

7.12.14 Abrasion

In the coastal environment, abrasion of wood is caused by wind- and water-driven sand as well as by the working or rubbing at joints in the wood structure. Abrasion can be from beneficial uses as from vehicular traffic on a pier or from rubbing of floats on anchor piles. The wearing away of wood in this manner will eventually reduce the integrity of the structure.

7.12.15 Seismic effects

Seismic activity can have a significant effect and in some locations a devastating effect. Ground shaking can stress structures to overload and cause destruction. Natural alluvial terraces or manmade landfills in the coastal zone are subject to liquefaction during severe earthquakes which cause the ground to slump and flow horizontally. Structures founded on such terraces in a severe seismic area are subject to destruction if liquefaction occurs. Under less severe conditions, however, wood performs very well in seismic events because of its resiliency. That characteristic of wood allows it to flex during ground shaking and reduce the stresses that might destroy structures made of more rigid materials.

7.12.16 Human activity

Human use of wood structures can eventually cause the wood to wear out; the worn wood parts can be replaced or the whole structure can be abandoned. On wharves and piers, ship moorings wear or break fender piles and vehicular traffic on the deck wears out the surface timbers. Human use engenders risks in the form of explosions, fires, and accidental impact loads, all of which can destroy wood members of coastal structures. Vandalism can cause serious damage to wood: Some wood may be sacrificed for firewood; wood may be destroyed by the target practice of shooters; and amateur wood carvers may cause deterioration.

7.13 Uses of Wood in Coastal Construction

Untreated dimension lumber can be used in temporary situations during the construction phase of a project or where the life of the wood is to be only a few months. It can also be used in any situations in which the wood can be protected by a covering, e.g., the interior framing of a building, or is to be painted and maintained in a painted state for the projected life of the installation. Untreated dimension lumber should not be used in direct contact with the ground or seawater. It is used in form work for concrete and also used in a variety of ways such as for dunnage of machinery supports.

Foundations and sill plates for frame buildings are usually pressure-treated with chromated copper arsenate. Whenever contact with the earth exposes the lumber to rot, fungus, or insect attack, treatment is necessary to obtain a satisfactory useful life. Specific treatment is determined by the conditions of service. Wood in exposed uses that subject it to severe weathering or prolonged (or periodic) immersion in seawater should be pressure-treated with coal-tar creosote; otherwise, one of the other treatments could be satisfactory.

7.13.1 Piles and poles

Nearly all wood piles and poles used in the coastal environment are pressure-treated with coal-tar creosote to resist attack by insects or, in water, marine borers and limnoria. Properly treated piles and poles will also withstand rotting and attack by fungus. Piles are used for building foundations and support for piers, wharves, trestles, jetties, groins, and bulkheads. Also, they are used in fender systems along the wharves and to anchor floating moorings for small boats.

It is unlikely that untreated piles or poles would be used in the coastal environment except for temporary uses during construction for falsework or to carry electrical power and telephone lines to the construction site.

7.13.2 Beams and stringers

Lumber classified as beams and stringers [having 125 mm (5 in) as their least dimensions] are seldom used untreated in the coastal environment. The principal uses are in protected spaces such as framing for buildings and covered structures such that covering or painting provides sufficient protection. Coal-tar creosote–treated beams and stringers are used extensively in the coastal environment.

7.13.3 Glued and laminated wood

Dry use plywood is seldom used in the coastal environment because of its extreme susceptibility to damage by the generally high humidity. Any use would have to be very temporary, or the plywood would have to be very well protected from the prevailing moisture.

Wet use plywood has many applications in the coastal environment. Diaphragms in buildings, roofs, walls, and floors are regularly sheathed with wet use plywood. Plywood is sometimes used for gussets in wood frames to join the members, and it is used extensively in making forms for concrete work. Signboards are frequently made of plywood. Covers, such as those for pits and valve boxes, can be made of plywood in light traffic or nontraffic areas. Wet use plywood can be further treated with preservatives to extend its useful life in extreme environments, such as immersion in seawater.

Wood in the glued and laminated category is generally referred to as glue-lam or, more properly, glued laminated wood. Because of better quality control, strength, and capability of being sized to suit the need, it can be the preferred material for many applications in which columns, posts, beams, and girders are used. In the coastal environment, wet use glue is absolutely essential, and glue-lam members must have preservative treatment in any use in which other wood forms would require it.

7.13.4   Miscellaneous wood forms

Although most people, when they think of wood, think of finished lumber and timbers cut to rectangular sizes from large trees or of piles peeled and trimmed, there are other useful forms of wood that can be used in the coastal zone. Small branches, saplings, brush, cane, bamboo, and reeds have all been used to make devices to control water currents, stabilize bottom sediments, or control dry sand buildup. When those wood forms are indigenous to the area or are readily available, they can be valuable materials.

7.13.5   Breakwaters and caissons

Wood is seldom used in offshore structures of the breakwater and caisson type, but dimension lumber and wet use plywood can be used for navigation aids or other incidental small structures that can be mounted on offshore breakwaters and caissons.

7.13.6   Pile dolphins

Wood is frequently used offshore for pile dolphins and other mooring or anchorage devices such as guide piles for floats and piles for channel markers. Pile dolphins are clusters of wood piles tied together as shown in Fig. 8.19.

7.13.7   Floats

Wood is used extensively in the construction of floating structures. Although the tendency is toward using synthetic materials for small flotation devices, wood remains the material most used for framing flotation units and providing platforms for access and mooring fastenings. Wood flotation units such as logs could be used to form floats or booms for the containment of surface debris.

7.13.8 Breakwaters and jetties

Wood uses in shore-connected breakwaters and jetties are the same as described for offshore structures.

7.13.9 Groins

Wood is frequently used in the construction of groins; Wakefield sheet piles are commonly used as shown in Fig. 7.9. The sheet piles are secured with timber wales at the top. Wood planks spanning between wood piles create another type of groin structure as shown in Fig. 7.22.

7.13.10 Bulkheads

Wood bulkheads are usually one of two kinds. Wakefield sheet piles are driven along the bulkhead line and tied back to the embankment by timber wales and tie-rods to imbedded anchors (deadmen). Otherwise, vertical piles acting as soldier beams are driven at regular in-

Figure 7.22 Beach protection with "billboard" groins, Ninilchik Harbor, Alaska.

tervals along the bulkhead line and wood planks are placed to span horizontally between them. Piles can be tied back to deadmen.

7.13.11 Revetments

Pile revetment. Slopes can be stabilized by using parallel piles laid along the slope as shown in Fig. 7.23. Piles used this way must be securely tied to headers or staked down.

Fascine mattresses. The word "fascine" comes from the Latin *fascina,* meaning a bundle of sticks. Fascine mattresses are used as submerged scour aprons and as filter blankets along revetments. There are many ways to construct them, but the blankets basically consist of sticks tied together in bundles and arranged in mattresses about 20 m (62 ft) wide up to 200 m (620 ft) long. The mattresses are made in a place that is normally dry but can be flooded (either at high tide or by removing a gate) so the mattress can be towed to its final location. The mattress is then loaded with stones, sunk into place, and covered with stones as needed to resist the expected currents. Examples of fascine mattresses are shown in Fig. 7.24. Finer material is placed on the bottom of the mattress, where the mattress contacts the sand, and coarser material is placed on top to support the stones. Fascine mattresses would deteriorate rapidly if they received only periodic wetting; they will work only if they are completely submerged all the time. When damage by marine borers is expected, preservative treatment could be

Figure 7.23 Piles laid along a slope to prevent beach erosion, Ninilchik Harbor, Alaska.

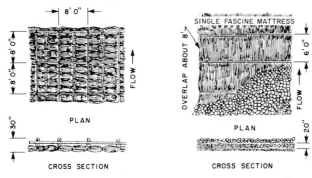

Figure 7.24   Two methods of making fascine mattresses. *(Courtesy of Fan Bendegon and Zanen)*

applied; but the cost of treatment may indicate that some other material be used (Van Bendegon and Zanen, 1960).

7.13.12 Submerged screens

Submerged screens for current control can be made of "wattles" (flexible branches woven around posts) as shown in Fig. 7.25, or they can be made by combining stones and small poles (or bamboo) into cribs. The cribs are formed by building a lattice work of poles to form a cage (crib) and filling it with stones for stability. Figure 7.26 shows some crib types.

7.13.13 Piers and wharves

Piers and wharves can be entirely of wood construction with only incidental use of metal fastenings and rock for slope protection. Piles,

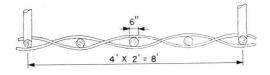

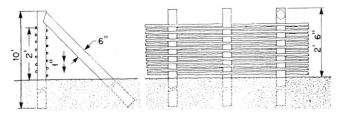

Figure 7.25 Submerged current control screen made of wattles. *(Courtesy of NEDECO)*

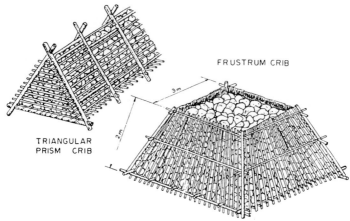

Figure 7.26   Examples of wood-formed cribs. *(Courtesy of Van Bendegon and Zanen)*

pile caps, stringers, and decking are treated and placed as discussed in earlier paragraphs of this section. These wood members can also be used in conjunction with other materials, such as concrete piles.

Mooring dolphins and fender piles for piers and wharves are frequently made of wood to take advantage of the energy-absorbing property of wood even when the remainder of the structure is of some other material such as concrete.

Figure 7.27   Sand fence using brush for filter material. *(Courtesy of Savage)*

7.13.14 Sand fences

Sand fences made of brush have proved more effective than fences made of boards in building sand dunes on the Outer Banks of North Carolina (Savage, 1963). In this installation, brush was held upright between pairs of wood plank rails as shown in Fig. 7.27. The filtering action of the brush apparently trapped more sand than the wind deflection action of wood slats.

in

8

Plastics

8.1 General

Chemically, the term "plastics" is applied to a large group of synthetic materials that are processed by molding or forming into final shapes. Plastics are composed of chainlike molecules of high molecular weight, called polymers, that usually have been built up from simpler monomers.

All plastics have many properties in common and, in general, have four things in common. First, at some stage of their production they are soft and pliable and can be formed, by the application of heat, pressure, or both, into definite desired shapes. Second, plastics are organic materials; i.e., they are based on a carbon structure. This distinguishes them from such materials as metals, ceramics, and concrete. Third, plastics are synthetic materials; they are products of chemical processes that alter the characteristics of the raw materials from which they are derived. Fourth, plastics are high polymers; they consist of monomer atoms joined together into molecular aggregations.

Different monomers are used to manufacture each different family of plastics. Each plastic has a particular combination of properties, processing requirements, and economics that make it ideally suited for certain applications yet unsuitable for many others.

8.1.1 Thermoplastics and thermosets

Plastics in general can be classified into distinct groups: the thermoplastics and the thermosetting plastics or thermosets. Thermoplastics soften repeatedly when heated and harden when cooled. At high tem-

---

This chapter was written by Robert J. Barrett and Lawrence L. Whiteneck.

peratures they melt, and at low temperatures they become brittle. The process of heating and softening and cooling and hardening may be repeated indefinitely for plastics such as polyethylene, polyvinyl chloride, acrylics, nylon, and polystyrene. Thermosetting plastics go through the soft plastic stage only once. When they are hardened, irreversible changes take place and they cannot be softened again by reheating. Some thermosetting plastics are polyesters, epoxies, phenol-formaldehydes, melamine-formaldehydes, and silicones.

8.1.2 Copolymers and composites

Like metal alloys, plastics can be combined to have the best qualities for particular end uses by selectively drawing from the best attributes of the blended components of the polymers. The process is referred to as copolymerization, and the products are called copolymers. Plastics used for structures, including those used in coastal zones, are most commonly composites or copolymers rather than pure forms. Reinforced plastics are a category of composites in which the plastic is strengthened and stiffened by combining it with high-strength fibers such as glass. Sandwich-type plastics contain a variety of strong, thin facings and lightweight cores. There are also the polymer concretes, which contain plastic matrix in place of or in addition to inorganic cement.

8.1.3 Structural properties from additives

Because all plastics are synthetic, various things can be done during their production to alter their characteristics. One is the introduction of additives: plasticizers, fillers, colorants, stabilizers, and impact-modifiers. For example, plastics which are hard and rigid or brittle at normal temperatures can be made pliable and flexible by the addition of plasticizers.

Fillers are normally added to both thermoplastics and thermosets to enhance their processing, performance, or economics. For example, phenolics, without the addition of fillers, are hard and brittle, shrink in molds, and may crack. The addition of finely ground wood flour makes them easier to mold and less costly. Powdered mica will enhance electrical resistance, and asbestos will improve heat resistance. Impact resistance can be improved by the addition of chopped natural fibers, tire cord, rayon, or glass. Colorants are easily added to plastics, although they are not usually necessary in coastal structures.

Stabilizers are an important group of additive materials used to increase the resistance of plastics to the deteriorating influences of weather, ultraviolet light, or radiation. Most plastics in their pure form do not have a great deal of resistance to those environments. Stabilizers retard thermodegradation and oxidation. Materials with stabilizer additives, in outdoor exposures, may have a design life of 30 to 40 years.

The impact-modifiers are another important category of additives. The inclusion of various fillers or plasticizers will increase the impact resistance of plastics which are normally very brittle.

8.1.4 Durability properties

In addition to structural qualities, plastics possess other desirable characteristics as a construction material. They are easily formed, corrosion-resistant, lightweight, wear-resistant, energy-absorbent, impact-resistant, flexible, and ductile, and they are used for insulation because of their thermal and electrical resistance. Energy absorption and impact resistance vary with the plastic. Rubber, a synthetic and not usually considered a plastic, can be formulated to have a high degree of impact resistance within a large range of stiffness characteristics.

Fire is a necessary consideration in the selection of all structural materials. Plastics will burn or disintegrate if they are exposed to fire or high temperatures. Some will burn easily, some slowly, others with great difficulty. Some will not support combustion in the absence of flame. Improved fire resistance can be achieved by incorporating flame-retardant chemicals into the molecular structure of the plastic materials. Phosphorus and halogens have been effectively used for that purpose.

8.2 Geotextile Filters

Because the most common use of plastics or geotextiles in coastal construction is as a filter, that use is the predominant topic of this chapter. The filters have been known as filter fabrics, construction fabrics, plastic filter cloth, geotechnical fabrics, and engineering fabrics. ASTM Joint Subcommittee D-18.19/D-13.61 developed test procedures for evaluating these fabrics and has adopted the name of "geotextiles." Geotextiles are used in engineering as filters, materials separators, and/or reinforcement for soils. The fabrics may be used in coastal structures in one or more of those roles, but they are most frequently used as filters which permit the passage of water but not soil or sand particles through the fabric. Geotextiles used as materials separators prevent the mixing of materials that should remain apart, such as poor subgrade soil and good subgrade gravel. Geotextiles have also been successfully used as reinforcement in the paving of roads and to restrain lateral movements of embankments built on soft soils.

The use of geotextiles has expanded rapidly since 1965, and many different kinds are available today. However, there are constraints that must be removed before geotextiles achieve unqualified acceptance. One of them is lack of standardization. Many fabrics are made by suppliers in different ways, out of different materials, and for different uses. In choosing a fabric for a project, it may be necessary to consider tensile, elongation, and puncture properties plus such factors as fabric elasticity, porosity, permeability, and resistance to abrasion, chemicals, light, weather, and temperature as well as resistance to biological attack.

Because geotextiles have many different uses in coastal structures, in drainage ditches, as riverbank protection, and in subgrade construction, no one fabric is right for all applications. It remains to be determined just what properties are important for each end use and what range of values for each property is acceptable. However, based on the successful use of many geotextile filters, the promise of longevity is exceedingly favorable. A prospective geotextile user should obtain advice and information from engineers experienced in geotextile use and also from more than one supplier.

The term "geotextile filter" as used in this book refers to a permeable fabric constructed of synthetic fibers designed to prevent piping (the passage of soil through it) and remain permeable to water without significant head loss and without permitting the development of excessive hydrostatic pressure.

8.2.1   Design properties

A geotextile filter must be sufficiently permeable to relieve the hydrostatic pressure differential between its sides by allowing the passage of groundwater flow without detrimental head loss, and it must prevent the passage, or piping, of adjacent granular or fine soil. A geotextile is used to replace all or part of a conventional filter system consisting of one or more layers of granular material. Figure 8.1 shows a geotextile replacing a layer of gravel beneath a revetment; it illustrates how the filter is designed to prevent protected soil from being washed through the overlying armor. It also demonstrates how a geotextile can be incorporated into a toe protection apron. To be effective, the geotextile must be designed to suit the grain size, groundwater, and wave conditions of each specific site, as well as the type of structure in which it is to be included.

In order to function satisfactorily, the geotextile filter must have

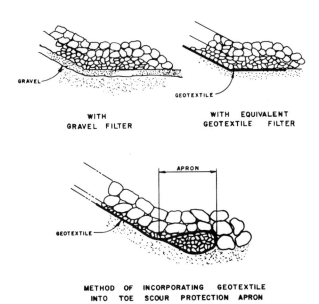

WITH
GRAVEL FILTER

WITH EQUIVALENT
GEOTEXTILE FILTER

METHOD OF INCORPORATING GEOTEXTILE
INTO TOE SCOUR PROTECTION APRON

Figure 8.1 Geotextile filters in revetments. *(Courtesy of Fabriform)*

the physical durability and filtering integrity to perform consistently throughout the design life of the structure. Durability depends on the chemical composition of the fibers, construction of the fabric, and physical properties of the fabric in its completed (finished) form. To ensure durability, specifications for fabric should describe the basic chemical composition, fabric construction, and additives. The 1977 Chief of Engineers Civil Works Construction Guide Specification CW 00215 (U.S. Army Chief of Engineers, 1977) states that "The plastic yarn shall consist of a long-chain synthetic polymer composed of at least 85 percent by weight of propylene, ethylene, ester, amide or vinylidene chloride." All geotextiles for coastal applications must meet that requirement. Filtering integrity depends on the fabric's ability to resist piping and clogging. Filtering integrity also depends on the fabric's resistance to loss of permeability because of distortion of the pores by elongation of the fibers under stress, a physical property of the fabric, or by melting in fire.

8.2.2 Chemical stability and resistance

The first extensive research and testing of geotextiles for filters was by Calhoun during the period 1969 to 1972 (Calhoun, 1972). In the course of the investigations, tests were conducted to determine the chemical stability and resistance of three types of synthetic polymers:

polyvinylidene chloride, polypropylene, and polyethylene. Bell and Hicks (1980) also investigated chemical properties of the three polymers plus polyester and polyamide. The results of the testing programs established that the synthetics tested have high chemical stability and resistance to chemical attack (acids and alkalies), and can be used in conventional soil applications with confidence. If the fabric is to be used in an environment containing petroleum products, it is recommended that the geotextile's resistance to those materials also be determined.

8.2.3 Fabric construction

Selection of a geotextile for a filter should be based on the filtering and physical properties as well as the chemical properties of the fabric consistent with the site-specific requirements. Bell and Hicks (1980) found that most fabric engineering properties are more strongly influenced by fabric construction than by the polymer. One of the most extensive and thorough fabric-strength-testing studies, involving 27 commercially available fabrics, was performed for the Army Engineers District, Mobile, by Haliburton, Anglin, and Lawmaster (1978). It was concluded that fabric construction had more influence than the type of synthetic fiber in the fabric. Because fabric construction is the predominant factor affecting physical properties and filtering performance, three general types of fabric construction are discussed: woven, nonwoven, and combination fabrics.

Woven fabrics. As the term implies, woven fabrics (commonly called cloths) are manufactured by weaving. Normally, the yarns cross at right angles, overlapped one over the other. The longer direction of the cloth, during weaving, is called the warp or machine direction; the shorter direction is called the fill. In geotextiles, the warp direction is normally stronger than the fill direction, although the cloth may be produced with equal strength in both directions or the fill may be stronger than the warp. Filters are woven with a variety of yarns as discussed below.

A monofilament yarn is a single filament of a polymer, which prohibits absorption of water by the yarn. This was the only type of yarn in geotextiles for coastal structure use in the United States from 1958 to the early or mid-1970s. Fabrics woven of monofilament yarns have relatively regular and uniform pore sizes. "Some engineers believe that because of their simple pore structure, the monofilament fabrics are more reliable filter materials and use them in critical installations, where their higher cost can be justified" (Bell and Hicks, 1980). The fabric is thin.

Multifilament cloths are woven of yarns containing many fine filaments; the exception is fibrillated yarns, which are formed of fibers cut from sheet plastic film. All multifilament fabrics can be produced with tensile strengths higher than those of monofilament fabrics. With the exception of fibrillated fabrics, multifilament cloths also have simple, relatively regular and uniform pore sizes, and they generally are thin. Fibrillated fabrics have slightly more irregular pore systems and generally are thicker.

Mono-multifilament combination fabrics contain monofilament yarn in one direction and multifilament yarn in the other. The pore sizes are consistent and are controlled by the weaving process. The openings are oblique to the plane of the fabric. The cloth is slightly thicker than monofilament cloth.

The term "slit film" as used herein refers to a fiber which has a width many times its thickness. Such a fiber is also called ribbon, split-film, slit-tape, and split-tape. Because of the poor distribution and uneven sizes of the pores, there is a great variation in retention and filtration capabilities. Slit-film fabrics are thin; they are not recommended for use as filters.

Nonwoven fabrics.   Nonwoven fabrics include all materials which are not woven or knitted. They consist of discrete fibers which may have a preferred orientation or may be placed in a random manner and do not form regular or simple patterns as wovens do. Nonwoven fabrics are composed of either continuous filament or staple filament fibers. Continuous filaments are extruded, drawn, and laid in the fabric as one continuous fiber. Staple filaments are cut to length before being laid in the fabric. The engineering properties of nonwoven fabrics are controlled by the fiber type, the geometric relations of the fibers, and the methods of bonding. Four methods of bonding are described below.

For the needle-punched type of nonwoven fabric, barbed needles are punched through the fabric web perpendicular to the plane of the fabric and then withdrawn, drawing filaments with them. This causes the fabric to become mechanically entangled. The needle-punched fabrics have a very complex pore structure and are compressible, so the nature of the pore structure changes. This results in a different in situ filtration performance than might be indicated by isolated permeability and particle retention tests. The fabrics are relatively thick and have the appearance of felt.

Heat-bonded fabric is subjected to a high temperature, which results in the filaments welding themselves together at the contact points. These fabrics have a relatively discrete and simple, though irregular, pore structure and are thin.

Resin bonding means that the fabric web is impregnated with a resin which coats and cements the fibers together. Pore structure and fabric thickness are intermediate between the two fabrics described above. Normally, they have less permeability and fewer voids.

A number of nonwoven fabrics are produced by a combination of two or more of the above methods to construct a finished product. Because of the variety and numerous combinations available, it is impossible to make statements regarding pore properties and thickness applicable to the entire nonwoven fabric classification.

Combination fabrics.   Fabrics have been produced by combining woven and nonwoven fabrics and using one of more of the bonding methods described above. Usually, the combination fabrics are produced to enhance particular property or performance requirements not found in either woven or nonwoven fabrics. Hundreds of such combinations are possible with an equal, or greater, number of finished fabric forms and properties. Each must be evaluated in view of the application being considered.

The 1977 CE Guide Specification CW 02215 (U.S. Army, Chief of Engineers, 1977) states that "The fabric should be fixed so that yarns will retain their relative positions with respect to each other. The edges of the fabric should be finished to prevent the outer yarn from pulling away from the fabric." Regardless of the fabric construction, this requirement is necessary to ensure continuous acceptable performance.

8.2.4   Physical property requirements

In few other applications is a filter exposed to so many damaging forces as in most types of coastal structures. Consequently, physical property requirements for geotextile filters are more stringent for coastal than for other applications. Suitable physical (mechanical) properties are necessary not only during the construction process but in the permanent structure as well. Because of a coastal structure's constant exposure to dynamic loading from waves and currents, armor and underlayer movement, earth and hydrostatic pressure, and rapid fluctuations, any geotextile used must have sufficient tensile and abrasive strength to retain its integrity throughout the life of the project.

The test methods used to determine a geotextile's physical property requirements are primarily textile tests. However, the test methods and requirements referred to herein can be related to the field performance of woven geotextile filters. Evaluation has confirmed successful performance of fabric(s) having a particular character. Many of the required test results have been verified by more than 20 years of field performance.

If test methods and/or results and specifications other than those required in this section are employed, they should be thoroughly evaluated to determine if the methods and results are applicable to the intended function or performance in the application of interest. Test methods for required physical property determination are described in Appendix B.

Tensile strength. Adequate fabric strength is necessary to (1) withstand dynamic forces, (2) prevent the movement of the geotextile filter through voids in the stone layer above the fabric, as often occurs with aggregate filters (Dunham and Barrett, 1976), and (3) permit the use of larger stones adjacent to the filter, thereby possibly reducing the overall thickness of the structure (Barrett, 1966; Dunham and Barrett, 1976). When armor is removed or rearranged, the fabric's independent strength should also retain the soil (Barrett, 1966; Fairley et al., 1970) and prevent cavity formation.

Elongation at failure. Elongation at failure is part of the tensile test described in Appendix B. Percent of elongation must be known because excessive elongation will distort and enlarge the pores and change the soil retention capabilities (piping resistance) (Steward, Williamson, and Mahoney, 1977). If excessive elongation is necessary to develop the fabric's ultimate strength, the fabric will probably never develop its required strength in situ.

Seam strength. It is advantageous to use geotextile filter sheets or panels in large lengths and widths in most applications. The larger panels reduce the number of overlaps required, which is the most probable cause of error during construction. Fabrics are manufactured in various widths, 1.8 to 5.2 m (6 to 17 ft), and then sewn together or bonded by cementing or by heat to form large panels as much as 25.6 m (84 ft) wide. When sections are sewn together, the yarn used must conform to the physical and chemical requirements of the fabric being joined. If the seam strengths are too low, the sheets may separate and permit piping to develop.

Puncture resistance. Puncture resistance is required to enable the geotextile filter to survive placement of other materials on it during the construction process and to prevent rupture or penetration by the overlying material when the structure is exposed to wave action.

Burst strength. Burst strength must be considered to assure the engineer that the fabric will retain its integrity when subjected to earth forces, especially when the material above it contains relatively large voids.

Abrasion resistance.   In all types of use, abrasion resistance is important during construction. In one case, fabric was damaged during construction of a French drain merely by placing small filter aggregate into the trench it lined. In coastal structures such as revetments, abrasion resistance is required not only during construction but also throughout the life of the structure; for these structures are subjected to continuous or intermittent wave attacks which result in movement of the overlying material adjacent to the fabric.

Requirements for special conditions.   In certain climatic conditions and geographic locations, it may be desirable to test the geotextile for freeze-thaw resistance, high-temperature survivability, and low-temperature survivability. When soils in the project site are contaminated or are subject to infiltration by high quantities of acids, alkalies, or JP-4 fuel, it may be advisable to test the geotextile's resistance to the specific contaminant. Test methods are described in Appendix B.

Properties for construction conditions.   The physical properties of geotextiles required for specific sites and structures vary with loadings as well as with functions. Loadings can be classified as follows:

- Severe dynamic loadings
- Dynamic and static loadings
- Most stringent placement and drainage

Severe dynamic loading is characterized by continued abrasive movement under wave action of materials adjacent to the fabric. Dynamic and static loadings are characterized by more restrictive placement procedures to limit abrasive movement and include gabion applications. The most stringent placement controls and drainage applications nearly eliminate abrasive movement of materials adjacent to the fabric. This category includes weep holes, linings of vertical walls, relief walls, linings for French and trench drains, and wrap collector pipe.

For each loading category, certain construction parameters and limitations must be met. In Tables 8.1 and 8.2 construction limitations for each category are listed with reference to three specific applications: quarrystone revetment, block revetment, and subaqueous applications. Block revetment includes precast cellular block (a cast or machine-produced concrete block having continuous voids through the vertical plane and, normally, smooth or near vertical sides) and interlocking concrete block (a cast or machine-produced concrete block having interengaging or overlapping edges). The subaqueous applications

TABLE 8.1   Construction Limitations: Quarrystone Revetment[a]

Parameter	Category		
	A	B	C
	Severe dynamic loading	Dynamic and static loading	Stringent placement and drainage
Steepest slope	1V on 2H	1V on 2.5H	1V on 3H
Min. gravel thickness above filter	None	None	20 cm
Stone adjacent to geotextile:			
Max. stone weight[c]	100 kg	71 kg	Gravel
Max. drop height	1 m	1 m	1.5 m
Max. stone weight	164 kg	118 kg	
Riprap weight range[d]	105 kg	20 kg	NA[b]
Max. drop height	0.61 m	0.61 m	
Max. stone weight	164 kg	118 kg	
Max. drop height	Placed	Placed	NA
Subsequent stone layer:			
Max. stone weight	NA	NA	0.67 kN
Max. drop height	1.2 m		
Max. stone weight	NA	NA	1.3 kN
Max. drop height	NA	NA	1 m
Max. stone weight	4000 kg	4000 kg	400 kg
Max. drop height	3 m	2.5	Placed
Max. stone weight	4000 kg	4000 kg	400 kg
Max. drop height	Placed	Placed	Placed

[a]This table can also be used for sand core breakwaters (a jetty, groin, or breakwater in which the core material consists of sand rather than stone).
[b]Not applicable.
[c]Weight of quarrystone armor units of nearly uniform size.
[d]Weight limits of riprap, quarrystone well graded within wide size limits.
NOTE 1: Stronger principal direction (SPD) and seams of the geotextile should be perpendicular to the shoreline.
NOTE 2: There is no limit to the number of underlayers between the armor and the geotextile.

include groins, jetties, and breakwaters, scour protection for piers, piles, and caissons, and toe aprons for bulkheads.

Minimum geotextile filter physical property requirements are shown in Table 8.3 based on the construction limitations in Tables 8.1 and 8.2. Test methods are described in Appendix B. The physical property requirements in Table 8.3 are not as stated in the current Civil Works Construction, Guide Specification, Plastic Filter Fabric, No. CW-02215 (U.S. Army, Chief of Engineers, 1977). Because this book is concerned only with coastal structures, the test methods and requirements set forth herein are based on field performance and verification in coastal structures and relevant laboratory research.

TABLE 8.2   Construction Limitations: Block Revetments and Subaqueous Applications

	Category		
	A	B	C
Block Revetment[a]			
Precast cellular block[b]			
Steepest slope, individual blocks	1V on 2H	1V on 3H	NA[c]
Steepest slope, cabled blocks[d]	1V on 1.5H	1V on 2H	NA
Max. block weight	> 210 kg	210 kg	NA
Interlocking concrete block[b]			
Steepest slope	NA	1V on 2H	1V on 2.5H
Min. gravel thickness above filter	NA	15.2 cm	15.2 cm
Max. block weight	NA	> 280 kg	280 kg
Subaqueous Applications[e]			
Steepest slope	1V on 15H	1V on 15H	1V on 15H
Stone adjacent to geotextile:			
Max. stone weight	610 kg	610 kg	226 kg
Min. drop through water	1.5 m	1.5 m	1.5 m
Max. stone weight	911 kg	911 kg	226 kg
Max. drop height	Placed	Placed	Placed
Subsequent stone layer(s)			
Max. stone weight	No limit	No limit	No limit
Max. drop height	NCP[f]	NCP	NCP

[a]Stronger principal direction (SPD) and seams of the geotextile should be perpendicular to the shoreline.
[b]With flat base.
[c]Not applicable.
[d]Precast cellular blocks are cabled together in a horizontal plane.
[e]No limit to the number of underlayers between the armor and the geotextile.
[f]As in normal construction practice, the geotextile does not require special limitations in these layers.

8.2.5   Filtering Integrity

To prevent piping, it is necessary to know the soil retention capability of the geotextile filter or granular filter when a filter system is considered. In his extensive and thorough research and development project with geotextiles for filters, Calhoun (1972) developed a special procedure for determining the piping resistance (soil retention capacity) of fabrics. The purpose of his procedure, described in Appendix B, was to determine the equivalent opening size (EOS) of the fabric.

Knowing that the soil retention ability of a fabric is directly related to the hydraulic pressures, flows, and forces it encounters, Calhoun (1972) conducted hydraulic filtration and clogging studies to develop a formula that related the EOS to required performance criteria. For geotextile filters adjacent to coarse-grained soils containing 50 percent or less particles by weight passing U.S. No. 200 sieve, the piping resistance is calculated by using this equation:

TABLE 8.3 Minimum Geotextile Filter Physical Property Requirements

| | | Category | | |
| | | A | B | C |
Property	Test method	Severe dynamic loading	Dynamic and static loadings	Stringent placement and drainage
Tensile strength[a]				
SPD[b]	App. B-1	1.56 kN		0.89 kN
BPD[c]			0.89 kN	
WPD[d]		0.98 kN		0.44 kN
Elongation at failure	App. B-1	< 36%	36%	36%
Seam strength[a]	App. B-2	0.87 kN	0.80 kN	0.36 kN
Puncture resistance	App. B-3	0.53 kN	0.53 kN	0.29 kN
Burst strength	App. B-4	3450 kPa (517 lb/in^2)	3790 kPa (569 lb/in^2)	1650 kPa (247 lb/in^2)
Abrasion resistance[a]				
SPD	App. B-5	0.44 kN		0.27 kN
BPD			0.29 kN	
WPD		0.29 kN		0.15 kN

| Optional Requirements | | |
Property	Test method	Percent of required strength
Freeze-thaw resistance	App. B-6	90
High-temperature survivability	App. B-7	80
Low-temperature survivability	App. B-8	85
Effects of acids	App. B-9	90
Effects of alkalies	App. B-10	90
Effects of JP-4 fuel	App. B-11	85

[a]In accordance with the specifications for the tests of these properties, the forces are applied over a width of 25.4 mm (1 in).
[b]SPD = stronger principal direction
[c]BPD = both principal directions
[d]WPD = weaker principal direction

$$\frac{D_{85} \text{ of protected soil}}{\text{EOS}} = 1$$

where D_{85} is the effective grain size, in millimeters, larger than the grain size of 85 percent of the sample by weight. For geotextile filters adjacent to fine-grained soils containing more than 50 percent particles by weight passing U.S. No. 200 sieve, the EOS should be no larger than a U.S. No. 70 sieve. Fabric with the largest possible EOS should be specified to promote drainage and reduce the likelihood of clogging. Geotextiles with EOSs smaller than the U.S. No. 100 sieve should not be used as filters.

When the protected soil contains particles ranging from 2.54 cm (1 in) in size to those passing the U.S. No. 200 sieve, only the gradation of soil passing a U.S. No. 4 sieve should be used in selecting the fabric. Whenever the protected soil is so sized or graded that a fabric cannot satisfy the above requirements and is to be protected by a multilayered granular filter, a geotextile filter will often satisfy the requirements of all but the filter layer immediately adjacent to the protected soil (primary filter layer).

There are additional restrictions on the percent of open area (POA) of the geotextile which must be considered in applying the piping resistance formula. The POA determination method is described in Appendix B. The criteria for determining piping resistance have been widely and successfully used by the engineering profession.

The original EOS determination method developed by Calhoun (1972) was based on the sieving of rounded to subrounded sands. Geotextiles rated by that method were used in the filtration and clogging tests which resulted in the piping resistance equation given above. The 1977 CE Civil Works Construction Guide Specification CW-02215 (U.S. Army, Chief of Engineers, 1977) modified the original EOS determination method by substituting glass beads for sand. The 1977 Guide Specification EOS determination method is described in Appendix B. Many geotechnical engineers and soil-testing laboratories who had experience with the sand determination method prior to 1977 indicate the sand behavior is more typical of the material to be protected. The objections to beads most often stated are:

- They develop static electricity.

- There is a size problem: many are not "true" when received from the manufacturer.

- A continuous breakdown of the beads occurs during sieving.

- Results obtained for the same geotextile when beads are used differ from the results obtained when sand is used. Generally, beads yield a larger apparent opening value (smaller sieve number) than sand (i.e., beads EOS equals No. 50 sieve vs. sand EOS equals No. 70 sieve), and on some occasions the discrepancy is reversed. [B.R. Christopher, P.E., Corporate Laboratory Director, STS Consultants Ltd. (formerly Soil Testing Services, Inc.), Northbrook, Illinois, personal communications, 1979 to 1982.]

Other workers have reported similar experiences and indicate that inconsistent results are obtained when sand is replaced by beads.

Geotextile filter selection criteria for piping resistance are the same for all applications in coastal structures. The criteria are based on the work of Calhoun (1972), the U.S. Forest Service criteria and experience with laboratory testing, field experiments, installations, and monitoring as presented by Steward, Williamson, and Mahoney (1977), and the authors' (of this chapter) personal experiences, performance records, and communications with users and researchers. Geotextile filters meeting the Calhoun EOS determination criteria have had field verification (service records) for more than 20 years.

Clogging criteria. In shore protection structures such as revetments, geotextile filters may be exposed to severe static and dynamic loading, turbulent flows, rapid fluctuations, high-pressure differentials, and sudden or regular drawdowns. Designers primarily concerned with subsurface drainage must recognize the necessity for more stringent property and performance criteria for fabrics being considered for this environment. It is especially true with respect to the filtration and clogging performance because, if the filter clogs, it could cause a more severe problem than if it had been omitted. Usually, underdrains have low rates of flow and relatively low hydraulic gradients. Because of the large-grained sand present in many coastal areas, filters adjacent to French and trench drains and surrounding collector pipes are often exposed to higher flow rates than would normally be expected in those applications. It is the responsibility of the designer to specify a geotextile filter that retains the soil being protected but has openings large enough to permit drainage and prevent clogging.

Many fabric suppliers provide fabric permeability and water flow rate at a specified head as fabric performance criteria, but the data are of little use in establishing filter-clogging criteria. Calhoun (1972) developed the most widely used filtration and clogging geotextile filter criteria in 1972 after an extensive 3-year research effort. Although the criteria were rather simple, they were based on numerous hydraulic-soil-fabric filtration and clogging tests. By using his method as described in Appendix B, Calhoun determined the EOS of the fabric and in addition the POA.

With certain restrictions for fine-grained soils, Calhoun's criteria allowed the ratio of the soil's D_{85} to the fabric's EOS to be equal to or greater than 1. The criteria had the added limitation that no woven fabric should have a POA less than 4 percent nor an EOS with openings smaller than U.S. No. 100 Standard Sieve. Calhoun established that the larger the POA, the less the fabric was susceptible to clogging.

The 1977 CE Guide Specification (U.S. Army, Chief of Engineers,

1977) introduced the gradient ratio (GR) based on Calhoun's original work. Determination of gradient ratio is set forth in Appendix B.

The U.S. Department of Agriculture Forest Service (USDAFS) has conducted numerous laboratory and field tests including an evaluation of geotextile filter performance in various types of structures. The Service's criteria for piping resistance (soil retention) and clogging are similar to Calhoun's, as discussed above and stated in Appendix B: the EOS-POA combined criteria. The service does have some disagreement with the current 1977 CE Guide Specification (U.S. Army, Chief of Engineers, 1977) as discussed by Steward, Williamson, and Mahoney (1977). The latter feel that the currently recommended GR test should be modified to represent the range of varying seepage rates and fabric strains accompanying the enlargement of openings in nonwoven fabrics due to stretching anticipated in the field and that intermittent flow should be added. They suggest that the GR test has not been confirmed by monitoring field performance. Concern is also expressed that, because of higher elongation, the EOS of nonwoven fabrics will be more variable and more subject to change under load than that of woven fabrics.

For all *critical* and *severe* filter applications the U.S. Forest Service indicates that only woven geotextiles should be used (Steward, Williamson, and Mahoney, 1977). The USDAFS definitions of these terms are quoted below:

Critical: Projects where failure of the filter could result in failure of an expensive or environmentally sensitive part of a project, such as:

- Rock blankets greater than or equal to a 3-ft horizontal thickness
- Retaining structures
- Road fills greater than 10 ft in height
- Underdrain trenches greater than 5 ft in depth
- Bridge repair

Severe: Conditions of moderate to high seepage out of erodible soils with a hydraulic gradient evident moving from soil toward the filter, such as:

- Spring areas
- Soils with flowing ground water
- Soils with high internal hydrostatic pressure

Both the above definitions seem applicable to most coastal applications. Calhoun (1972) also concluded that only woven fabrics should be used in coastal projects. The Forest Service indicates a preference for the woven filters and says that the sometimes "lower material cost of the lightweight non-woven fabrics for critical or severe seepage conditions appears to be outweighed by the risk and consequence of possible failure at this time." The Service also states that, in similar installations, graded aggregate filters have 50 percent chance of functioning properly, whereas woven geotextile filters have a nearly 100 percent chance.

In order to develop geotextile selection criteria for filtration-clogging properties verified by laboratory and field performance, the authors of this chapter relied on the experience of knowledgeable users, their own personal experiences, and a combination of parts of reports citing criteria relevant to coastal applications by Calhoun (1972), Steward, Williamson, and Mahoney (1977), CE 1977 Guide Specification (U.S. Army, Chief of Engineers, 1977), and Haliburton, Lawmaster, and McGuffey.

To achieve desired clogging resistance, woven geotextile filters adjacent to soils containing 50 percent or less particles by weight passing through a U.S. No. 200 sieve should have an effective POA equal to or greater than 4 percent. (When overlaid with stone, the POA of the geotextile is the effective percent open area. If half of the geotextile is covered by flat-based concrete blocks without a gravel layer between the fabric and the blocks, a POA equal to or greater than 8.0 percent is required to yield an effective POA equal to or greater than 4.0 percent.)

Nonwoven geotextiles in the same application should have a gradient ratio equal to or less than 3.0. This same gradient ratio is used as the criterion for selection of all geotextiles adjacent to soils with more than 50 percent particles by weight passing the U.S. No. 2090 sieve, soils with a very slight gradation curve, or soils that are skip-graded (gap-graded). Geotextiles with the largest possible POA available in the required EOS sieve number should be specified.

As stated previously, ASTM Subcommittee D-18/D-13.61 is developing test methods for geotextiles. When these or other test determination methods and/or formulas are submitted to the specifier, they should be evaluated to determine if their results meet the requirements discussed above. It is recommended that both sets of tests be conducted and the results be correlated to the requirements stated in this book.

8.2.6 Placement

The geotextile filter must be laid loosely, not in a stretched condition, but free of wrinkles, creases, and folds for all applications on slopes and beneath jetties. When the slope continues above and beyond the structure, the filter should be keyed in by being placed in a trench at the upper terminus of the structure. When a gravel layer is placed on the geotextile, it must have permeability sufficient that it does not reduce the flow from the filter. The largest-size sheets available should be used to reduce the number of overlaps required. Overlaps of adjoining sheets should be a minimum of 46 cm (18 in) and staggered for installations in the dry. For underwater applications, the overlaps should be 1 m (3 ft). Strict inspection and enforcement in regard to drip height limitations and overlaps are required.

On slopes, construction begins at the toe and then proceeds up the slope. Horizontal underwater placement (such as groins, jetties, and scour protection for vertical walls and piers) starts at the shoreward end and proceeds away from the shore or starts adjacent to the protected structure and proceeds to the outer limits of the scour protection.

When securing pins are required to prevent the geotextile from slipping during construction, they should be 3/16-in in diameter, of steel, pointed at one end, and fabricated with a head to retain a steel washer having an outside diameter of no less than 3.8 m (1.5 in) when used in soils having a medium to high density. In loose soils, longer pins should be used. They should be inserted through both strips of overlapped fabric at the midpoint of the overlap. The maximum pin spacing along overlaps should be 0.6 m (2 ft) for slopes steeper than 1V on 3H, 1 m (3 ft) for slopes of 1V on 3H to 1V on 4H, and 1.5 m (5 ft) for slopes flatter than 1V on 4H. Regardless of location, additional pins should be installed as necessary to prevent any slippage of the geotextile.

8.2.7 Repair method

If the geotextile filter is damaged during the placement of the fabric or of the stone (or blocks) on the fabric, it should be repaired as follows: Cut the damaged part of the fabric out of the sheet and position an undamaged piece of geotextile filter, 1.2 m (4 ft) longer in each direction, where the fabric has been removed. Extend the edges of the new fabric 0.6 m (2 ft) beyond and under the edges of the undamaged original filter.

8.2.8 History of uses in coastal construction

The first use of geotextiles was as a filter beneath an interlocking concrete block revetment of the Atlantic coast in South Palm Beach, Florida, in 1958 (Dallaire, 1977). The fabric used was woven of monofilament yarns of polyvinylidene chloride (Saran) containing stabilizers to make the filaments resistant to ultraviolet and heat deterioration. The equivalent opening size (EOS) was equal to a U.S. Standard Sieve No. 100, and the percent of open area (POA) was 4.6 percent. Physical properties were as follows: tensile strength approximately 350 newtons per centimeter (N/cm) [200 pounds per inch (lb/in)] (warp), 175 N/cm (100 lb/in) (fill), elongation at failure less than 33 percent; burst, 455 N/cm (260 lb/in); puncture, 310 N (70 lb); abraded strength, 250 N (57 lb) (warp), 85 N (19 lb) (fill).

In the 4 years following 1958, geotextile filters were used in a number of coastal structures on the east coast of the United States. In every instance, the fabric was the same as in the first use at South Palm Beach. Although the fabric performed satisfactorily in those installations, field observations during construction led to the conclusion that construction would be simplified and a superior structure would result if a filter could be developed with higher tensile strength and burst, puncture, and abrasion resistance for use in conjunction with quarrystone construction materials. Development of a geotextile woven of polypropylene monofilament yarns consisting of at least 85 percent propylene and containing stabilizers and inhibitors to make the filament resistant to ultraviolet and heat deterioration was completed in 1963. The new fabric had an EOS equal to a No. 70 U.S. Standard Sieve and a POA of 5.2 percent. Physical properties were as follows: tensile strength approximately 665 N/cm (380 lb/in) (warp); 385 N/cm (220 lb/in) (fill); elongation at failure less than 30 percent; burst, 946 N/cm (540 lb/in); puncture, 620 N (140 lb); abraded strength, 440 N (100 lb) (warp); 310 N (70 lb) (fill).

In 1969 the U.S. Army Engineer, District Memphis, inspected three bridge abutments protected by geotextile filters overlaid with 276-kg (125-lb) stone. In one abutment built in 1962, the fabric, similar to the 350-N/cm tensile strength fabric referred to above, had numerous holes attributed to abrasion and could be easily torn by hand. The other two abutments, built in 1964 with the stronger, 6645-N/cm tensile strength fabric, were in excellent condition and no evidence of loss of strength was apparent (Fairley, et al., 1970).

The first uses of geotextile filters in coastal structures by U.S. government departments and agencies were as follows:

1961   U.S. Navy, U.S. Naval Station, Mayport, Florida. Beneath stone revetment.

1962   U.S. Army Corps of Engineers, Coastal Engineering Research Center, Fort Belvoir, Virginia. Beneath interlocking concrete block revetment (Hall and Jachowski, 1964).

1962   U.S. Department of the Interior, National Park Service, Colonial National Historical Park, Yorktown, Virginia. Beneath stone revetment and repair of damaged shoreline riprap.

1963   U.S. Department of Agriculture, Forest Service, Lake Winnibigoshish Chippewa National Forest, Minnesota. Erosion control beneath gabions (first filter application with gabions).

1964   U.S. Air Force, Capehard Marina, Tyndall Air Force Base, Florida. Beneath stone breakwaters.

By 1966, woven geotextile filters had been included in the following types of coastal structures in North America: filters beneath stone and interlocking concrete block revetments; linings for the interiors of vertical seawalls (bulkheads) to permit the relief of water through weep holes and the joints (tongue and groove, king pile and panel, T-pile and panel); wrapping for collector pipes and French drains; beneath stone jetties, groins, and breakwaters; security for the slopes of sand core jetties; linings for the interior of steel cells; and scour protection around steel cells and piers of drilling platforms down to a 46-m (150-ft) depth as in the North Sea (Barrett, 1966).

Lake Texarkana, Texas, was the location of the first installation of a nonwoven textile in a coastal structure in the United States. Construction in 1976 consisted of a precast cellular block revetment lying directly on the fabric. The geotextile filter was composed of 100 percent polyester continuous fiber, and the filaments were mechanically interlocked by needle punching. The EOS equaled a No. 100 U.S. Standard Sieve, and other properties were tensile strength, 525 N/cm (300 lb/in); elongation, 65 percent; burst, 876 N/cm (500 lb/in); abraded strength, 289 N/cm (165 lb/in).

8.3   Other Forms of Plastics

Although geotextiles are the most common plastic materials used in coastal construction, other forms of plastic materials have found special uses.

8.3.1   Flexible forms for concrete

High-strength fabric such as nylon can be used in conjunction with concrete to control erosion. To form slabs, the fabric is put down as a double layer along a bank or shoreline and acts as a mold form for concrete that is injected into it. Figure 8.2 shows two types of double-layer fabric forms. Figure 8.3 shows installation of a concrete-filled form.

Grout-filled fabric tubes can be arranged in various configurations along the shoreline. They are useful as groins, dikes, breakwaters, or weirs. Figure 8.4 shows an arrangement of tubes and filling point; Fig. 8.5 shows the use of fabric pillows as concrete forms for erosion control.

8.3.2   Sheet forms

Synthetic sheet materials made of polyethylene, vinyl, or rubber are used as linings and as covers for controlling water seepage and pre-

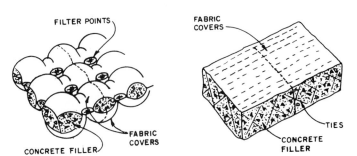

Figure 8.2 Two types of double-layer fabric forms. *(Courtesy of Fabriform)*

venting pollution. These liner-type materials can be bonded together to form a large continuous sheet useful in lining storage ponds qr pits where coastal pollution is a problem. Synthetic sheet materials are also utilized in harbors to control and to clean up oil spills. Flexible plastic sheeting is held between floats so that it passes through the surface layer as a pollution dike (Fig. 8.6). The polluting material is retained within the flexible floating dike. Such diked areas can be set up in advance of operations that might result in a spill. For example, the regular off-loading of oil at a cargo terminal would require that a containment boom be on standby or be deployed in a particularly en-

Figure 8.3 Double-layer fabric forms being installed.

Figure 8.4 Longard tubes being filled with sand for beach nourishment project, North Sea coast, Germany. *(Courtesy of Langeoog)*

Figure 8.5 Concrete-filled synthetic fiber bags used for shore protection. *(Courtesy of Fabricast)*

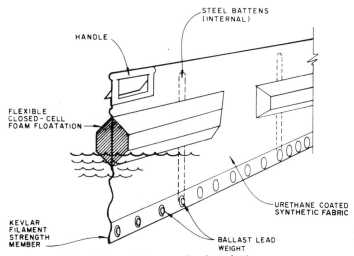

STEEL BATTENS
(INTERNAL)

HANDLE

FLEXIBLE
CLOSED-CELL
FOAM FLOATATION

URETHANE COATED
SYNTHETIC FABRIC

KEVLAR
FILAMENT
STRENGTH
MEMBER

BALLAST LEAD
WEIGHT

Figure 8.6   Oil containment boom made of synthetics.

vironmentally sensitive location. The confined retained pollutant can then be cleaned up by a simple surface-skimming operation.

8.3.3 Molded forms

Fenders or guards are frequently molded or fabricated from rubber and high-strength synthetic plastics or plastic-fabric combinations. Examples of these structures are shown in Fig. 8.7.

Fenders and bumpers.   Rubber and high-density polyethylene (HDPE) are excellent and widely used materials for fenders and bumpers. High-density polyethylene has excellent properties for such marine applications as sliding fenders. It has excellent low-friction properties,

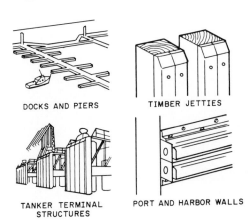

DOCKS AND PIERS

TIMBER JETTIES

TANKER TERMINAL
STRUCTURES

PORT AND HARBOR WALLS

Figure 8.7 Typical uses of molded high-density polyethylene. *(Courtesy of Schlegel Corporation)*

good toughness, and resistance to abrasion and impact damage. High-density polyethylene can be cross-linked to form a three-dimensional structure to make it even higher in strength. However, its elongation and flexibility properties are reduced as its strength is increased.

Polyethylene outlasts wood rubbing strips on fenders 4 to 5 times, is easily machinable or extrudable, and requires little maintenance. It also has greater fire resistance than wood, and the resistance can be enhanced by certain formulation modifications of the polyethylene.

Rubber, in the form of tires and molded shapes, is utilized with excellent success as rubbing bumpers. Old tires are frequently found in harbors as bumpers for small craft, but their energy-absorption capacity is unpredictable and is not relied upon for larger vessels. For larger vessels, a chain net of tires over a rubber fender or HDPE cushion block may be used to provide energy absorption. Figure 8.8 shows such an application.

8.3.4 Pipe forms

Fiber-reinforced plastic (FRP) is a special pipe in wide usage. It is often referred to as RTRP (reinforced thermosetting resin pipe). As mentioned above, thermosetting resins such as epoxy or polyester can be used in combination with glass fiber to manufacture a tough corrosion-resistant pipe. This pipe is utilized in and around the waterfront to avoid the necessity of other corrosion prevention procedures.

The material costs of FRP pipe are higher than those of steel pipe, but installation costs for FRP pipe can be significantly lower than those of steel pipe because of FRP's light weight, ease of handling, and ease of making field joints. In fact, the total installed cost of FRP piping is usually lower than that for steel pipe in the same size range.

Figure 8.8 Sea cushion with tire chain net. *(Courtesy of Seaward International)*

Nonreinforced plastic pipe in the smaller size range, to about 25-cm (10-in) diameter, is used in many construction projects. It does not have the structural strength of FRP in large diameters, but it is sufficiently strong in most small-diameter applications. Nonreinforced plastic pipe is produced by the extrusion process. The present technology for pipe extrusion is to use PVC powder compounds produced by cold-blending techniques that do not require high-cost intensive mixing equipment. Nonreinforced plastic pipe has a large use in the electrical industry as electrical conduit, and because of its lower cost and excellent corrosion resistance, it is finding expanded use as water and drainage pipe.

8.3.5 Epoxy grouts

Epoxy resins, when mixed with sand, form a chemical grout which has excellent chemical and physical properties. Epoxy grouts can be used to patch cement construction such as roadways or to patch certain worn or corroded metal parts. They have superior adhesion properties, high strength, and high corrosion resistance. Many other grout types, e.g., silicates, acrylics, and lignin, are available.

8.4 Environmental Considerations

The environment interacts with sustained stress and strain to further alter material response and strength of plastics. Chemical environments [e.g., ultraviolet (UV) exposure, contact with petroleum products, and sustained elevated temperature] can have profound influences on performance and hence must be considered in the design and use of plastics. However, many synthetics have high chemical stability and resistance to chemical attack by acids and alkalies. The failure to design for environmental effects as they interact with sustained stress or strain has been the chief cause of failure of plastic products.

Fillers and plasticizers alter the basic response and strength of plastic materials. Fillers (e.g., clay, limestone, carbon black, and other inert materials) introduced to increase stiffness, improve processing characteristics, or lower costs may also be used to improve UV resistance or heat resistance. Plasticizers change the physical properties of plastics such as impact resistance, flexibility, and toughness and abrasion resistance. The introduction of strong fibers, such as glass, will improve strength, stiffness, and dimensional stability.

Stabilizers are an important group of additive materials used to increase the resistance of plastics to the deteriorating influences of weather, UV light, or radiation. Stabilizers are also used to retard degradation by heat. Flame-retardant elements, such as phosphorus and halogens, can be incorporated in the molecular structures to improve fire resistance.

Plastics in common use in the coastal environment, such as the epoxies, polyester, polyethylene, polypropylene, and polyvinyl chloride, are generally not considered to be biodegradable. However, these plastics in virtually all structural forms (such as tanks, pipes, buoys, or geotextile fabrics) have high corrosion deterioration in the coastal environment and when exposed to chemicals, except for some of the aliphatic solvents such as ketones.

8.4.1 Geotechnical fabrics

Atmospheric temperature, chemicals (in concentrations normally found in soils), and wetting and drying are factors having little or no effect on geotextiles conforming to the chemical and physical requirements stated in the test and Table 8.3. Trees may grow through the fabric. In the few cases where this has been observed, there was no detrimental effect on the functioning of the filter because the geotextile was sealed tightly against the tree trunk at the point of penetration.

No standard test method has been developed to determine the biological resistance of geotextiles. However, all investigators have concluded that fabrics composed of the synthetic polymers described earlier are inert to biological attack, with the possible exception of polyamide (nylon). Research has shown, however, that bacterial activity in the fabric interstices can clog a fabric and thereby reduce its permeability. B. C. Beville, U.S. Department of Agriculture, Soil Conservation Service, Orlando, Florida, in 1968 performed the following test (Calhoun, 1972): Two slotted collector pipes were installed in separate trenches. Each was wrapped with a geotextile filter of different physical type and chemical composition. In a matter of weeks, the nonwoven glass fiber fabric on one pipe became clogged with an iron sludge. The sludge was formed by iron bacteria that oxidized and precipitated iron into the water. There was no sludge buildup on the woven polyvinylidene chloride cloth on the other pipe.

Other factors discussed in detail below can also have an adverse effect on the performance or physical properties of geotextiles.

8.4.2 Ultraviolet (UV) radiation

Without UV stabilizers, all synthetics discussed previously are subject to degradation when exposed to UV radiation. The fabric will be exposed to UV rays during construction. The length of time of exposure will vary with the size of the project and the construction sequence. In a drainage installation, when completed, there is no concern for the effects of UV radiation, but in certain types of coastal structures, such as revetments, UV resistance must be considered. Continuous or intermittent UV exposure may result from any one or a combination of the following:

- The stone armoring may be relatively thin, and rays may penetrate to the fabric through voids in the armor.

- The armor may be precast cellular blocks with holes through the concrete from the tops to the bottoms of the blocks that permit daily exposure to UV rays.

- Armoring materials may be rearranged or removed by storm or other occurrences exposing the previously shielded fabric.

- A construction oversight may have permitted the geotextile filter to be exposed after the structure was completed.

The 1977 CE Guide Specification (U.S. Army, Chief of Engineers, 1977) requires that the fabric "contain stabilizers and/or inhibitors added to the bare plastic if necessary to make the filaments resistant to deterioration due to ultraviolet and/or heat exposure." For most coastal installations the phrase "if necessary" should be eliminated.

Of the synthetics discussed, in an untreated state (no stabilizers or inhibitors added to the polymer) polyester has the greatest resistance and polypropylene and polyethylene the least resistance to UV degradation. Steward, Williamson, and Mahoney (1977) report that untreated nonwoven polypropylene and polyester fabric samples completely disintegrated within 18 months when left exposed in the field. Bell and Hichs (1980) indicate that, when constructed of the same fiber and an equal amount of UV stabilizers, woven monofilament fabrics would be the most resistant to UV radiation, multifilament woven and the nonwoven fabrics would have intermediate resistance, and slit-film woven would be the least resistant.

One monofilament woven polypropylene fabric in which carbon black was incorporated in the filament during the extrusion process retained satisfactory strength properties after 11 years of exposure (without cover material) in a coastal environment (Soil Testing Services, 1980).

As yet, no standard test has been developed to measure the length of time over which fabric, either untreated or treated, can be exposed to UV before harmful degradation takes place. ASTM Subcommittee D-

18.19/D-18.61 is charged with developing such a test method. As stated above, there are numerous design reasons for UV stabilizers being necessary in geotextiles for coastal structures, especially in the absence of a method for determining fabric life.

8.4.3 Fire

The melting point varies with the polymer used in the fabrics from 135 to 260°C (274 to 500°F). If fire generates heat beyond the fabric melting point, it will alter the geotextile filter's piping and permeability performance. Some polymers will burn (support combustion); others will only melt.

8.4.4 Ice

Ice formation within the structure of thick nonwoven fabrics will enlarge the pore openings. Depending on the polymer and fabric construction, some recovery (of unknown extent) may take place. The soil retention capability will be reduced.

8.4.5 Kinetic energy

Kinetic energy in the form of direct wave attack on unarmored geotextiles may cause rupture. Wave energy transmitted by armor stone may damage covered geotextiles if requirements given in Tables 8.1 and 8.2 are not met.

8.4.6 Abrasion

Abrasion can tear fibers and weaken the fabric as mentioned earlier. Fabric can be abraded by overlying material during construction and storms and by waterborne debris if the cover is removed.

8.4.7 Vandalism

If the fabric is not protected by a cover of earth or armor, it can be damaged by vandalism.

9

Recycled and Other Materials

9.1 General

In emergency situations or when funds are lacking and the need is great, almost any material with a specific gravity greater than water has been used as either a temporary or a permanent device to protect against damaging waves or currents. Even for temporary protection, however, materials with a specific gravity of less than 1.5 are of little value. Other materials that may provide emergency or short-term protection may be so difficult to recover, or remove, as to be undesirable. Others may be environmentally undesirable because of hazards to bathers or visual or chemical pollution, or there may be the possibility that waves or currents will transport the material to an undesirable location or cause undue scour.

New materials are continually being offered for coastal installations, but durability and resistance to fatigue and chemical breakdown are vital to the economic life of a coastal structure. All too frequently, only over a long period of time can it be determined how the untested new material will function.

9.2 Concrete

Salvaged concrete can be used as found in its original form so long as lifting and transport equipment can handle it. It can be reinforced or unreinforced concrete broken into sizes more easily handled, or it can be crushed and reduced to sizes ranging from sand to cobbles.

9.2.1 Concrete rubble

Concrete broken into sizes capable of being transported and handled by conventional rock-placing equipment can be used in the same man-

---

This chapter was written by William J. Herron.

ner as rock rubble for the armor and underlayers of rubble mound structures. It is generally one of two types: unreinforced or reinforced concrete. Unreinforced concrete is preferred. If reinforcing steel is protruding, it is unsightly and dangerous to bathers or recreationalists climbing on it. If possible, protruding sections of steel should be cut off. In salt water the exposed steel will corrode rapidly and will split the pieces of concrete into smaller units. That will reduce the effectiveness of the units in resisting the forces of waves or currents.

Revetment. The most common use of concrete rubble is in a revetment. The concrete can be used for armor stone or for the underlayer, and design slope and sizing should follow the same procedures as for rock. It must be kept in mind that concrete has a specific gravity of only about 2.3 to 2.5 and does not have the hardness of most stone, so it will have a limited life. In place, it generally presents a somewhat unattractive appearance as compared to rock; but if it is in the wave wash area, it will be abraded into less angular shapes and will be similar to water-worn rocks in appearance. It is acceptable as an underlayer beneath armor rock, and near urban areas, it may be less costly than rock.

The primary revetment use of salvaged concrete or concrete rubble is in emergency, low-cost, or temporary revetment of an eroding bank or bluff. The concrete is generally available for the cost of hauling; in many areas, like the coastline of the Gulf of Mexico, the southeastern United States, and parts of the shorelines of the Great Lakes, it may be the only protective material immediately available. Sources of salvaged concrete for use in revetments are broken highway or landing strip paving, foundations for structures, broken piles or light standards, manholes, and large sewer or water pipe. Thin slabs, particularly those reinforced with wire mesh, should not be used because (1) they tend to form flat planes upon which other materials slide into deeper water, (2) the exposed mesh is hazardous to recreationalists, and (3) the rusting of the mesh causes rapid deterioration of the concrete.

Groins. The application of salvaged concrete in groins is the same as in revetments; but because groins are usually located in recreational areas, the concrete is much less desirable from the standpoints of esthetics and safety.

Jetties. Concrete rubble is generally not desirable as armor material for jetties unless the jetties are located in small bays or lakes where the waves are of limited height. The several thousand pieces of armor material required are usually not available in design size from broken concrete. Jetties are also usually designed for long lives, and the concrete blocks tend to wear rapidly or disintegrate. Concrete rubble may, however, be used in jetties as graded core material.

Breakwaters. Concrete rubble is generally undesirable for the armor or underlayers of a breakwater unless the structure is in a small bay or lake. Wave action may be too severe, and it is not feasible to obtain an adequate quantity of design size pieces. Concrete rubble is acceptable as a filler within the core section as long as it is well mixed with the remaining core material to avoid excessive voids.

9.2.2 Crushed concrete

Equipment is available to economically crush unreinforced concrete to most any size and gradation desired. Because of the need to completely rebuild miles of highways, crushed concrete will, in some areas, compete with the cost of crushed stone.

Protective structures. The principal use for crushed concrete in breakwaters, groins, and jetties is for the core and the bedding layer. Here density is not as important as for the cover layers, but the crushed concrete should satisfy the same size, gradation, and durability requirements as those established in the SPM [U.S. Army, Corps of Engineers, Coastal Engineering Research Center (CERC), 1977] for stone. The same criteria apply to crushed concrete for use as filter layers in revetments, seawalls, and bulkheads, and special measures may be required to control the amount of dirt in a crushed concrete mix.

Roads, parking lots, and storage areas. Crushed concrete is frequently used as a base course for roads, parking, or storage areas, particularly in commercial harbors. When it is properly graded, it is as effective as crushed stone and in urban areas may be less costly.

9.2.3 Unbroken concrete

Unbroken concrete is normally so large as to be a supplement to cover stone. It consists of foundation blocks, light standards, concrete piles, manholes, and other reasonably compact concrete structures weighing between 0.44 and 88 kN (100 and 20,000 lb). The pieces are generally of an awkward shape to place and, if made of reinforced concrete, can cause severe handling and corrosion problems, particularly if they are used in salt water. They are generally unsightly and can be difficult to recover when no longer needed.

Concrete barges and concrete-hulled ships have, on a few occasions, been used in breakwater, groin, or jetty construction. They can be incorporated into the mass of a breakwater or jetty or be used singly or in a line to act as a groin. In general, they have not served well because their smooth sides and bottoms and large surface areas compared to masses permit them to slide or tip out of position. The interiors of these barges and ships may be difficult to fill with ballast, and the vessels are subject to the same deterioration problems as slabs of concrete. The hulls can be extremely difficult and costly to remove or salvage.

Temporary or emergency protection. The most prevalent use of unbroken concrete units is temporary or emergency protection, and often it does more harm than good. Unbroken concrete units are usually large enough that they should have an underlayer, or bedding layer, of smaller rocks, as in the design of a rubble mound stone structure. Shapes such as light standards and pilings are usually too long and rigid to act as flexible and effective units, and in general there is little variety in size or shape to provide a well-graded section such as that available with stone. Without a bedding layer, these units tend to work into the sand bottom and can be very difficult to remove. However, in emergencies they may be the only medium- to high-density material available, and so they can be used pending later availability of properly graded stone.

Supplement to armor stone. Generally, for rubble construction, unbroken concrete units are not available in a sufficient range of sizes to be used as armor stone. However, if they fit some of the design sizes for armor stone, and esthetics and the safety of recreationalists are not a controlling condition, they can be used in combination with armor stone to reduce costs.

Supplement to core stone. Unbroken concrete units are frequently used in conjunction with core stone simply to provide bulk and reduce costs. Care must be taken to fill all voids inside and around the units so that they become effective parts of the core. Problems may be encountered if their dimensions do not allow them to fit within the core boundaries, and they should not be so placed that large flat surfaces are nearly horizontal and thus encourage the sliding of other stones across them.

9.3 Asphalt

The primary types of asphalt mix used as a salvaged material in shoreline structures are the asphalt and concrete mixes and the asphalt and sand mixes. Both are used in much the same manner as salvaged concrete, i.e., as a substitute for stone. Depending on the proportion of asphalt, both mixes tend to be of lower specific gravity than stone or concrete and so are less desirable. Both are less durable than stone or concrete. Because of its black-to-brown coloration, salvaged asphalt mix is less desirable from an esthetic point of view than concrete or rock.

It is possible through heat processing to melt asphalt down and reuse it as a mixing agent, but the cost is prohibitive and the material is of no value for shoreline structures.

9.3.1 Asphalt in rubble structures

Broken asphalt concrete can be used as a cover layer or in the underlayer if the design criteria for stone are followed. As a cover layer it is not desirable because it is of low specific gravity, is not as durable as stone or broken concrete, and its color and appearance make it unappealing esthetically. Asphalt-sand has little durability, a characteristic that makes it undesirable in the cover layer.

9.3.2 Crushed asphalt concrete

Crushed asphalt concrete or asphalt-sand is an acceptable substitute for stone as core or bedding material in coastal structures, but it must be used with care as filter material because of the difficulties in obtaining and maintaining an adequately graded mix to act properly as a filter. The most common use is as a base material for roads, parking lots, and storage areas in the same manner as discussed for broken concrete. Either broken or crushed asphalt concrete or asphalt-sand can be used as fill material so long as the voids can be filled.

9.4 Concrete Block and Bricks

Salvage material, consisting generally of hollow concrete blocks, cinder blocks, and bricks, may be used in random placement. These materials are seldom used except for emergency protection out of economic necessity.

9.4.1 Concrete blocks

Concrete blocks are generally hollow blocks salvaged from dismantled buildings or broken in production during the curing period. In the western United States such blocks are made of concrete, have a specific gravity of about 2.3 in seawater, and are not particularly durable. In the eastern United States they are frequently fabricated by using cinders or slag for aggregate and are known as cinder blocks. They are even less durable than the concrete block and have a very low specific gravity (about 1.5 to 2.0).

Rubble structures. Concrete blocks are frequently used for emergency protection because they are often readily available for only the cost of transporting. They generally break down through handling and are of no value in the armor layer or in the underlayer. They are not durable, nor are they aesthetically acceptable. However, they can be used as temporary protection during an emergency in isolated, nonrecreational areas as long as it is realized that they must be covered with designed layers of stone that will act as a protective material.

Crushed concrete block. Concrete blocks will generally crush easier than broken concrete or stone and can be used in the same manner for bedding layers, filter layers, or as part of a base course for roads, parking lots, and storage areas.

9.4.2 Bricks

Common building bricks have a specific gravity of about 1.9. They are not generally durable when subjected to abrasion, nor are they esthetically desirable. Salvaged bricks may be rejects from the kilns, or they may have been obtained from dismantled buildings or structures. They may be whole bricks, broken bricks, or clusters of bricks still bound together by mortar.

Rubble structures. Salvaged bricks are not desirable for either the armor layer or the underlayer of a breakwater, jetty, groin, or revetment because of their low specific gravity, lack of durability, and lack of esthetic appeal. Also, they tend to create slip planes. They may be used as core material as long as they are mixed with other materials and satisfy the general design criteria for core stone.

Crushed bricks. When bricks are crushed, they can be used as base course material for roads, parking areas, and storage areas in the same manner as crushed stone. They have limited value as bedding material because of their low specific gravity and tendency to reduce to sand and clay sizes. They are of almost no value as filter materials because they lack durability and tend to break down and fill the voids necessary for a material to act as a filter.

9.5 Salvaged Ships, Barges, Railroad Cars, Automobile Bodies, Refrigerators, and Other Objects

Because of their size and weight, it is always a temptation to use no-longer-functioning objects to achieve "instant structures." In particular, some salvaged ships and barges might provide up to 300 m (1000 ft) in length for a protective structure. The weight may be several thousand tons. But there is the difficulty of placement and perhaps an even greater difficulty if the objects must be removed.

9.5.1 Salvaged ships and barges

Salvaged ships are generally steel-hulled, although there may be a few concrete-hulled ships from World War II, and they may be from 15 to 300 m (50 to 1000 ft) in length. The ships are generally 20 to 50 years old, and much of the internal equipment, particularly the heavy engines, usually has been removed. The curved cross section of a typical ship's hull and the partially streamlined bow and stern make the ship difficult to place precisely and not shift in position or capsize. The general concept is to maneuver the ship into a predesigned location, sink it in place, and fill it with sand, gravel, or rock to provide stability.

The steel hull plates of these ships originally vary from 6 to 19 mm (¼ to ¾ in) in thickness, have already been exposed to 20 to 40 years of corrosion, and have a very limited life as a partially sunken ship. In shallow water the plates can be subjected to abrasion by the sand that is constantly in motion, and the force of breaking storm waves in shallow water can be considerable. Once the hull opens, the disintegration of the ship can be very rapid, and the structure will soon lose its effectiveness as a protective measure.

A steel ship is of questionable value even as a temporary protective structure. If it settles too deeply into a sand bottom, rolls over, partially disintegrates, or is no longer floatable, it can be extremely difficult and costly to remove. It is obvious that, as the ship disintegrates and is moved by waves or currents, it can become a hazard to navigation, be a danger to recreationalists, and be very unsightly from an esthetic point of view.

Salvaged barges. Barges may be of wood, steel, or concrete. They are usually smaller than ships, more rectangular in shape, and easier to place in position. They are usually flat-bottomed and have a tendency to slide out of position under the force of storm waves. If positioned on sand, they are also subject to severe scouring action, and they may tilt or slide out of position. Like ships, they are usually many years old and seriously deteriorated before being salvaged. Wooden barges are the least desirable. They lack the deadweight of steel or concrete; and once they start to disintegrate, the process is more rapid. Also, if they are filled with sand or gravel for stability, they may even attain partial buoyancy as the waves remove the material.

Concrete barges are unsightly and dangerous because of the mass of reinforcing bars exposed as they deteriorate and break up. Steel barges, like ships, have relatively thin hull plates. Through corrosion or wave forces, they deteriorate very rapidly, especially in seawater.

Breakwaters and jetties. Neither salvaged ships nor barges are recommended for breakwater or jetty construction. Wave action is generally too severe and, in the case of breakwaters, is usually broadside, the most unstable position. Even for temporary or emergency protection, salvaged ships and barges are not recommended because of the difficulty of removal.

Groins. There has been some limited success in groin construction with the use of several barges in tandem. They must be securely fastened to each other and well seated on the bottom. Even so, because of deterioration, provisions must be made to either remove them or cover them with rock at a later date.

Revetment. Salvaged ships and barges are not recommended for revetment use. They are rigid structures; and in an area of breaking waves, they will generate so much scouring action that more erosion may result with the structures than without them.

9.5.2 Salvaged railroad cars, automobile bodies, refrigerators, and other objects

Miscellaneous salvaged objects are used mostly for bank or shore protection in nonrecreational areas. All of them are unsightly, and, although made of metal, they generally do not have much weight compared to their total dimensions and are easily moved by waves or currents. Because of their rigidity and numerous flat surfaces, they can cause accelerated scouring, and they may even accelerate the erosion process. Like other metals, they are normally corroded before salvage; and if they are used in seawater, the corrosion process is accelerated. Railroad cars and automobile bodies, in particular, will disintegrate within a few years. Their use is not recommended except in a nonrecreational area and only then as an emergency temporary measure if they are to be removed as soon as a long-term protective system can be implemented.

9.6 Rubber Tires

About 2 million rubber tires, too worn for further use on trucks and automobiles and not suitable for being recapped or retreaded, are available annually throughout the United States. Although, as a material, rubber tires are strong and durable, they have almost no salvage value. Hence, they are generally available at very low cost or just for the cost of hauling. They have been used for years as fenders on barges, work boats, and docks, but it is only since about 1983 that they have been seriously considered as a low-cost, and readily available, material for protection structures.

9.6.1 Characteristics

Salvaged rubber tires have a specific gravity of about 1.2. They are tough, flexible, durable, and they are almost inert to chemical reaction in either fresh or salt water. In fact, the critical strength factor of a scrap rubber tire system is not the tire portion, but the fastenings and the mooring system.

9.6.2 Uses

Salvaged rubber tires have been used primarily to form floating breakwaters, but they also have potential use for revetments, groins, bottom stabilizers, and fishing reefs. Experiments are underway to use them as an additive to asphalt concrete paving.

Floating breakwater. Like all floating breakwater systems, the floating rubber tire system is most successful where the need is to protect a basin area against short-period waves as in bays, harbors, and lakes. Several different arrangements of tires have been used and model-tested, but the basic principles are the same.

Flotation is provided either by entrapped air or by urethane foam filling a part of the tire. The air system works only for tires held in a vertical position. A regular schedule of adding air to replace lost air or compensate for added weight due to sea growth or the entrapment of silt must be established. Tires are bundled together in modules of a workable size and weight, and then the modules are assembled into a floating breakwater of design length, width, and depth. Other design factors are the density of the tire assembly and the allowable load stresses on the fasteners and the mooring system.

The scrap tire assemblies have been secured by steel cable, galvanized iron chain, nylon rope, or (one of the most successful) scrap cuttings from conveyor belt material fastened together with nylon bolts, nuts, and washers. The floating breakwater is a dynamic system in constant motion, so it is imperative that an adequate inspection and maintenance schedule for the fasteners and their hardware be established. Salvaged telephone or power poles can be used as inexpensive spreaders to frame the assemblage of scrap tire modules. Poles and spreader frame work can be specially treated to extend the useful life of fasteners; but in view of the low-cost aspect of salvage systems generally, the expense may not be justified.

Standard mooring systems of steel cable or galvanized iron chain with anchors or anchor blocks are generally used to hold the breakwater units in place. Mooring stresses will depend on wave forces, density of the tire modules, width of the system, and depth of the water as related to the depth of the system. As with fasteners, a prudent inspection and maintenance schedule is mandatory.

Revetments. If the tires or modules of tires can be securely anchored to the bottom of the natural slope, they can serve as revetments. With a specific gravity of only 1.2 they cannot be expected to stay in place of their own weight. They do not act as a revetment in the same manner as rock rubble, i.e., by completely absorbing or reflecting wave energy before it reaches the native bank material. The rubber tire revetment will only partially reduce the energy of the waves; and if it is under persistent attack, increased turbulence may even accelerate the erosion of the native material. It may be feasible to use the rubber tires in conjunction with underlayers of rock to act as a revetment, provided the tires are securely anchored in place.

Bottom stabilizers. Rubber tires are being used in parts of the Chesapeake Bay where waves are small and erosion is slow or intermittent, and they apparently function by encouraging the growth of marsh grasses and the resultant increase in an accumulation of mud to stabilize the entire mass.

Some success in use of tires to control littoral drift has been reported in areas of low turbulence. The technique is to simply anchor rubber tires to the sea bed to slow the bedload movement of sand either by currents or wave-induced movements in the littoral zone. This is obviously not feasible in the breaker zone of the open coast, and the entrapment of sand outside the breaker zone would be slow and minor in quantity. This application may have particular merit in lakes, reservoirs, or bays where surf is not severe.

Fishing reefs. Modules of rubber tires placed in deep water will, because of added surface and the many voids and crevasses, encourage the growth of marine plants and animals. The modules must be anchored in place, but the anchoring is not as difficult as when they are used for floating breakwaters or revetments.

9.7 Use of Recycled Materials in Coastal Construction

The use of recycled materials is very sensitive to the degree of emergency or lack of availability of new materials, the availability of the recycled materials, and the suitability of the materials to accomplish the desired design objectives. In some cases such as floating breakwaters, the recycled material, in this case rubber tires, may actually be the preferred material. In every case, the conditions peculiar to the project determine the usability of recycled materials. When suitable, recycled materials can be used in the types of coastal structures discussed in the following subsections.

9.7.1 Breakwaters

Concrete rubble and salvaged asphaltic concrete can be used as a substitute for stone underlayers or cores, but the lower densities must be considered when making the substitution. Crushed concrete can be an effective core material. Concrete blocks and brick can also be used as rubble in place of stone underlayer or core if allowance is made for the lower densities. When crushed, these materials can be used in cores. Rubber tires are used to make floating breakwaters capable of attenuating short-period waves. Salvaged ships and barges should not be used as breakwaters in the ocean or in large lakes. They have limited usefulness as a type of caisson in that they are generally located at remote stations and towed to the sites where they are sunk and perhaps filled with sand or rock to serve as breakwaters or reefs.

9.7.2 Reefs

In addition to barges, automobiles, and railroad car bodies, broken concrete and rubber tires can be formed to trap sand for beach contouring and to encourage the growth of marine biota. Broken asphaltic cement, concrete blocks, and bricks can be used to create rubble mound reefs. Below the wave and breaker zone, such materials are more durable and are more likely to serve successfully.

9.7.3 Seawalls and bulkheads

Toe protection and backfill materials, where rock would normally be used, can be of salvaged concrete, asphaltic cement, concrete blocks, or brick. These materials could also serve as aggregate for preplaced-aggregate concrete.

9.7.4 Revetments

All the materials discussed in this section except the ships, barges, and car bodies can be used to stabilize revetments. The finer particles of crushed materials are suitable for filter blankets and the rest as rubble or cover material.

9.7.5 Piers and wharves

Salvaged rubber tires make good bumpers and fenders for small craft using piers and wharves. Tires are also good buffers between independent floats or structures that might otherwise bump or scuff each other in moving with tides, waves, or currents. Broken concrete, asphaltic concrete, and broken concrete blocks or bricks can be used as revetment to protect the shoreside embankment of piers and wharves.

10

Protective Systems for Materials

10.1 The Corrosion Process

To understand the reason for using coatings or applying cathodic protection, a brief review of corrosion fundamentals is necessary. The principles apply to any metal structure. Corrosion is defined as the deterioration of a material, usually a metal, or of its properties because of a reaction with the environment. Three conditions are necessary for metallic corrosion to occur:

- There must be an electrical potential difference between two metallic electrodes, anode and cathode. This can exist because of metallic composition differences, metallic surface condition differences, or differences in the environment contacting the electrodes.
- The contacting environment (electrolyte) must be electrically conductive with positive and negative ions present.
- There must be a metallic connection between the electrodes.

Corrosion is a natural process involving electrochemical reactions with a resulting flow of direct current from anodic areas of the substructure (corroding areas) to cathodic areas of the substructure through the surrounding and contacting electrolyte (soil or water environment). A simplified diagram of the corrosion process on iron or steel in water is shown in Fig. 10.1. The circuit is completed through the metallic connection between anode and cathode, both of which may be part of the same structure. The current flow is called galvanic current, and it is usually in microampere or milliampere quantities.

This chapter was written by Dr. Albert L. Roebuck.

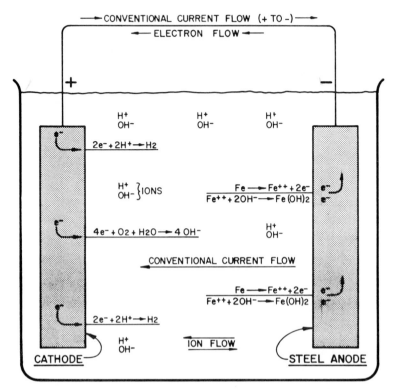

Figure 10.1 Corrosion process.

Corrosion (loss of metal) of steel substructures takes place only at anodic areas as the result of current flow into the electrolyte from anodic areas.

In the case of iron or steel, metal loss amounts to about 9 kg (20 lb) per ampere-year of current flowing from the metal into the contacting electrolyte. Loss of metal is directly proportional to the amount of current. One milliampere of current leaving the substructure from one point into the electrolyte will cause penetration of 9.5 mm (⅜ in) steel plate in less than 1 year. See Table 10.1 for corrosion rates of other metals, including anode materials. In iron or steel substructures the electrochemical reaction prevents corrosion of areas where current flows from the electrolyte into cathodic areas of the substructure. Anodes, cathodes, and corrosion current as related to steel in water are shown in Fig. 10.2.

10.1.1 Corrosivity of the environment

Corrosivity of the environment on a structure will depend on the temperature (ranges and mean averages), relative humidity, wind condi-

TABLE 10.1   Corrosion Rates of Some Common Metals

Metal	Consumption rate, kg/(A • year)
Lead	33.36
Copper	20.46
Tin	19.13
Zinc[a]	10.23
Iron	8.90
Magnesium[a]	7.56
Aluminum[a]	2.67(seawater)
Carbon[a]	0.89
High-silicon iron[b]	Less than 0.44
Magnetite, Fe_3O_4[b]	Less than 0.0445
Lead-silver[b]	Less than 0.0445 (seawater)
Platinized titanium[b]	Less than 0.000044 (seawater)

[a]Galvanic anode material
[b]Impressed current anode material

tions, proximity to the water, rainfall, and chemical fumes (from cargo or nearby plants). Corrosivity generally increases with increases in temperature, relative humidity, wind velocity (particularly off-water), closeness to the water, rainfall, and concentrations of chemical fumes.

If the structure is to be immersed, the corrosivity will depend on the temperature of the water, chemical composition (salt content, dissolved oxygen, and the presence of other chemicals), velocity of the water (movement by tides, waves, gravity, etc.), and splash zone effects (a combination of mechanical and corrosive effects). Corrosivity on immersed or partially immersed structures generally increases with increases in temperature, corrosive chemical content, and fluid

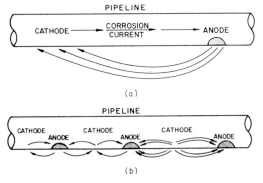

Figure 10.2   Corrosion process in pipelines. (a) Single cell; (b) multiple cells.

velocity and in the splash zone. It should be noted that high-purity water such as distilled or deionized water is a special condition that can cause coating to blister or delaminate.

10.1.2 Corrosion prevention

On a given structure, the corrosion process can be prevented or stopped if any one of the three conditions necessary for corrosion can be eliminated. The principal methods for preventing or mitigating corrosion are described below.

Coatings. A perfect coating will electrically insulate the anode and cathode areas from contact with the electrolyte and thereby prevent flow of corrosion current. As is well known, a perfect coating is an impossibility for anything other than a laboratory-scale project. Also, all coatings disintegrate in time and impose an ever-increasing area of poorly coated or bare metal. Cathodic protection is an ideal way to deal with coating discontinuities (holidays) and poor coating in general. Coatings (with cathodic protection) are feasible for practically any subsurface structure. Exceptions are the underwater parts of off-shore oil-production platforms, where coating repair or replacement is not possible.

Insulated joints. Insulating joints between metal plates or piping joints will minimize stray current or galvanic corrosion by interrupting corrosion current flow. In this application, the metallic connection between widely separated anode and cathode areas is broken by the insulating joint. Insulated joints also serve to separate dissimilar metal areas, as well as to separate cathodically protected areas from unprotected areas.

Cathodic protection. On a given subsurface structure the corrosion process can be prevented or stopped by supplying an excess of electrons to all subsurface parts of the structure. The result is that the structure becomes all-cathode because of the electrons provided by forcing direct current to flow through the contacting electrolyte (water or soil) from a nearby subsurface source (anode) onto all subsurface parts of the structure. Hence the name, cathodic protection. When the current is adjusted properly, it will counteract corrosion current flowing into the electrolyte from the substructure with an opposing and slightly more than equal flow of current from the electrolyte into the substructure. Loss of metal has been transferred from the protected substructure to the external anode, which will require occasional replacement.

10.2 Coatings

Environmental conditions affecting coastal structures range from mild to severely corrosive. To provide suitable service life for coastal structures, protective coatings are usually required; they range from little or none (other than decorative painting) to complex and extensive multicoat systems. Specific coating demands depend upon the type of the substrate to be coated and its environment.

This section will cover the basic design considerations that must be given to the structures to be protected and to the selection, application, and inspection procedures necessary to provide a protective coating system with years of dependable service life. Repair, rehabilitation, proper maintenance procedures, and other important facets, including economics, also are described.

Protective coatings are designed primarily to isolate metal surfaces from exposure to corrosive elements. Coating thickness will vary from 50 to 75 micrometers (μm) (2 to 3 mils) for simple alkyd coating systems to 380 to 760 μm (15 to 30 mils for certain high-build coal tars, coal-tar epoxies, or urethanes. (1 mil equals 25.4 μm or 0.025 mm. When any conversion in a discussion results in a number less than 0.1, micrometers and mils will be used.)

Conventional paints, surface preparations, and methods of application should not be used in corrosive areas of coastal structures. Only high-performance protection coatings such as epoxies, zinc-rich, chlorinated rubber, and polyurethanes should be considered. Products selected for use should be resistant to the environment and be capable of serving as a barrier between the substrate and the environment. Writing appropriate specifications will play an important role in obtaining the right kind of protective coating including, particularly, cost per unit area per year of service life.

10.2.1 Design and specifications

In the design of new structures, it is important to consider surface requirements for ease of application and effectiveness of coating. The design should (1) provide smooth, flat, easily curved surfaces and (2) avoid overlapping surfaces, edges, back-to-back structures (brackets, beams, or L's), riveted surfaces, sharp protrusions, and weld splatter (see Fig. 10.3).

Like other personnel in any design and construction organization who are engaged in translating an owner's requirement into design

SPECIFICATION   CONSIDERATIONS

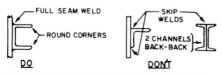

POCKETS OR CREVICES. ALL CONSTRUCTION INVOLVING POCKETS
OR CREVICES THAT WILL NOT DRAIN OR THAT CANNOT BE PROPERLY
BLAST-CLEANED SHOULD BE AVOIDED.

CONTINUOUSLY WELDED JOINTS. ALL WELDED JOINTS SHOULD BE CONTINUOUSLY
WELDED. ALL WELDS SHOULD BE SMOOTH, WITH NO POROSITY, HOLES, HIGH
SPOTS, LUMPS OR POCKETS. GRINDING SHOULD BE USED TO ELIMINATE POROSITY,
SHARP EDGES AND HIGH SPOTS THAT DO OCCUR.

REMOVE WELD SPATTER. ALL WELD SPATTER SHOULD BE
REMOVED.

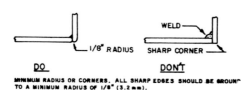

MINIMUM RADIUS OR CORNERS. ALL SHARP EDGES SHOULD BE GROUND
TO A MINIMUM RADIUS OF 1/8" (3.2 mm).

Figure 10.3   Examples of design details that facilitate
the application of coatings.

and the design into a structure, the protective coating specifier must
give careful consideration to all aspects of the coating work requirements. The specification must detail coating selection, surface preparation, coating application, coating inspection, touchup, and repair to
ensure a successful job.

10.2.2   Generic coating classes

Numerous coastal marine coating tests and surveys of field applications have been made. Table 10.2 lists some of the coating systems
used for various surface substrates. In the discussion which follows,
various uses of and precautions for each coating system are presented,
together with a brief description of chemical composition and properties.

TABLE 10.2 Types of Coatings Commonly Used on Different Substrates

Substrate	Paint	Comments
Interior wood	Oil	Generally slow drying and relatively soft.
	Alkyd	May be hard or soft.
	Latex (vinyl or acrylic)	Can be applied over oil, alkyd, or latex primer.
Exterior wood	Oil	Good wetting of weathered wood and paint chalk; slow drying; soft.
	Alkyd	Good wetting; variations give variety of properties.
	Silicone alkyd	Good wetting; good gloss.
	Latex (vinyl or acrylic)	Poor wetting of weathered wood and paint chalk; easily applied.
Interior masonry, plaster, and wallboard	Acrylic latex	Easily applied; brushing is good on coarse surfaces; must remove all loose chalk.
	Vinyl latex	Same as acrylic latex.
	Chlorinated rubber	Good for waterproofing.
Exterior concrete and masonry	Acrylic latex	Fill coats of these materials will reduce water penetration.
	Vinyl latex	Same as acrylic latex.
	Chlorinated rubber	Good for waterproofing.
	Vinyl	For concrete in very corrosive environments.
Interior iron and steel	Alkyd	Never on continuously damp or immersed environments.
	Vinyl	Good resistance to water and poor to strong solvents.
	Epoxy	Good durability and chemical resistance.
	Urethane	Good durability and chemical resistance.
Exterior iron and steel	Oil	For mild environments only.
	Alkyd	For mild environments only.
	Silicone alkyd	For mild environments only; good gloss.
	Inorganic zinc	Very abrasion resistant; limited life in seawater without topcoat.
	Vinyl	Good durability; easily touched up.
	Epoxy	Good durability and chemical resistance, but chalks in sunlight.
	Urethane	Aliphatic type has good weathering over epoxy primer.
	Chlorinated rubber	Good ductility and fair corrosion resistance.

Coatings are composed of many raw materials that can be divided into three categories—vehicle, pigment, and additives—as shown in Table 10.3. The vehicle, or liquid part of a coating, is composed of resin, solvent, and plasticizer. The vehicle resin contributes to many of the basic properties of a coating including water resistance, chemical resistance, cure time, elongation, toughness, and adhesion to substrate.

Coatings are usually named or designated by their resin systems. For example, an alkyd coating contains an alkyd resin, an epoxy coating contains an epoxy resin, and a silicone alkyd contains an alkyd resin as the primary resin system with additions of silicone resins. Plasticizers are added to modify the properties of the resin. Solvents are used to dissolve or disperse the resin for manufacture. Solvents are also useful for developing and controlling application properties of a coating. For example, in a cold climate, the solvent must be volatile enough to evaporate at low temperatures. Conversely, in a tropical or desert climate, the solvent must be slow enough (less volatile) to allow the coating to flow out and properly cover a hot surface before it volatilizes.

The pigments are the finely ground solids which are added to give the coating body, color, and corrosion-inhibitive properties. Special additives are added, usually in small amounts, to give the coatings many special properties.

Table 10.4 lists generic coating types suitable for use on concrete and steel (both immersed and nonimmersed). Selection of a recommended coating system for a given service condition is determined by specific properties of coating types, which are described below.

Alkyds. Alkyds are formed when oil is combined chemically with glycerol phthalate, and they cure by reaction with oxygen. They have excellent wetting properties, fair-to-good weather resistance, poor resistance to acids and alkalies, and only fair abrasion resistance. They

TABLE 10.3   Coating Components

Resin	Vehicle plasticizer	Solvent	Pigment	Additives
Alkyd	Glycols	Prime diluents	Prime extenders	Surfactants, drying agents
Epoxy			Color	
Coal tar		Mineral spirits	Corrosion inhibitor	Antiskinning agents
Vinyl		MEK toluene		

TABLE 10.4   Recommended Coating Systems

Type of surface	Coating type
Concrete	Coal tar
	Coal tar–epoxy
	Epoxy
Steel, not immersed	Alkyd
	Silicone alkyd
	Silicone acrylic
	Coal tar (good)
	Coal tar–epoxy (good–excellent)
	Epoxy (good–excellent)
	Urethane (good–excellent)
	Vinyl
	Zinc-rich
Steel, immersed	Coal tar
	Coal tar–epoxy
	Epoxy
	Epoxy phenolic
	Phenolic
	Vinyl

should be used only for mild environments in selected harbor locations, and they should not be used where surfaces are continuously damp or immersed in water.

Silicone alkyds.   Silicone alkyds possess properties similar to those of alkyds as defined above but have somewhat better heat resistance, weather resistance, and gloss retention. They cannot be used for immersion service. Also, they are softer and are less resistant to abrasion than alkyds.

Modified alkyds.   Modified alkyds have additional resins (such as ester gum, phenolic, styrene, vinyl, acrylic, and chlorinated rubber) to improve properties such as weather resistance and corrosion resistance. They can be used for some coastal structures; but because of the limited corrosion resistance of alkyds, immersion service or exposure to corrosive environments is not recommended.

Acrylic (solvent base).   Acrylic coatings can be formulated as solvent-base or as water-base materials. The solvent-base coatings are composed of copolymer acrylic resins in an aliphatic hydrocarbon solvent. They cure by evaporation of solvents, and in some respects they possess the toughness and corrosion resistance of a baked enamel. Acrylics have good resistance to general weathering, and they have excellent gloss retention and color stability. Their resistance to marine atmospheres and corrosive environments is good. They have good impact resistance and fair heat resistance. Acrylics can be used as topcoats for epoxy, modified alkyd, zinc-rich, and universal primer. They are not recommended for immersion.

Acryl' 'ater base). The water-based acrylic products are widely used for maso₁ stucco, or wood for both interior and exterior application. They are also used to a limited extent for metals. Because they are water-based, they are nonflammable and can be used in fire hazard areas. They are quick-drying and can be easily cleaned with water. Acrylic water-base coatings have excellent color retention and good flexibility and toughness, and they are easy to repair. They have only fair corrosion resistance. Acrylic resins are used in conjunction with other resins to improve color stability and general weathering resistance.

Chlorinated rubber. Chlorinated rubber coatings are composed of rubber polymers chemically treated with chlorine in a blend of solvents. They cure rapidly by solvent evaporation, have fair-to-good corrosion resistance to marine and chemical corrosion. They resist most dilute acids, alkalies, and salts, and have fair weathering resistance and color stability. They are suitable for immersion in salt and fresh water to 49°C (120°F). Chlorinated rubber coatings have fair abrasion and impact resistance but possess only poor resistance to organic solvents.

Coal-tar coatings. Coal tar is a byproduct of the coke industry. It has outstanding water resistance, and for that reason it has been found to be an excellent coating material for many coastal structures. Its high resistance to moisture makes it useful for immersion or for splash zone service. It is one of the finest coating materials for resistance to water or moisture, and for that reason it is frequently used for protecting steel or concrete in immersion. Coal-tar products are usually applied by spray because of their heavy consistency, but touchup is frequently handled by brush. Additional thinner can be added when the coating is applied by brush.

Coal-tar coatings are frequently modified with other materials such as epoxy or polyurethane. The modified products have the excellent water resistance of the coal tar with the improved toughness and solvent resistance of the modifier. Properties of coal tar and the modified materials are as follows:

- Coal-tar properties: low cost, excellent resistance to water, low moisture vapor transmission, poor resistance to organic solvents, and poor abrasion resistance and toughness

- Epoxy properties: tough and resistant to chemicals and solvents
- Urethane properties: elastomeric and abrasion-resistant

Two-component coal tar–epoxy and coal tar–urethane coatings are discussed. Coal tar–epoxy systems contain coal tar, pigments, solvents, epoxy resin, and either an aliphatic or aromatic amine or polyamide curing agent. Coal tar–urethanes contain coal tar, pigments, solvent, and reactive urethanes. These "chemically cured" coal-tar coatings can be applied in one coat to dry film of 0.13 to 0.64 mm (5 to 25 mils). Most coal tar–epoxies have excellent adhesion, abrasion, and impact resistance. Their resistance to immersion in salt and fresh water is excellent, and they withstand a wide range of chemical corrodents. Proper measurement of two components is mandatory for proper cure and coating properties. It is recommended that these coatings be prepared by using power mixers.

Subsequent coatings must be applied within manufacturers' specified time and temperature limitations to avoid delamination between coats. The coal tar–epoxies have variable pot lives when mixed depending on the temperature. Repairability is a major problem, and color is limited to black. They tend to chalk lightly in sunlight and weather and lose their original black gloss within 6 months to 1 year.

Epoxies (catalyzed). Epoxy coatings are based on epoxy resins and are supplied as two components. One component contains the epoxy resin, the other the curing agent. The two components must be thoroughly mixed before application. A broad spectrum of service conditions is covered by a wide variety of materials based on these versatile resins. Amine- or polyamide-type curing agents are most common. Epoxies have excellent durability and toughness and possess high chemical and moisture resistance. They resist strong alkalies and have excellent adhesion properties and abrasion resistance. They can be used in marine environments and are not damaged by moisture or immersion in water.

Epoxies can be used in conjunction with special fillers to make mastic-like materials which find applications as concrete grouts, surfacers, or "bug hole" fillers. They can be applied as heavy coats by trowel or in some cases by spray. They are used in marine applications to protect concrete or steel in the splash zone. Heavy layers of the mastic are built up in the splash and tidal zones. Layers of glass fabric can be introduced between coats of mastics to add strength to the splash zone coating system. Application is sometimes by hand in a process known as palming (Fig. 10.4). Gloves should be worn when

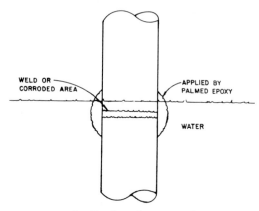

WELD OR CORRODED AREA

APPLIED BY PALMED EPOXY

WATER

Figure 10.4 Application of epoxy underwater coating by palming.

handling epoxy materials because many people have toxic reactions to epoxies.

Certain epoxy polyamide coatings and mastics will cure on moist surfaces or even under water. It is this property that allows the materials to be utilized for coating and for repairing wet or submerged structures which require protection. Epoxy coatings have excellent corrosion resistance to most chemicals. They perform well in alkaline media but are only fair in contact with acids. Because of their inertness they are sometimes difficult to repair or topcoat. Very few coating materials will adhere to a well-cured epoxy. Epoxy coatings tend to chalk, and for that reason colors fade, particularly dark colors. The lighter colors are most frequently specified for use where the coatings will be exposed to sunlight and weather. To summarize, epoxy coatings have excellent corrosion resistance and toughness, but they are fair to poor in weather resistance.

Epoxy-phenolics. Epoxy-phenolic coatings are epoxy-type coatings which have been modified by the addition of phenolic resins. The phenolic resin adds corrosion resistance because it promotes cross linking and bonding of the epoxy resin as illustrated in Fig. 10.5. Because they are resistant to attack by corrosive chemicals or environments, the epoxy phenolic coatings find application as linings for tanks, barges, ships and piping. The disadvantages of epoxy-phenolics as compared to epoxies are:

- Increased brittleness of coating

- Lack of toughness and thus greater susceptibility to damage upon impact (cracking, shattering, delamination)

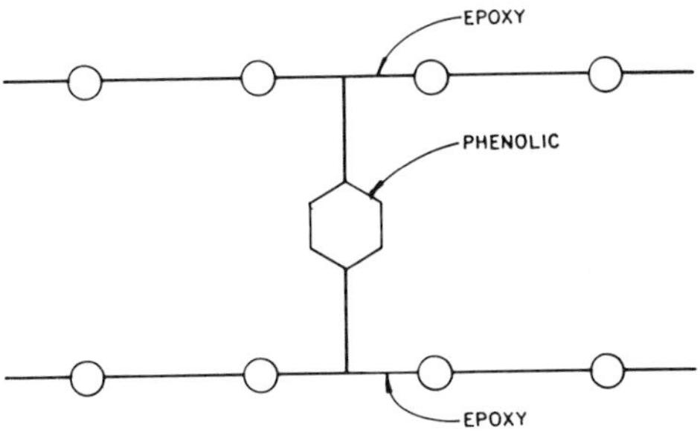

Figure 10.5 Crossed-linked epoxy phenolic.

- Requirement for heat in curing (not true of all epoxy phenolics)
- Lack of availability in high-build formulations
- Slightly greater difficulty in repairing and topcoating because of adhesion problems discussed under Epoxies, catalyzed.

Phenolics. Phenolic coatings include baking phenolics of the alcohol-soluble type and phenolics of the oil-soluble type in combination with drying oils. The baking phenolics have excellent resistance to corrosive chemicals, acids, and caustics. They require baking for curing (cross-linking). The baking phenolics form tightly cross-linked coating films with exceptionally high corrosion resistance, which is higher than that of the epoxy-phenolics. Because of their tight cross-linking, they are brittle and are subject to damage and delamination by impact. These materials are used as linings for tanks, barges, and piping.

The phenolic coatings designed for atmospheric service are made from phenolic resin combinations and drying oils. At one time, they were considered to be the best air-drying coatings for water resistance and weak chemical resistance, and they were used extensively as marine maintenance coatings. They are still considered superior to most alkyd coatings in hardness, abrasion resistance, and chemical resistance. Their limitations include somewhat less weather resistance than certain alkyd coatings have. They possess only fair gloss retention and are applied at comparatively low film thicknesses. They also require good surface preparation for best performance. Many phenolics become brittle with age.

Polyurethanes. The polyurethanes are generic products, more commonly called urethanes, comprised of moisture-cured single-package systems, single-package systems which require heat for curing, or two-package systems cured by agents such as hydroxyl-bearing polyol. All urethanes contain isocyanate.

Because of the wide variety of formulations available, the selection of the proper urethane for a specific job is often difficult. The chemical resistance (to fumes, splash, and spills) of polyurethanes, especially those cured by hydroxyl-bearing polyols, is very good. Others may be somewhat limited in this respect. The coatings are best known for their toughness and resistance to abrasion and impact. The urethanes based on aliphatic resins possess outstanding resistance to sunlight and weather including excellent gloss retention. They are used extensively on ships and aircraft where color retention is important. Formulations based on aromatic urethane resins have poor weather resistance. They should not be used as exterior coatings.

Adhesion to properly primed metal or direct to masonry is fairly good. Urethanes also generally adhere well when directly applied to glass fiber materials. Some formulations have high-build properties. Many urethanes have excellent low-temperature-cure characteristics and can be used at temperatures as low as 2°C (35°F). Urethanes have excellent flexibility and elongation properties. Because of the inertness of the two-component polyurethanes, repairability and recoating can be made difficult by adhesion problems.

Vinyls. The coatings known as vinyls include all polymers and copolymers of vinyl chloride with vinyl acetate and vinylidene chloride. They are single-package coatings that cure by solvent evaporation. Vinyls are also processed with oil-base materials such as alkyds, phenolics, and acrylics, most of which offer good overall corrosion resistance. They are somewhat similar in that respect to chlorinated rubber.

Vinyls withstand high humidity, and in their resistance to salt atmospheres and immersion in water they are comparable to epoxies in many respects. They have good abrasion and impact resistance. Flexibility and elongation properties of vinyl coatings are good. Vinyls are among the best coatings from the standpoint of resistance to oxidation. Many have good gloss retention. Their repairability is very good, and they usually present very few problems on recoating. Their resistance to most solvents, however, is poor, and they have a limited heat resistance.

Inorganic zinc. The inorganic zinc products are formulated with metallic zinc dust, at relatively high pigment volume concentration, and inorganic binders. Metallic zinc content is 60 to 90 percent by volume in the dried film. As single-coat systems, they have outstanding corrosion resistance on coastal structures for protecting metal exposed to aggressive marine atmospheres. They can be topcoated to produce coating systems with even greater anticipated service lives. Inorganic zinc coatings can also be used in industrial environments; in mild industrial environments they can be used without topcoats. One application which has found wide usage in the past is lining hydrocarbon tanks both on ships and shore.

Inorganic zinc has excellent abrasion resistance and is often used as preconstruction primer. It must be topcoated for use in aqueous immersion. Zinc coatings offer excellent resistance in the splash zone to salt water and to fresh water. Zinc-rich primers are often used as primers for organic topcoats such as chlorinated rubber, vinyls, epoxies, urethanes, and certain acrylics in more aggressive chemical atmospheres. Inorganic zinc must also be topcoated for continuous immersion in water. Surface preparation requirement is stringent, and application requires skilled workers.

Organic zinc.   The organic zinc coatings are composed of zinc dust with organic binders such as epoxy resins. Zinc dust content will vary between 45 and 80 percent by volume in the dry film. The performance of a coating is related to the percent of zinc in the final dry film; generally, the more metallic the zinc, the better the performance. Organic zinc coatings also show excellent resistance to high-humidity splash and spray conditions of both fresh and salt water. Their abrasion resistance is somewhat less than that of the inorganic zinc primers. They will tolerate mildly alkaline atmospheres, and they are often preferred to inorganic zinc primers for chemical atmospheres because surface preparation and application procedures are less critical.

Underwater-curing coatings and mastics.   The underwater-curing materials are two-component, 100 percent solids, polyamide-cured, epoxy nonshrinking compounds (sometimes combined with appropriate fillers) designed to chemically displace water on the surface and form a tight bond. Extensive and careful laboratory and field tests, combined with experience information, have shown that good protection can be provided to in-place underwater structures with these mastics.

The mastics are applied by gloved hand (palming) or trowel to a dry film thickness of 3.2 to 6.4 mm (125 to 250 mils), and they cure after several days while under water. Tensile strength, adhesion, impact strength, and abrasion resistance are fair to good. They can be used as patching compounds to seal metal, concrete, wood, fiberglass, and many other substrates specifically in situations where the surface is damp, wet, or under water. The surface preparation recommended is sandblasting for best results. Costs of material and application are very high—$54 to $162 per square meter ($5 to $15 per square foot).

Other.   Other special coatings can be used from time to time for certain applications in coastal structures. They are combinations and modifications of the generic coating classes discussed above. They will not be discussed in detail in this book. They include vinyl polyurethanes, epoxy–fiber-reinforced mortars and coatings, polyester–glass flake combinations, glass flake–coal tar–epoxy, 100 percent solids urethane-elastomeric membranes, catalyzed hypalon–coal tar and others. The qualities of various types of coatings are compared in Table 10.5.

10.2.3 Surface preparation

Proper preparation of the surface to receive the coating may be the most important activity in a coating job. Regardless of its chemical and physical properties (such as chemical resistance, moisture resistance, impermeability, abrasion resistance, and weather resistance), a coating cannot fulfill its function properly unless it adheres to the substrate. Proper surface preparation consists of getting the surface to the proper degree of cleanliness and roughness (surface profile) to receive a specific coating. A number of methods are used for cleaning and roughening a surface.

Coatings designed for use in corrosive environments have been found to give best performance when applied to clean, freshly blasted surfaces. The more resistance or durability expected of a coating system, the better the surface preparation must be. The better the surface preparation, the better the performance. Methods of surface preparation have been defined and described by the Steel Structures Painting Council (SSPC) (1975) and by the National Association of Corrosion Engineers (NACE) of Houston, Texas.

Methods of cleaning include hand cleaning with brushes, mechanical cleaning with brushes, and blast cleaning with abrasives. Hand and mechanical cleaning procedures are defined and described by SSPC specifications: SSPC-SP-2, Hand Cleaning, and SSPC-SP-3, Mechanical Cleaning. The procedures are usually carried out by using wire brushes.

Blast-cleaning procedures are preferred because they provide surface roughness (surface profile) as well as clean the surface. The procedures, described and defined by the SSPC and NACE, are based on the degree of cleanliness of the surface. All loose surface contamination must be removed. The percents of firmly adhering residues allowed to remain are indicated in Table 10.6.

Immersed zones. For immersed zones or areas of severe chemical exposures, the surface preparation recommended is white metal (SSPC-

TABLE 10.5 Qualities of Various Coatings

Coating	Gloss and color retention	Abrasion	Weather	Resistance to moisture		Resistance to chemical attack				Time to handle	Cost
				Conden-sation	Immer-sion	Acid	Alkali	Aliphatic	Aromatic		
Alkyds	F	P	F–G	F	P	P	P	F	P	F–P	Low
Silicone alkyds	G–E	P	G–E	F	P	P	P	F	P	F–P	Medium
Chlorinated rubber	F	F–G	F–G	E	E	G	G–E	F	P	F–G	Low–medium
Coal-tar coatings	F	F	F	E	E	G	F	F	P	F	Low
Coal tar–epoxies	F	G	F	E	E	G–E	E	G	F	F	Medium
Epoxies	F	G–E	F–G	E	E	G	E	E	G	F	Medium
Epoxy phenolics	P	F–G	F	E	E	F–G	F	E	E	F	Medium–high
Phenolics	F	F	F–G	E	E	F–G	F	E	E	F	Medium–high
Polyurethanes	E	E	E	G	F–G	F–G	F–G	G	G	G	High
Vinyls	G	G	G	G–E	E	E	G–E	G	P	F–G	Medium
Zinc-rich coatings	NA	E	E	E	F–P	P	P	E	E	E	High

E = excellent; G = good; F = fair; P = poor; NA = not applicable

TABLE 10.6   Residues Permitted to Remain on Blast-Cleaned Surfaces

Surface preparation		Degree of cleanliness	Firmly adhering material remaining on surface, %
SSPC-SP	NACE		
5	1	White metal	0
10	2	Near white	5
6	3	Commercial	33.3
7	4	Brush blast	NA

SP-5; NACE-1). Near white (SSPC-SP-10; NACE-2) is sometimes specified for the same applications, but it is not as satisfactory and the life expectancy of the coating system will be decreased. The near-white surface preparation blast clean is usually reserved for slightly less corrosive exposures such as nonimmersion marine. Commercial blast cleaning surface preparation (SSPC-SP-6 or NACE-3) is utilized for still less corrosive exposure such as mild marine or industrial exposure. Table 10.7 shows the recommended surface preparations for maximum results with specific generic classes of coating.

Abrasives. In addition to the degree of surface cleanliness recommended in Table 10.6, it is necessary that close attention be given to the surface roughness (profile) as illustrated in Fig. 10.6. If the pattern of peaks and valleys is too shallow, proper adhesion may not be obtained; and if the pattern is too deep and irregular, pinpoint rusting can occur because the prime or first coat may not cover the peaks. Generally, it is believed that the anchor pattern should not exceed one-half of the dry film thickness of the first coating applied (the prime coat). That, however, is not always true for high-build primers or mastics. In general, manufacturers' recommendations with respect to the degree of cleanliness and the depth of the anchor pattern should always be carefully followed.

Table 10.8 indicates the average height of profile produced by different abrasives. The profiles will vary, to some extent, with the angle and velocity of the abrasive particles and the hardness of the steel surface being blasted. Under most conditions, it is possible to protect the surface with a coating system which provides a total dry film thickness of 0.13 to 0.25 mm (5 to 10 mils).

In normal sandblasting, the anchor pattern should run about 38 to 50 μm (1.5 to 2.0 mils) deep. Most manufacturers recommend a 16 to 30 (16/30) mesh silica sandblasted at the surface at 690 kPa (100 lb/in^2) to produce that profile pattern. An 8/30 mesh sand may be needed to remove tightly adhering rust and paint. Seldom are anchor patterns of more than 75- to 100-μm (3- to 4-mil) depths recommended

TABLE 10.7   Recommended Surface Preparation for Specified Coatings

Coating type	Surface preparation
Alkyds	Commercial blast[a] SSPC-SP-6 NACE-3
Acrylic enamel	Commercial blast[a] SSPC-SP-6 NACE-3
Acrylic latex	Commercial blast[a] SSPC-SP-6 NACE-3
Chlorinated rubber	Commercial blast[b] SSPC-SP-6 NACE-3
Coal-tar coatings	Commercial blast[b] SSPC-SP-6 NACE-3
Epoxies	Near-white blast[c] SSPC-SP-10 NACE-2
Epoxy-phenolic	Commercial blast[a] SSPC-SP-6 NACE-3
Phenolics	Near-white blast[c] SSPC-SP-10 NACE-2
Modified phenolics	Commercial blast[a] SSPC-SP-6 NACE-3
Polyurethane	Commercial blast[c] SSPC-SP-6 NACE-3
Vinyls	Commercial blast[c] SSPC-SP-6 NACE-3
Zinc (inorganic)	White metal blast[d] SSPC-SP-5 NACE-1
Zinc (organic)	Near-white blast[c] SSPC-SP-10 NACE-2

[a]When a compromise or alternate is necessary, brush-off blast (SSPC-SP-7) can be used.
[b]Use near-white blast (SSPC-SP-10) for immersion or other severe exposure.
[c]Use commercial blast (SSPC-SP-6) when exposure is not severe.
[d]For severe exposure, surface preparation should be improved to an SSPC-SP-10 or NACE-2.

even for high-build mastic coatings up to 5-mm (200-mil) dry film. That profile can usually be achieved by use of 8/30 mesh silica sand blasted at the surface with a nozzle pressure of 690 kPa (100 lb/in^2). Zinc-rich coatings, on the other hand, require a blast by 30/60 mesh sand to produce the required profile of 25 to 40 μm (1 to 1.5 mils).

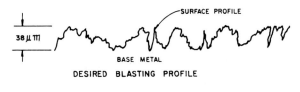

38 μm

SURFACE PROFILE

BASE METAL

DESIRED BLASTING PROFILE

12.7 μm

SURFACE PROFILE

BASE METAL

BLASTING PROFILE NOT DESIRED

Figure 10.6   Surface-blasting profiles.

TABLE 10.8   Types, Sizes, and Resulting Profile of Abrasives Used in Air Blast Equipment

Abrasive	Maximum particle size passing through ASTM mesh	Average height of profile	
		mils	m
Sand, very fine	80	1.5	40
Sand, fine	30	2.0	50
Sand, medium	18	2.5	65
Sand, large	12	2.8	70
Steel grit G-80	40	1.3–3.0	30–75
Iron grit 3G-50	25	3.3	85
Iron grit G-40	18	3.6	90
Iron grit G-25	16	4.0	100
Iron grit G-16	12	8.0	200
Steel shot S-170	20	1.8–2.8	45–70
Iron shot S-230	18	3.0	75
Iron shot S-330	16	3.3	85
Iron shot S-390	14	3.6	90

Other abrasives include garnet, flint, steel grit, steel shot, and aluminum oxide. Costly abrasives, such as steel shot and steel grit, are primarily used where they can be recovered and reused, as in cabinets and blast rooms. Special equipment such as surface profile comparator furnished by Zorelco and/or NACE TM-01-70 visual standard should be used to verify the surface roughness (anchor pattern) produced by various abrasives.

Sandblasting. No coating system can perform better than the surface to which it is applied; it must be able to reach and adhere to that surface in order to perform its function. In general, as previously stated, sandblasting steel for use in harbor facilities provides the best and often the only suitable foundation for the majority of materials which should be used in those environments. In addition to removing surface contaminants, sandblasting produces the surface roughness known as the anchor pattern which enhances coating adhesion. Sandblasting is, therefore, one of the most important steps to consider in any protective coating program.

The compressed air supply is perhaps the most critical part of a sandblasting operation. Speed of work and quality of results will be in direct proportion to the volume and pressure of air passing through the nozzle. To achieve good economics and a good profile, sandblasting steel requires a high nozzle pressure of 620 to 690 kPa (90 to 100 lb/in^2) and high volumes of air at 2.5 to 10 m^3/min (80 to 350 ft^3/min). The larger the nozzle, the faster the work will be completed, assuming proper air pressure.

It is generally recommended that an additional 35 kPa (5 lb/in^2) compressor pressure should be used for each additional 15-m (50-ft) length of blast hose. When only low nozzle pressure is available (as with long hose), the blasting rate will be low, which will make removal of mild scale, old rust, paint, or heavy marine growth expensive or incomplete.

The efficiency of blasting at 655 kPa (95 lb/in^2) is only 50 percent of that of 690 kPa (100 lb/in^2). Lower pressures often result in the need to compensate by using larger-sized abrasives with their attendant deep anchor pattern.

The air supply and couplings should be related to the size of the job; generally, they should be as large as possible in order to supply a constant high volume of air at proper blasting pressure. For large jobs, large air lines and couplings should be used to minimize friction loss in the hose. Universal-type couplings are recommended for use on air lines.

For successful low-cost blasting, the sand pot also must be given careful consideration. Several types are available. All of them furnish regulated amounts of abrasives to high-pressure airstreams for the blasting function. However, pots with only one chamber are normally used for intermittent blasting, whereas a two-chamber pot is better adapted to continuous work. Many sizes as well as designs are available. The proper size is determined by the abrasive capacity desired. All sand pots used for abrasive blasting should be equipped with separators to remove any contaminating moisture or oil. The separator trays must be emptied and maintained on a regular basis.

A moisture trap prevents condensation from forming on the lines and flowing onto the work surface. It should be placed as close as possible to the sandblast pot. An oil separator should be placed on the air line between the compressor and the blast machine to prevent oil contamination of the blasted surface. Although these precautions are frequently disregarded, they are important in obtaining good coating jobs. If moisture or oil is allowed to remain in the compressed air, it will contaminate the surface and degrade the quality of the coating. Water will oxidize and corrode the surface and reduce the quality of the final coating system. Oil will lower the adhesion properties of the final coating system, and it can cause delamination.

Sandblast hoses are often too small. The inside diameter of the hose should be 3 to 4 times the size of the orifice in the nozzle. Avoid any coupling or pipe-fitting connection that fits internally into the hose; it can reduce the inside diameter enough to lower the air-carrying capacity by more than 50 percent. Use only externally fitted quick couplings.

Consumption of abrasive and volume of air used are directly related to the size of nozzle being used. Most commonly used nozzle sizes are 6.4, 8, and 19 mm (¼, 5/16, and ¾ in) inside diameter. The larger the nozzle the larger the area cleaned in a given time. For example, at 690 kPa (100 lb/in^2) a 9.5-mm (⅜-in) nozzle will clean at 2.25 times the rate of a 6.4-mm (¼-in) nozzle.

Sand is the most widely used abrasive, and for that reason it will be used as the basis for consumption rates presented. Other abrasives, infrequently used, include aluminum oxide, walnut shells, steel shot, and steel grit. In general, rates for other materials are related to sand consumption rates. The specific consumption rate of each material will depend on the specific density and velocity of the abrasive.

Upon impingement on a surface, the energy released by an abrasive particle will be directly related to the cleaning rate. That is, the more energy released, the faster the cleaning rate. The energy of a particle in motion is determined by mass and velocity according to the equation $E = \frac{1}{2}MV^2$, where E is the energy of a moving abrasive particle (energy is released when the particle strikes the surface), M is the mass of abrasive particle, and V is the velocity. Thus, the cleaning rate increases as the mass and velocity of the particle increase.

Sand consumption rates and airflow consumption are presented in Table 10.9. Increased consumption of sand and air results in a greater area being cleaned per hour. The actual production per hour is affected by the condition of the surface to be prepared for coating and the cleanliness or surface preparation to be achieved, as shown in Table 10.10.

TABLE 10.9   Sand and Air Consumption

Nozzle diameter, mm	Component	Nozzle pressure, kPa					
		344.7	413.7	482.6	551.6	620.5	689.5
3.2	Air, m^3/min	0.32	0.37	0.43	0.48	0.52	0.57
	Sand, N/hr	298	342.5	391.4	449.2	498.2	547.1
4.8	Air	0.74	0.85	0.93	1.08	1.16	1.27
	Sand	667.2	760.6	871.8	960.8	1058.6	1174.3
6.3	Air	1.33	1.53	1.73	1.92	2.09	2.29
	Sand	1192	1387.8	1574.6	1814.8	1992.8	2197.3
7.9	Air	2.18	2.52	2.86	3.2	3.57	3.88
	Sand	2081.7	2375.2	2686.6	2989	3291.5	3611.8
9.5	Air	3.06	3.57	4.05	4.56	4.9	5.55
	Sand	2971.3	3398.3	3843.1	4270.1	4679.3	5124.1
11.1	Air	4.16	4.81	5.49	6.14	6.79	7.19
	Sand	3985.4	4590.3	5230.8	5835.8	6440.7	7045.6
12.7	Air	5.52	6.34	7.13	7.92	8.74	9.57
	Sand	5159.7	5942.5	6725.4	7472.6	8255.5	9002.8

TABLE 10.10   Approximate Sand Usage and Labor Rates on Sandblasting

Condition of surface to be prepared	Surface preparation to be achieved							
	SSPC-SP-5 white		SSPC-SP-10 near-white		SSPC-SP-6 commercial		SSPC-SP-7 brush	
	$S,^a$ kg	$L,^b$ m^2	$S,^a$ kg	$L,^b$ m^2	$S,^a$ kg	$L,^b$ m^2	$S,^a$ kg	$L,^b$ m^2
Loose mill, fine powder rusting	43	11.6	35	14.4	17	23.2	9	37.2
Tight mill scale and little rust	52	8.4	44	10.7	26	18.6	9	33.4
Existing coating which is failing	87	6.5	70	8.4	35	13	7	29.7
Badly pitted steel	109	4.6	87	6.5	52	9.3	7	27.9

^{a}S denotes abrasive usage per square meter.
^{b}L denotes labor production per hour.
NOTE: These figures are approximate and reflect work on large areas such as tanks or large I-beams.

Sandblasting operations should be carried out by following careful procedures. In particular:

- Strict adherence to the manufacturer's operating instructions
- Careful observance of all safety regulations and proper use of safety equipment
- Nozzle held at proper altitude and angle for surface area being cleaned
- Removal of any and all dust after blasting, with vacuum or brush, to avoid coating adhesion problems
- Careful attention to inside corners, pits, edges, bolt threads, nuts and bolts, welds, and overblasting ("overblasting" means blasting near an area already coated and allowing the abrasive to hit the coated areas)

10.2.4   Coating application

The goal in applying a coating is to place a continuous, uniform film of sufficient thickness that is securely bonded to the substrate. Coating application is initiated after surface preparation has been satisfactorily completed and inspected. The prime coat should be applied promptly after the surface is cleaned, because cleaned steel is easily rusted if it is not protected. The coating must be applied under good dry conditions. If the temperature is low enough that drops can occur and cause condensation on the surface, the application should not be undertaken.

Coating specifications usually designate that coating work should not be undertaken unless the temperature is 3 to 6°C (5 to 11°F) above the dewpoint (the temperature at which condensation will occur at a given pressure and relative humidity). Application can be by brush, roller, spray, or dip.

Brush.   Application of coatings by brush is one of the oldest and most widely used procedures. Maintenance and repair work are frequently done by brush, particularly if the job is small. Most alkyds and a limited number of other products with good wetting properties can be applied by brush. Some manufacturers recommend that the prime coat be brushed on to ensure that it has been worked well into the surface. This is a desirable procedure when the surface is rough or when the surface preparation is poor.

When brushing is called for, the primer is often thinned to improve flow-out on the substrate. If the surface is rough or pitted, brushing will help work the coating into the roughened surface. The ability of a primer to protect is directly related to surface wetting. If areas of the surface are not wetted by the primer, they will be subject to corrosion. When conditions dictate that a brush be used for touchup work, tacky materials should be applied by filling the brush liberally and quickly applying the material to the surface with a minimum amount of brush-out. One advantage of application by brush is that there is only 4 to 8 percent loss of material as opposed to 20 to 40 percent by conventional spray.

Brushes must be of best quality and appropriate style. Brushing should be done in such a manner that a smooth coat as nearly uniform as possible is formed. The coating must be worked well into all crevices and corners and be applied without runs or sags and with a minimum of brush marks.

In short, brushing is used by professionals only when it is specified or recommended by the coating manufacturer, when surface preparation is relatively poor, and for touchup work where spraying is unsuitable because of plant or area regulations.

Roller.   Roller application, like brushing, is fine for some materials and poor for others. It is generally used with alkyds and water-based (latex) materials over large areas such as tanks and buildings. Rollers are also called into play because of regulations against any spray fog. Fast-dry lacquer-type products will dry on the roller and do not dissolve in dipping. Others also show poor flow-out properties. Pickup of the undercoat is another problem which must be considered a possibility with rolling.

Roller application is faster than brushing, and it involves the same low material loss as contrasted with conventional spray. Particular care must be exercised when a roller is used on rough spots, pits, rivet heads, edges, corners, welds, and the like to make certain they are properly covered. Like brush application, roller coating is especially useful where spraying cannot be used because of spray fog and possible flammability of solvent.

Spray. Protective coatings are applied by either conventional or airless spray in almost all industrial work including many relatively small jobs. Wastage of materials is higher because of the fine atomization of the coating. The advantages of spray application are coverage speed and application uniformity. Newer spray methods include hot airless spraying and electrostatic spraying, which combine speed with low material loss. Hot airless, as the name implies, is simply a heated airless spray application. Electrostatic techniques involve putting a charge on the spray particulate and an opposite charge on the workpiece being coated so that the coating is electrostatically attracted to the part. Spray coating has the obvious advantage of relatively easy application in corners, cracks, and crevices. The newer spray methods practically eliminate overspray and spray fog.

The spray gun, in simple terms, is a tool for using compressed air to atomize products and apply them to a substrate. Air and coating enter the gun through separate passages, which brings the two together and mixes them at the air cap. Most spray guns have two adjustments. One is for regulating the amount of fluid which passes through the gun when the trigger is pulled; the other is to control the amount of air passing through the gun. The latter determines the fan spray width. The internal-mix gun mixes air and material inside the cap before expelling them. It is generally used when only low air pressure is available or when slow-drying products which do not contain abrasive particles are to be applied.

The external-mix gun, the type more commonly preferred, mixes and atomizes material and air outside the air cap and uses no air for atomization. Hydraulic pressure is used to atomize the fluid by pumping it at high pressures through an accurately designed small orifice in the spray nozzle, which controls the fluid flow and spray pattern. The air-operated hydraulic fluid pump multiplies many times the air input pressure to deliver material at desired spraying pressures. Various sizes of caps are required, since coatings do not atomize in the same way because of variations in coating pigment grind, viscosity, and other formulation features. Other types of spray guns include automatic, extension, and special.

Not all coating materials can be applied by airless spray. Many heavy, pigmented, and fiber-filled abrasive materials cause some problems and are difficult to spray. However, the method has considerable advantages; a major one is the almost complete absence of overspray. There is less tendency for the coating material to bounce, so there is less air contamination and coating porosity. Airless spray is ideal for large flat areas such as tanks and walls, where speed is an asset and good coverage is required.

The maneuverability and convenience of airless spray equipment, especially when the work is in a confined area, are other advantages. When heavy coating thickness is needed, airless spray is the desirable procedure. Table 10.11 shows material losses by various application methods.

Dip. A coating can be applied by dipping; most galvanizing is done by dipping. Parts which have a high surface area–to–weight ratio, and which are relatively small, such as gratings, ladders, and fencing, can be dip-coated. The quality of a dip coating depends on the surface preparation and the application. The surface of the part being coated must be clean and free of contamination, and it must have the right profile for the coating being applied. Most coating systems can be applied by dip; they include alkyds, coating systems which cure by evaporation (e.g., vinyls and acrylics), and coating systems which cure by reaction (e.g., epoxies and urethanes).

As stated above, the quality of dip coatings depends not only on the surface preparation but also on the application. The dip bath must be properly and continuously stirred and agitated so that the pigments and other coating components are uniformly dispersed in the bath. The temperatures of the part and the bath should be controlled to lie within the range recommended by the coating manufacturer. Coatings which cure by reaction have a limited application life that is dependent upon the temperature: the higher the temperature, the shorter the life. Care must be exercised not to apply coatings which have exceeded their effective application life.

Care must also be exercised in the application of coatings which cure by evaporation. Solvent must be added from time to time to make up for the solvent lost to the atmosphere by evaporation. Dipping work should be stopped during the application of solvents, and it

TABLE 10.11 Estimated Loss of Coating Materials During Application

Method	Loss, %
Conventional air	20–40
Hot-air spray	15–30
Airless spray	5–20
Electrostatic spray	5–15
Brush and roller	4–8

should not be commenced again until the added solvent is thoroughly and uniformly dispersed. Parts which are dipped are often "turned" while curing in order to avoid drips and runoff "strings" from the bottom edges. Heat and air movement may be used to increase the cure rate (shorten the cure time).

10.2.5 Coating coverage

The volume of a 3.785-liter (L) [1-gallon (gal)] container of paint is 0.0038 m^3 (231 in^3) according to the U.S. Bureau of Standards. If the paint is 50 percent solvent which volatilizes as the coating cures, the effective remaining paint for surface coverage (paint film) is 50 percent, or 0.0019 m^3 (115.5 in^3). Some paint products contain more than 50 percent solvent; others contain less; and there are a few that contain no solvent, i.e., are essentially 100 percent solids. If 0.0038 m^3 (230 in^3) of nonvolatile material is spread on a surface as a coating film $25.4 \text{ }\mu\text{m}$ (1 mil) thick, it will cover 150 m^2 (1604 ft^2).

If the coating contains 50 percent volatile solvent by volume, its surface-covering capabilities are reduced 50 percent. At 25.4-μm (1-mil) dry film coating thickness, it will cover 75 m^2 (802 ft^2). The coating as applied, containing 50 percent solvent by volume, will be "wet." It will be $50.8 \text{ }\mu\text{m}$ (2 mils) thick. As the solvent evaporates, its thickness will decrease. When all the solvent has volatilized (vaporized), the dry film thickness will be $25.4 \text{ }\mu\text{m}$ (1 mil). As mentioned earlier, coating thickness will vary from 38 to 76 μm (1.5 to 3 mils) for alkyds and up to 250 to 760 μm (10 to 30 mils) for heavy coal-tar coatings or mastics.

10.2.6 Inspection

Coating inspection work starts with the surface preparation. Details relating the degree of surface cleanliness to the surface preparation specified are found in the Steel Structures Painting Council standards and will not be discussed here. Coating application must be in strict accordance with the coating manufacturer's recommendations. Inspection of coating application, including touchup, repair, and maintenance work, will be centered around compliance with these recommendations.

Inspection includes such items as making sure that coatings with dense pigments (such as zinc-rich coatings and red lead primers) are continuously stirred during application, measuring and recording both air and surface temperatures and the humidity, determining the shelf life of the coating and using the coating within its shelf life, and becoming thoroughly familiar with the coating product being used and the manufacturer's application procedures.

10.2.7 Coating repair and maintenance

During construction there will be damage to the coating system. Also, after erection, welding areas (weld lanes) will require surface preparation and coating. This section will consider touchup and repair of such areas before going into maintenance coating work. Priming is frequently done before erection. After erection, the weld lanes are cleaned and primed and the coated areas damaged during erection are repaired.

Inorganic zinc is frequently used as a preconstruction primer because of its toughness (resistance to damage) and its excellent corrosion resistance. Organic zinc-rich primers (such as epoxy–zinc-riches) are often used for weld lanes and repair work after erection. The organic zinc-rich coatings have better adhesion qualities and are more tolerant of imperfect surface preparation than the inorganic zinc coatings are. Other primers utilized for weld lane touchup work include those from all of the generic coating classes: alkyds, vinyls, epoxies, and universal (a primer that usually contains a phenolic resin and can be covered with any topcoat without lifting).

Final topcoating after erection is considered good practice. Topcoating before erection is infrequently done because of the problem of damage during erection. Damaged coatings should be removed by abrasive grinding or wire brush procedures. The edges of the adjacent nondamaged coated areas should be feathered and slightly roughened for good adherence of the repair coating. The repair coating system should be applied and should overlap the previous coating about an inch or two.

Touchup and repair topcoats are usually semigloss; in most industrial coating systems, a semigloss topcoat rather than a high-gloss topcoat is used. Semigloss colors are much easier to match (in texture and color). High-gloss products tend to show up any slight variations of color or texture.

Maintenance of a coating system is essential for long life and minimum cost. In general, all critical areas, including splash zones and the like, should be inspected at least once a year. Regardless of type and service protective coatings will eventually wear or erode away. They require periodic touchup and restoration. The effective service life of a coating system will vary widely with the service condition, coating type, surface preparation, and application. No fixed maintenance program can be developed for good coating life unless accurate periodic inspection records are kept. Determining the most effective time for maintenance work will require careful analysis of the periodic examination records.

The expense of corrective maintenance work can be greatly reduced by proper timing, as well as by the choice of the proper repair materials. A good maintenance program will initiate repair work before coating failure. It is simple and inexpensive to apply a fresh topcoat; it is costly and time-consuming to remove a failed coating, clean the surface, and reapply a coating system (primer and topcoat). In choosing a repair topcoat, it is important to select a product which will adhere to and is compatible with the old coating. It is recommended that product suppliers be consulted when such choices must be made. When the coating failure is greater than 10 percent or there is widespread delamination, cracking, or incipient surface rusting, it is best to remove the old coating completely and recoat. The best maintenance program is a good inspection program coupled with the application of a fresh topcoat before excessive failure starts. A good weathering topcoat, such as an acrylic, can add many years of life to a coating system at minimum cost.

10.2.8 Specifications

A satisfactory coating system and protective coating program begin with the specifications, yet many coating specifications are vague and wordy. All too frequently, the importance of developing a structure is given so much attention that the architect-engineer underemphasizes the necessity for clear-cut coating specifications to protect the structure. Writing appropriate coating specifications will do more than almost any other phase of coating work to achieve long-life economical protection in any harbor facility. The specifications play conspicuous roles in both initial cost and cost per square meter per year of service life.

Many specifications, primarily because of efforts to streamline the job or put specifications in the category of routine work, open loopholes for contractors and result in jobs that are poorer because of misinterpretation. The coating specifications should be as brief as possible without forfeiting any of the primary objectives. They should be written in such a manner that the contractor can present his bid wisely, efficiently, and competitively and the coatings will be applied in strict accordance with all requirements. A good coating specification should contain the following:

- Scope of the work (a general description of the job, its location, and work to be done)

- A detailed list of the products, preferably by brand name (including those that may have been evaluated), or military (MIL) specifications when required

- A full description of the methods and equipment required to do all phases of the work (surface preparation, application, and inspection)
- Identification of the material supplier's responsibility to supply quality coatings that meet specific performance requirements or are manufactured to meet specific formulation requirements
- Safety requirements to ensure proper procedures in sandblasting and application work
- The quality of workmanship requirements spelled out to include cleanup work, work logs expected, inspection acceptance, and prevention of contamination of nearby structures
- A coating schedule that shows exactly what is to be applied, where, when, and how
- Inspection requirements to show what inspection measurements and reports are required

10.2.9 Safety

The combustible and toxic nature of protective coatings must be considered during both application and the cure period. Special precautions are required when coatings are applied in confined or enclosed areas. Containers should always be opened in a well-ventilated area and kept away from open flames or sparks. Proper ventilation must be provided to prevent possible buildup of explosive or toxic materials. Inhalation of toxic fumes and dusts, whether inside or outside, must be avoided. Suitably designed air masks should be worn while spraying is in progress. The manufacturers' safety requirements must be studied and followed.

10.3   Cathodic Protection

The simplest way to make the subsurface structure all-cathode is to install galvanic or sacrificial anodes in the electrolyte and electrically connect them to the substructure. The anodes may be of magnesium alloy (usually used in soil), aluminum alloy (usually used in seawater), or zinc alloy (used in either soil or water). Alloys are used because, in the cases of the three metals mentioned, they result in greater efficiency (i.e., one or more of higher output current, higher output voltage, or longer life) than is possible when pure unalloyed metals are used.

10.3.1   Types of anodes

Aluminum or zinc anodes are used in seawater or fresh water. Magnesium also can be used, but its cost is higher and its life per newton is shorter. A zinc or magnesium anode for use in soil is installed with a chemical backfill completely surrounding it. The backfill material should be uniform to provide current efficiency. When the backfill is not uniform, the anode will supply increased current where the backfill has low resistivity and therefore waste more rapidly in that area. Also, a uniform backfill will aid in keeping the anode continuously moist and will prevent anode contact with adverse materials that may be in the soil. Low backfill resistivity has the same effect as increasing the anode size and thereby decreasing effective anode resistance to earth.

Magnesium anodes in prepackaged backfill are available in many sizes; the anode with surrounding backfill is contained in a cloth sack or porous tube. The backfill commonly supplied is 75 percent gypsum, 20 percent bentonite alloy, and 5 percent sodium sulfate, but other mixtures usually are available on request. Magnesium anodes are normally furnished with a 3-m (10-ft), No. 10 or No. 12, AWG copper lead wire with type TW insulation.

Zinc anodes are normally not sold in prepackaged backfill or with leads attached because of the difficulty of handling the much heavier anodes. If they are not provided with prepackaged backfill, galvanic anodes in soil should be completely surrounded by well-tamped chemical backfill to ensure best utilization of anode material. Chemical backfill for either magnesium or zinc anodes is shown in Table 10.12. No backfill of any kind is used with anodes in water.

Zinc and aluminum anodes, without backfill, are particularly useful in seawater where the low-resistivity electrolyte permits good current output in spite of the relatively low anode-driving potential. Magnesium is not suitable for seawater use because of the low efficiency caused by the tendency of magnesium to self-corrode in the low-resistivity electrolyte.

Whether or not a galvanic anode system will work depends on the electric circuit resistance and on the current required for protection. The circuit resistance is determined almost entirely by the resistivity of the electrolyte environment. Galvanic anodes work best in low-resistivity electrolytes such as seawater with a resistivity of 16 to 20 ohm-centimeters ($\Omega \cdot cm$). Good performance usually is obtained in electrolytes up to 1000 $\Omega \cdot cm$. Anodes have been made to work in resistivities as high as 2000 $\Omega \cdot cm$ when conditions permit use of the limited output current. The output current is necessarily limited by the relatively low driving potentials of all types of galvanic anodes. Table 10.13 shows approximate data for some common anode materials.

Current required for protection is based on current density requirements, which are usually stated in milliamperes per unit of area. Bare

TABLE 10.12   Chemical Backfill for Galvanic Anodes in Soil

Backfill mix	Gypsum (CaSO$_4$) Hydrated, %	Molding plaster (plaster of paris), %	Bentonite clay, %	Sodium sulfate, %	Approx. resistivity, $\Omega \cdot cm$
—^{a}	25	—	75	—	250
—^{b}	50	—	50	—	250
—^{c}	—	50	50	—	250
—^{d}	75	—	20	5	50

^{a}Useful in low-soil-moisture areas. Utilizes moisture-holding characteristic of bentonite clay.
^{b}Usually used with zinc anodes.
^{c}Useful with zinc or magnesium in wet or marshy soils. Prevents rapid migration of backfill from anode surface.
^{d}Low resistivity; useful in high soil resistivity to reduce anode resistance to earth. Type usually furnished with prepackaged magnesium anodes.

TABLE 10.13   Approximate Data for Common Galvanic Anode Alloys

	Zinc alloy in soil^a or seawater (Mil Spec. A-18001)	Magnesium alloys in soil^a Standard (H-1 alloy)	Magnesium alloys in soil^a High potential (Galvalum alloy)	Aluminum alloy in seawater (Galvalum alloy)
Specific gravity	7.14	1.94	1.94	2.77
Kilonewtons per cubic meter	69.1	19.0	19.0	—
Theoretical ampere-hours per newton	84	225	225	302
Current efficiency, %	95^b	50^c	50^c	95^b
Actual ampere-hours per newton	79^b	112^c	112^c	288^b
Actual newtons per ampere-year	111	76^c	76^c	31
Solution potential, V	− 1.1^d	− 1.55^d	− 1.80^d	− 1.1^e
Driving potential, V	− 0.25^f	− 0.60^g	− 0.85^g	− 0.25^f

^{a}Anodes in suitable chemical backfill.
^{b}Current efficiency of zinc and aluminum reasonably constant over wide range of output.
^{c}Current efficiency of magnesium varies with current density. Figures given are for approximately 300 mA/m^2 of anode surface.
^{d}To CuSO$_4$ reference cell in neutral soil.
^{e}To CuSO$_4$ reference cell in seawater.
^{f}Solution potential, the equilibrium potential of a metal exposed in a given solution, minus the polarized potential, the potential of a metal after the flow of sufficient current to come to equilibrium (− 0.85 V to CuSO$_4$ reference electrode).
^{g}Solution potential minus polarized potential of protected structure minus approximately 0.1 V for anode polarization in service.

steel in average soil or quiescent seawater usually is considered to be about 10 milliamperes per square meter (mA/m^2) [1 milliampere per square foot (mA/ft^2)]. Any coating present, regardless of quality, can produce a drastic reduction in the current density requirement. Actual current required for a given substructure depends, of course, on the subsurface area exposed to the contacting soil or water. In the case of bare steel, 93 m^2 (1000 ft^2) probably would require at least 1 ampere (A) for protection. Approximate current density requirements for wrap-coated steel pipe can be estimated by reference to Table 10.14. The values are for guidance estimates only and may vary widely depending on specific conditions such as moving seawater.

The actual current requires that a calculation of the total area, in square feet, be made. Based on the current requirement, a decision on whether a galvanic system or an impressed current system should be used can be made.

Impressed current (inert) anodes. The driving potential of galvanic anodes may not be high enough to provide sufficient current for effective cathodic protection. That is particularly true when the subsurface structure to be protected is surrounded by an electrolyte of high resistivity such as that frequently encountered with steel in fresh water or soil. Also, there are applications requiring higher current density than can be delivered economically by galvanic anodes; examples are protection of the external bottom surface of a large oil storage tank and either soil or water sides of a steel sheet pile bulkhead. To be free from the limitations of galvanic anodes, an external power source can be used to provide "impressed" current for protection. Here the required direct-current driving voltage and current are limited only by the availability of an external source of power. Various power sources can be used; the most common one is commercial alternating-current power connected to a low-voltage transformer-rectifier combination (usually called simply a rectifier) to provide the required direct current with easily adjustable output voltage. That provides the means to

TABLE 10.14   Current Requirements for Coated Steel or Wrapped Pipe

Age of wrap or coating	Approx. current density required, mA/m^2
1930–1957	1.1
1950–1960	0.54
1960–1970	0.32
1970–present	0.11

counteract the flow of corrosion current to make the substructure "all cathode." A simplified diagram as applied to a pipeline is shown in Fig. 10.7.

Impressed current anodes differ from galvanic anodes in two important respects: (1) The galvanic potential difference between anode and protected substructures is of no importance. (2) The impressed current anode should be as inert as possible, i.e., have a very low consumption rate for long life. Consumption rates for various metals, including those commonly used for impressed current anodes, are shown in Table 10.1. Carbon or graphite, high-silicon cast iron, magnetite, lead-silver, and platinum all have critical current densities above which rapid consumption may occur. The consumption rates listed for these materials assume operation below critical current densities.

As with galvanic anodes in soil, impressed current anodes in soil should be completely surrounded by backfill material. Impressed current anodes in water do not require backfill; those in soil require a carbonaceous backfill well-tamped to eliminate air pockets and to provide the best possible electrical contact with both the anode and the soil. Two functions are thereby served:

- The very low resistivity of the backfill material has the effect of increasing the anode size and consequently reducing resistance to the surrounding soil.

- Most of the current flows to the backfill from the anode by direct electrical contact, so that most of the electrolytic consumption is at the soil contact with the outer surface of the backfill column.

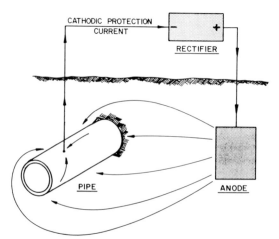

Figure 10.7 Diagram of simple impressed current cathodic protection.

Resistivity of carbonaceous backfill should not exceed 50 $\Omega \cdot$ cm. There are at least three materials in present use: coal coke breeze, calcined petroleum coke breeze, and manmade or natural graphite flakes or particles. All of these are basically carbon in low-resistivity form. The term "breeze" is loosely defined as being a finely divided material. For backfill use, specific sizes are obtainable. Coke breeze should be procured by specification; particle size and resistivity are most important. Particle size should not exceed 9.5 mm (0.375 in) and not more than 10 percent dust should be included. Petroleum coke breeze must be calcined to produce resistivity of 50 $\Omega \cdot$ cm or less. Graphite flakes should not be used because of possible gas blockage problems (accumulation of gas around the cathode from the cathode reaction).

10.3.2 Comparison of anode types

Cathodic protection for a given subsurface structure can be provided by either galvanic or impressed current anodes as long as respective limitations are recognized. Some of the more important characteristics of each method are listed in Table 10.15.

10.3.3 Connections between anodes and structures

In the case of galvanic anodes, the connecting wire from the anode is part of the cathode and bare areas resulting from cut or broken wire insulation or poorly insulated splices will be part of the protected structure and suffer no electrolytic damage. The opposite is true of impressed current anode systems. If any of the insulation is less than

TABLE 10.15 Comparison of Galvanic and Impressed Current Protection Systems

Galvanic	Impressed current
No external power required	External power source required
Relatively low installation and maintenance costs	Higher installation and costs
Frequently requires no additional right of way	Applied voltage and current can easily be varied
Adjusts current output as structure potential varies (especially zinc anodes)	Protects larger and more extensive structures
Severely limited current output	Suitable for high-resistivity electrolytes
Useful primarily in low-resistivity electrolytes	Monthly power bill
Interference with foreign structures usually nonexistent	Can cause interference problems with foreign structures

perfect, current discharge into the surrounding electrolyte (soil or water) will occur and the wire will corrode too quickly, thereby producing failure of part or all of the system. This means that top-quality insulation must be used for all buried or immersed anode leads and header cables. Insulation should be the 600-volt (V) type and be suitable for direct-burial service. High-molecular-weight, high-density polyethylene has a good record for satisfactory use.

Insulation quality of all subsurface electrical connections is equally important for the same reasons. Satisfactory electrical connection methods for copper wire include soft soldering, powder welding (such as Cadweld), silver soldering, phoscopper brazing, crimp-type couplings, and split-bolt couplings. The first four methods, if carried out properly, will provide metallurgical joining and will be permanently low in resistance. Mechanical methods such as the last two mentioned, again if carried out properly, will also be satisfactory. Joint insulation should be of such quality as to at least equal the electrical insulating qualities of the wire insulation.

Acceptable insulation methods include cast epoxy as well as various tapes. There are several manufacturers of cast joint insulation. Details of joining and insulating can be obtained from cathodic protection material supply house catalogs. In all cases, satisfactory performance life depends on the proficiency of the people doing the work. Careful inspection of all phases of impressed current anode installation is mandatory.

10.3.4 Criteria for cathodic protection

Earlier it was stated that, if electric current is caused to flow from an external source through a common and contacting electrolyte into a subsurface structure, corrosion current and attendant corrosion will be stopped provided the external current flow counteracts the corrosion current at all parts of the substructure surface. In this case the substructure becomes all-cathode and is protected against corrosion. That is, if all areas on a corroding substructure are polarized (or made equal) to the same open-circuit potential, corrosion will be impossible because there will be no potential difference between anode and cathode areas and no corrosion current can flow. In practice, the potential applied to the surface, called the polarization potential, must equal or exceed the open-circuit potential of the most anodic area in order to stop all corrosion.

To determine when or if the open-circuit condition exists, there must be some way to measure the potential between the protected substructure and its contacting electrolyte environment. Based on this concept, potentials should be measured directly across the interface between the substructure and its environment. That is relatively simple with marine substructures in seawater, but it is seldom feasible when the work is with substructures in soil such as pipelines or buried cylindrical storage tanks. Common practice with buried substructures is to measure the potential between the substructure and the soil at the surface directly above the substructure. The measured potential includes polarization potential plus a potential, usually called *IR* drop, caused by current flowing through a part of the resistance between the structure and the external anode installation.

Potential measurement apparatus. Since potential measurement is a useful approach to determining if corrosion is present or absent, potential measurement methods should be considered. The connection to the substructure usually is easily made by direct contact or by a suitable test wire. The actual measured potential will vary (sometimes drastically) with the method used to contact the environment. To get reproducible results, this contact must be made through some stable and reliable reference. The method in common use for measurements of substructures in soil (and frequently in water) environments involves a copper–copper sulfate half-cell reference electrode contacting the electrolyte. The potential may be referred to as above or as "copper sulfate electrode," "copper sulfate reference," "copper sulfate half cell," "$CuSO_4$ electrode," or "CSE." The working parts of a copper sulfate electrode are shown in Fig. 10.8, along with an equivalent circuit to illustrate the half-cell concept. A silver–silver chloride electrode is similar: Silver metal is in equilibrium with a 0.1 normal (*N*) solution of silver chloride with a porous plug contacting the electrolyte.

There are a few precautions to be observed when any reference electrode is used. Care must be taken to prevent contamination of the fluid in the electrode if the potential is to remain constant. Such contamination is possible when making potential measurements in a fluid electrolyte such as seawater. With a copper sulfate electrode, the observed potential will vary slightly with temperature; there will be a positive gradient of 0.9 millivolt per degree Celsius (mV/°C) [0.5 millivolts per degree Fahrenheit (mV/°F)] up to about 50°C (120°F) where hydrated copper sulfate begins to change structure. One authority advises correction of all copper sulfate electrode potential readings to 25°C (77°F). Such precision usually is not required for normal fieldwork, but it is good practice, before measuring a potential, to invert the electrode two or three times to equalize internal temperature.

All reference electrodes are subject to polarization to some degree when they are placed in an electric circuit with current flowing. This means that any appreciable current drawn by the potential measuring circuit (voltmeter) would result in lower potential readings. In general, the accuracy of potential measurements will increase as the mea-

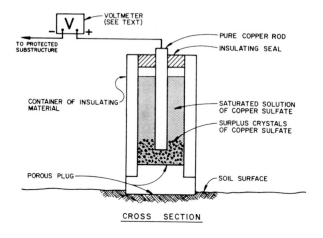

CROSS   SECTION

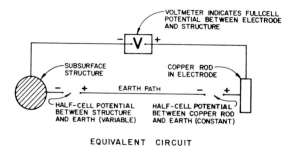

EQUIVALENT   CIRCUIT

Figure 10.8   Copper sulfate electrode test circuit.

suring circuit current approaches zero. The most accurate voltmeter suitable for field or laboratory use is the standard-cell potentiometer with a suspension-type galvanometer. When balanced for reading of potential, the reference cell current is zero. The most easily handled and easily operated instrument for general field use, ashore or afloat, is the electronic type, preferably with an input resistance (sensitivity) of 10 megohms (MΩ) or more. This ensures a reference cell current of a fraction of a microampere. Conventional voltmeters of 1000 or 5000 ohms per volt (Ω/V) are not suitable for these measurements.

When a suitable voltmeter is connected between the substructure and the reference electrode, the measured potential is a combination of the potential between the reference electrode and electrolyte (soil or water) and the potential between the substructure and electrolyte. The substructure-to-earth half-cell potential is the variable of interest; the reference half-cell potential is considered constant for practical engineering purposes. The copper sulfate electrode has been found to be practical for field use because its half-cell potential is reasonably constant over a wide range of conditions. The actual value of the half-cell potential is not important; the total cell potential, as measured, is the one used in engineering practice. It is the structure-to-electrolyte potential. Several discussions of the copper sulfate electrode, covering theory, history, and development, are available in the literature (National Association of Corrosion Engineers, 1979).

It has been established that the most anodic areas to be expected on a freely corroding steel structure in moist soils and waters will show a potential of about -0.8 V measured to a copper sulfate electrode contacting the electrolyte as close as possible to the anodic area. Allowing for some variation in potential of the most anodic areas, the value of -0.85 V to a copper sulfate electrode contacting the electrolyte has been adopted as a practical indication of satisfactory protection. Wide and varied experience in many environments has indicated the accuracy of this criterion for practical field use.

Although the copper sulfate electrode is by far the most common, other electrodes are in use; they include calomel (usually the saturated type), silver–silver chloride, and occasionally pure zinc. Other common metals are not sufficiently stable for reference electrode use. The calomel electrode, although very stable, is more adapted to laboratory work than field use because of its largely glass construction. The silver–silver chloride electrode also is quite stable, and it may be encountered frequently in marine operations. Pure zinc (special high grade, 99.99 percent pure) is occasionally used as a reference electrode but is subject to variations of as much as 50 millivolts (mV), making it suitable only for approximate values. Table 10.16 shows the potential readings for these common reference electrodes compared to the reading for copper sulfate electrodes at 25°C (77°F).

Measurement techniques. NACE Standard RP-01-69 (1976 Revision), "Control of External Corrosion on Underground or Submerged Metallic Piping Systems," applies equally to any buried or immersed struc-

TABLE 10.16 Potential Readings for Various Reference Electrodes

Type of reference electrode	Structure-to-electrode reading equivalent to -0.85 V to copper sulfate electrodes, V	To correct structure-to-electrode reading to equivalent reading to copper sulfate electrodes, V
Calomel (saturated)	-0.778	Add -0.072
Silver–silver chloride (0.1 N Kel solution)	-0.840	Add -0.010
Pure zinc (special high grade, 99.99% pure)	$+0.25^a$	Add -1.10^a

[a]Based on zinc open-circuit potential of -1.10 V to copper sulfate electrode.

ture. The following information and criteria for cathodic protection are quoted from the above publication:

6.2.3 The criteria in Section 6.3 have been developed through laboratory experiment or empirically determined by evaluating data obtained from successfully operated cathodic protection systems. It is not intended that the corrosion engineer be limited to these criteria if it can be demonstrated by other means that the control of corrosion has been achieved.

6.2.4 Voltage measurements on pipelines are to be made with the reference electrode located on the surface as close as practicable to the pipeline. Such measurements on all other structures are to be made with the reference electrode positioned as close as feasible to the structure surface being investigated. The corrosion engineer shall consider voltage (IR) drops other than those across the structure-electrolyte boundary, the presence of dissimilar metals, and the influence of other structures for valid interpretation of his voltage measurements.

6.2.5 No one criterion for evaluating the effectiveness of cathodic protection has proven to be satisfactory for all conditions. Often a combination of criteria is needed for a single structure.

6.3.1.1 A negative (cathodic) voltage of at least 0.85 V as measured between the structure and a saturated copper–copper sulfate reference electrode contacting the electrolyte. Determination of this voltage is to be made with the protective current applied.

6.3.1.2 A minimum negative (cathodic) voltage shift of 300 mV, produced by the application of protective current. The voltage shift is measured between the structure surface and a stable reference electrode contacting the electrolyte. This criterion of voltage shift applies to structures not in contact with dissimilar metals.

6.3.1.3 A minimum negative (cathodic) voltage shift of 100 mV measured between the structure surface and a stable reference electrode contacting the electrolyte. This polarization voltage shift is to be determined by interrupting the protective current and measuring the polarization decay. When the current is initially interrupted, an immediate voltage shift will occur. The voltage reading after the immediate shift shall be used as the base reading from which to measure polarization decay.

Paragraphs 6.3.1.4 and 6.3.1.5 are not quoted, since they apply only to specific situations not generally encountered.

The U.S. Department of Transportation has issued "Regulations for the Transmission of Natural and Other Gas by Pipeline, Part 192, Title 49," provisions of which are now in effect. Subpart I contains requirements for corrosion control. Criteria for protection are, in effect, identical with those in the NACE publication quoted above.

The voltage drop between two points in a medium is equal to the current I flowing between the points multiplied by the resistance R of the medium. For this reason, voltage or potential drops caused by current flowing through resistive electrolytes are called *IR* drops.

The NACE Standard RP-01-69, 1976 Revision, quoted above states, in paragraph 6.2.4, that *IR* drops are to be "considered." Unfortunately, there are still many workers in the field who ignore the *IR* drop contribution to potential measurements and continue to record structure-to-electrolyte potentials with no consideration of *IR* drop. In such cases the recorded potential will always be higher (more negative) than it actually is. If a reading of −0.85 V was recorded, indicating protection according to the criterion given in paragraph 6.3.1.1, the true potential, after deducting *IR* drop, may well be substantially below protective levels. The amount of error will depend primarily on electrolyte resistivity. Examples are shown in Table 10.17.

As can be seen in Table 10.17, the only time that *IR* drop can be ignored safely is when potentials are measured in seawater or a similar electrolyte with a resistivity less than 50 Ω • cm. In seawater the *IR* drop is usually negligible if the reference cell is within 1.5 m (4.9 ft) of the substructure.

If the cathodic protection current ceases (is turned off or is otherwise interrupted), the *IR* drop ceases instantly. The potential of the protected structure (polarization potential) decays at a relatively slow rate compared to the *IR* drop, depending on several factors, and, of course, it is free from *IR* drop. If this potential can be measured while the current is momentarily interrupted, it will be close enough to the true polarization potential for normal engineering purposes. This potential is usually known as the OFF or INSTANT OFF potential as con-

TABLE 10.17   Electrolyte *IR* Drop, mV, 1.5 m from Bare Pipe

Electrolyte resistivity, Ω • cm[a]	Nominal pipe diameter, in	*IR* drop, mV
50	1	0
50	5	1
50	15	2
50	30	3
5,000	1	33
5,000	5	110
5,000	15	225
5,000	30	331
50,000	1	328
50,000	5	1100
50,000	15	2250
50,000	30	3309

[a]Resistivity of seawater is generally about 20 cm. Current density = 10 mA/m^2.

trasted with the ON potential, which is read while protection current is flowing.

The relations among the various potentials involved are shown in Fig. 10.9. Prior to time of current OFF, the only potential measurable is the ON potential which includes the *IR* drop. After that point, the measurable potential is the rapidly decaying polarization one. The meter response curve shown is typical of a conventional high-resistance voltmeter.

In view of the preceding background on criteria, the criterion of −0.85 V (− 850 mV) with the current momentarily interrupted is recommended. With this criterion in use, the NACE criterion in paragraph 6.3.1.1 will be met. The criterion in paragraph 6.3.1.3 usually will be met also.

There are a few considerations. There should be provisions for reading the potential as soon as possible after interruption of the current and, to minimize loss of polarization, for keeping the interruption period as short as possible. A simple way to do that is by means of the circuit shown in Fig. 10.10. A single-pole double-throw microswitch is

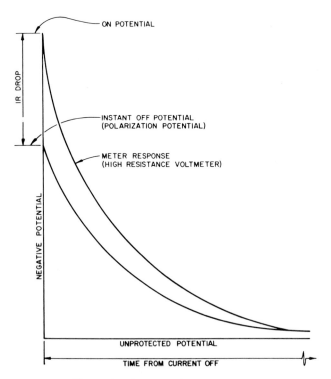

Figure 10.9 Time-related potentials upon interruption of cathodic protection current.

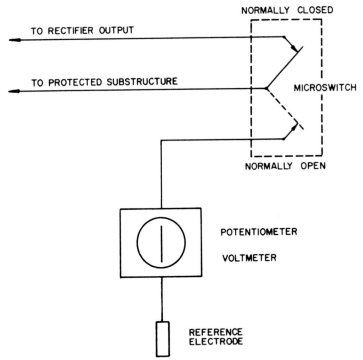

Figure 10.10   Circuit for OFF potential measurement.

used to interrupt the cathodic protection current and at the same time connect the potentiometer voltmeter and reference electrode to the structure under protection. The voltmeter is adjusted to some expected value, and the switch is cycled as rapidly as possible. This will interrupt the current for about a tenth of a second. The indicating meter needle will kick up or down the scale depending on the potentiometer setting. The voltmeter is adjusted in the direction of needle kick, and the switch is again cycled. This is continued until a narrow band, probably about 5 mV wide, in which no needle movement is observed, is found. The center of this band is very close to the true OFF potential.

Care should be taken that the added resistance of the microswitch circuit does not appreciably reduce the rectifier output current. If necessary, the microswitch circuit can be used to drive a low-resistance relay-switching circuit. This system is not feasible for general pipeline use because of the length of required leads and switching of one or more distant rectifiers. The system works well where the protected structure is relatively compact such as subsurface tanks, exterior tank bottoms, interiors of water tanks, oil or water well casings, short pipeline sections, and most marine installations.

As mentioned before, the time between interruption of current and measurement of potential should be as short as possible. Actual time before significant loss of polarization can vary from fractions of a second to hours. Any factor that quickly removes products of the cathodic reaction will accelerate depolarization. All things considered, current interruption, if done properly, gives the simplest and most reasonably accurate correction for *IR* drop. At least one supplier provides an electronic system for current interruption that uses a very short interruption time interval.

For general pipeline application where two or more rectifiers are used, synchronous interrupters are required. Two such interrupters for adjacent rectifiers will usually provide sufficiently accurate OFF potential data. Interrupters that use quartz timing and provide for accurate synchronization are available. A timing interval of 25 s ON and 5 s OFF will usually be satisfactory for pipeline work.

A possible source of error when measuring substructure potentials could be *IR* drops caused by the presence of stray currents of unknown magnitude in the contacting electrolyte. The best protection against such an error is to place the reference electrode as close as possible to the structure under test.

To summarize briefly, a cathodically protected substructure potential to copper sulfate electrode of -0.85 V or more indicates that protection exists under the following conditions:

- If the substructure is immersed in seawater or a similar electrolyte, with the rectifier on and the reference electrode contacting the electrolyte within 1.5 m of the substructure.

- If the substructure is buried in soil with the rectifier output momentarily interrupted and the reference electrode contacting soil above or immediately adjacent to the substructure.

10.3.5   Design

Ideally, design for cathodic protection should be a part of the original design of the substructure. Placing an existing substructure under cathodic protection can be expensive if the substructure was installed without any consideration of cathodic protection. Some measures to ensure low cathodic protection current requirements that should be included at the design stage of subsurface structures are:

- If coating is feasible, use a high-quality coating and apply it properly with minimal bare areas or discontinuities (holidays).

- Provide electrical isolation for large bare metal areas not to be protected, such as steel bulkheads, tank bottoms, electrical grounding systems, and local utility piping.

■ Be sure that there are no metallic contacts with other subsurface structures such as pipelines, pipeline casings, or cables.

A very important part of cathodic protection design is to check carefully for the presence of other subsurface structures in the area that are not to be protected or are owned by others. These are usually known as foreign substructures, and they may suffer from a side effect of cathodic protection known as interference. The point to emphasize here is that a proposed anode installation should not be placed closer than 90 m from an existing foreign substructure in either soil or water. This applies particularly to a proposed impressed current anode installation. Also check carefully for locations of other cathodic protection systems in the area, especially anode beds.

Anode choice. Whether to use galvanic anodes or impressed current anodes to protect a given substructure normally will depend on three factors taken singly or in combination:

■ Amount of protective current required

■ Resistivity of the contacting electrolyte (soil or water)

■ Electric power availability

Current requirement can be calculated or, preferably, determined by actual test. If it is calculated, the total exposed area, in square meters, must be determined. If the substructure is bare steel in normal soil or normal quiescent fresh or brackish water, a current density requirement of 10 mA/m^2 (0.9 mA/ft^2) can be assumed. Total current, in amperes, required would equal total exposed area in square meters multiplied by 0.01. If all or part of the exposed steel is coated, an assumption of coating quality must be made. For good-quality, relatively undamaged coating, it is customary to assume that the substructure is 1 percent bare for the coated part and figure the total current requirement accordingly. Thus, a coated substructure will require far less current for protection. The main difficulty with calculating current requirements is that the 10 mA/m^2 (0.9 mA/ft^2) figure for bare steel can vary over a wide range, especially in moving water, where it is almost always higher, sometimes by orders of magnitude. That is why an actual test for current requirement is preferable to reliance on calculated values.

Resistivity of the electrolyte (soil or water), in ohm-centimeters, should be determined by using the Wenner four-point method. Resistivity is a measure of electrolyte resistance to current flow. The higher the resistivity, the lower the output from a galvanic anode will be and the higher the voltage required to produce necessary current from an impressed current anode will be. Reported resistivity of an area should be an average of a number of measurements made at various locations in the area. Measurements should be made by a person qualified by experience in resistivity measurement for cathodic protection work. Table 10.17 shows electrolyte resistivity and *IR* drop for bare pipe.

Electric power may be from one of several sources. Commercial alternating-current power is the most widely used and is lowest in cost. Other sources are solar power (increasing in use as costs come down), thermoelectric, engine-driven generator, and wind. Along with resistivity, power availability may be a deciding factor in an anode-type choice.

As discussed in an earlier section, galvanic anodes installed in a soil environment should normally be completely surrounded by chemical backfill to provide uniform low resistance to the soil. Galvanic anodes installed in water do not require any backfill because the water will provide uniform contact with the anode. Impressed current anodes in soil also require backfill, but for different reasons. Here the backfill material is carbonaceous in nature. The anode makes electrical contact with the backfill, and ion transfer, with resulting loss of material, occurs at the backfill interface with the soil. The result is longer anode life and lower contact resistance because the backfill is of low resistivity and has much greater surface contact area with the soil than would be possible with the anode alone. Like galvanic anodes in water, impressed current anodes in water do not use backfill.

Calculating anode resistance.   Anode installation design requires that the effective resistance of the anode (or anode bed) to its environment be known or be calculated. Dwight's equation is generally used for single galvanic or impressed current anode resistance determinations in either water or soil. The equation is

$$R = \frac{\rho}{2L} (\log_e \frac{4L}{a}) - 1 \qquad (10.1)$$

where R = resistance of vertical anode or backfill to ground or water, Ω
L = length of anode, cm
a = radius of anode, cm
ρ = electrolyte resistivity, cm

Equation (10.1) can be simplified to:

$$R_v = \frac{0.012\rho}{L} \log \frac{35.3L}{D} \qquad (10.2)$$

where R_v = resistance of vertical anode or backfill column to ground
    (soil or water), Ω

$\quad L$ = length of anode or backfill column, ft

$\quad D$ = diameter of anode or backfill column, in

$\quad \rho$ = electrolyte resistivity (backfill, soil, or water), $\Omega \cdot cm$

Anode installed in backfill soil.   If a single anode is to be used in soil, the internal resistance (anode to backfill) should be considered. Internal resistance is not a factor when the anode is installed without backfill, as in seawater. The resistance will depend on the type of backfill used, which in turn will depend on whether the anode is to be the galvanic or the impressed current type. For an impressed current anode with carbonaceous backfill, a backfill resistivity of 50 $\Omega \cdot cm$ can be used.

Assume that a graphite anode 76 cm (3 in) in diameter and 1.5 m (5 ft) in length is to be centered in a vertical backfill column 203 cm (8 in) in diameter and 2.1 m (7 ft) in length (Fig. 10.11). Resistances are calculated for the anode and for the backfill column by using Eq. (10.2) with ρ = 50 $\Omega \cdot cm$. The difference between the two values represents the internal resistance of the anode (0.213 − 0.128 = 0.085 Ω). For most conventional impressed current anodes used singly, a figure of 0.1 Ω may be safely used.

When two or more anodes are to be connected in parallel, the internal anode resistance of the group becomes the single-anode internal resistance divided by the number of anodes. If the number of anodes to be parallel-connected is more than three or four, the internal resistance becomes negligible. The same method is used to calculate the internal resistance of a single galvanic anode (Fig. 10.12). Here the backfill resistivity and also the internal resistance will be higher.

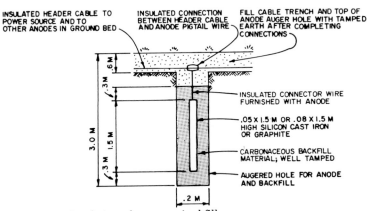

Figure 10.11   Anode in carbonaceous backfill.

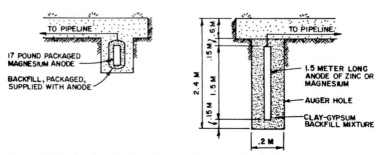

Figure 10.12   Anodes in other than carbonaceous backfill materials.

Anode installed in water.   For suspended vertical-anode installations in water, the anode should be so installed that its top is never less than 1.5 m (5 ft) below the water surface. Refer to tide data. The bottom of the anode should be 1.5 m (5 ft) above the channel or marine bottom. Header cables should be far enough above the water surface to ensure no water contact. In protected areas that is a minimum of 3 m (10 ft); in open-sea areas it is much greater. That requires anodes with leads large enough to permit connections to the header cable with no underwater splices. In some cases anodes may have to be installed in perforated nonmetallic pipe to avoid damage by water movement.

Resistance of anode groups.   Usually, anodes are used in groups, installed in a line, and connected in parallel to a header cable which in turn is connected to the substructure to be protected (galvanic anodes) or to the positive output of the power source (impressed current anodes). A calculation of the overall resistance of the parallel-connected group (usually termed "anode bed" or "ground bed") will be required. The effective resistance of the group, differing from normal parallel electric circuits, will not be equal to the resistance of one anode divided by the number of anodes in the group. That applies to marine installation also. Because of mutual interference between anodes, the resistance of the group will always be higher than the resistance determined by parallel electric circuit calculations; it will vary with the number of anodes, anode spacing, and electrolyte resistivity. Several methods have been used for calculating parallel anode resistance. One method utilizes the following equation:

$$R_v = \frac{0.001566\rho}{NL} \, 2.303 \log \frac{423.6L}{D}$$

$$+ \frac{2L}{S} \, 2.303 \log 0.656N \tag{10.3}$$

where R_v = resistance to electrolyte (soil or water) of the vertical anodes in parallel, Ω

ρ = resistivity of electrolyte, $\Omega \cdot cm$

N = number of anodes in parallel

L = length of anode (or backfill column), m

D = diameter of anode (or backfill column), m

S = anode spacing, m

The resistance of the anode group is the sum of R_v and the internal resistance of the group, i.e., the internal resistance of a single anode divided by the number of anodes in the group.

Equation (10.3) can be used to construct a chart of anode size and backfill size for a particular project. Such a chart, based on impressed current anodes 0.05 m (2 in) in diameter and 1.5 m (5 ft) long in 0.2-m (8-in) by 2.1-m (7-ft) backfill columns of 50 $\Omega \cdot cm$ resistivity, is shown in Fig. 10.13. A similar typical design chart for galvanic anodes is shown in Fig. 10.14. Note that both charts are based on electrolyte resistivity of 1000 $\Omega \cdot cm$.

Anode (or backfill column) resistance to electrolyte is directly proportional to electrolyte resistivity. For example, consider 15 anodes in parallel at 7.6-m (25-ft) spacing in 2200 $\Omega \cdot cm$ soil. Anode (in backfill) resistance in 1000 $\Omega \cdot cm$ soil, shown on the chart in Fig. 10.14, is 0.233 Ω. Resistance in 2200 $\Omega \cdot cm$ soil = 0.233 × 2200/1000 = 0.513 Ω. To this add the internal resistance of the group. From use of Eq. (10.2), the internal resistance of one electrode is 0.106 Ω, making the internal resistance of the group (0.106/15) = 0.007 Ω, a negligible amount. The total resistance is 0.520 Ω, but 0.513 Ω could be used safely.

Calculating resistance in cables. Cathodic protection design also requires a knowledge of the resistance of various sizes of copper wire or

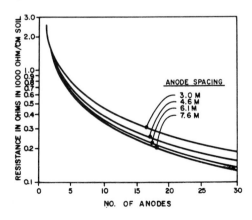

Figure 10.13 Chart of anode spacing (impressed current method).

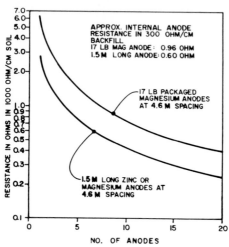

Figure 10.14 Chart of anode spacing (galvanic anode method).

cable most often used in anode installations. Resistance data and the common uses of some of the most common sizes are given in Table 10.18.

Calculating anode life. If the current output of a galvanic anode of any given weight is known, the approximate useful life can be calculated. The calculation is based on the theoretical ampere-hours per newton and the current efficiency of the anode material (see Table 10.13). Also involved is a utilization factor, which may be taken as 85 percent. That means the anode will require replacement when it is 85 percent consumed. The reason is that insufficient anode material remains to maintain a reasonable percent of the original current output.

Expressions for determining individual anode life for different materials are presented below. Efficiency and utilization factors are expressed as decimals.

- For magnesium:

$$\text{Life, years} = \frac{0.026 \text{ anode weight, N} \times \text{efficiency} \times \text{utilization factor}}{\text{anode current, A}}$$

- For zinc:

$$\text{Life, years} = \frac{0.0095 \text{ anode weight, N} \times \text{efficiency} \times \text{utilization factor}}{\text{anode current, A}}$$

TABLE 10.18 Copper Wire Resistance

Common use	Resistance of standard copper wires, $m\Omega/m,^a$ at 25°C (77°F)	
	Wire size (AWG)	$m\Omega/m$
Impressed current anode installations	4/0	0.167
	3/0	0.211
	2/0	0.266
	1/0	0.335
	1	0.423
	2	0.531
	4	0.850
	6	1.35
Galvanic anode installation	8	2.15
	10	3.41
Substructure test stations	12	5.41
	14	8.60
Instrument test leads	16	13.7
	18	21.9
	20	34.8
	22	55.8

Temperature Correction Factors		
Temperature		Multiply resistance at 250°C by:
°C	°F	
0	32	0.901
10	50	0.941
20	68	0.980
25	77	1.000
30	86	1.020
40	104	1.059

^{a}1 m = ¹⁄₁₀₀₀

- For aluminum (Galvalum II alloy) in seawater:

$$\text{Life, years} = \frac{0.035 \text{ anode weight, N} \times \text{efficiency} \times \text{utilization factor}}{\text{anode current, A}}$$

By using the values from Table 10.13 for theoretical ampere-hours per newton and current efficiency, along with an 85 percent utilization factor for the three anode materials, the above expressions can be simplified to:

Magnesium
$$L = \frac{1.105W}{I} \qquad (10.4)$$

Zinc
$$L = \frac{0.767W}{I} \qquad (10.5)$$

Aluminum
$$L = \frac{2.826W}{I} \qquad (10.6)$$

where L is the anode life in years, W is the anode weight in newtons, and I is the anode current in amperes.

As may be noted, Eqs. (10.4) to (10.6) can also be used for calculating anode bed life, where L is the anode bed life in years, W is the total anode weight in newtons (all anodes), and I is the anode bed current in amperes.

Deep-well anode beds. Some mention should be made of deep-well anode beds because in recent years they have attracted much interest in impressed current systems, primarily on pipelines. Such installations can be very useful if conditions permit. In the case of pipelines, the well may be in the pipeline right of way, which will make unnecessary any additional right of way for conventional surface-type anode beds. A deep-well anode bed, usually 60 to 120 m (200 to 400 ft) in depth, can be described as one in which the anodes are placed in remote earth by drilling straight down or by using an existing hole such as an abandoned water well. For a pipeline this accomplishes the same general result as that obtained by locating a conventional surface-type anode bed laterally several hundred feet from the pipeline. Advantages of a deep-well anode bed include small surface space needed (little or no additional right of way), probably fewer interference problems, and frequently lower anode-to-soil resistance than with conventional anodes. Disadvantages include the great difficulty or impossibility of repair, the need to prevent contamination of underground potable water sources, difficulty in determining soil resistivity at depths of several hundred feet, and expense of installation.

Application of calculation methods. Given the preceding background on design considerations, some examples follow to show how designs for several types of subsurface structures can be worked out. Professional consultation is advisable before plans for any cathodic protection installation are finalized. Each location has specific problems which must be recognized and considered if the installation is to be effective and reasonably troublefree.

10.3.6 Example project

A part of a waterfront facility consists of a steel sheet pile bulkhead (see Fig. 10.15 for the cross section). It has a typical seawater cross

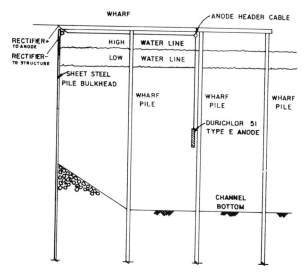

Figure 10.15 Cross section of the wharf in the example project.

section with the various zones of exposure. The water side of the bulkhead should be provided with cathodic protection as soon as possible to prevent further loss of steel caused by the corrosive action of the contacting seawater. Average water resistivity is 20 Ω • cm. The soil side of the bulkhead also is to be provided with cathodic protection at an early date. Average soil resistivity is 500 Ω • cm. A minimum of 20-year life for the waterfront facility is anticipated.

Conditions. The steel sheet pile bulkhead is 300 m (980 ft) in length. There is approximately 9 m (30 ft) of water-exposed steel or 2800 m^2 (30,000 ft^2), including a 0.6-m (2-ft) splash zone and about 3000 m^2 (32,000 ft^2) embedded in the sandy clay soil. The steel surface contacting the water is expected to require about 54 mA/m^2 (5 mA/ft^2) for protection. The embedded part, 3000 square meters, is expected to require about 22 mA/m^2 (2 mA/ft^2) for protection.

Calculations. Current requirement:

$$2800 \text{ m}^2 \text{ at } 54 \text{ mA/m}^2 = 150 \text{ A}$$

$$3000 \text{ m}^2 \text{ at } 22 \text{ mA/m}^2 = 65 \text{ A}$$

$$\text{Total current required} = 215 \text{ A}$$

Resistance. Anodes used will be Durichlor 51 Type E, 7.6 cm (3 in) in diameter, 1.5 m (60 in) long, to be suspended in the water under the wharf by using polypropylene rope. These anodes are rated at 4 A per anode for long-service life.

$$\frac{215 \text{ A required}}{4 \text{ A per anode}} = 54 \text{ anodes}$$

For safety margin use 60 anodes spaced evenly under the wharf about 5 m (16 ft) apart. By using Eq. (10.3), the anode resistance of each component of the example project is calculated to determine the total anode resistance:

60 anodes, averaging 20 Ω•cm resistivity	2 mΩ
Rectifier negative to center of header cable under wharf: 30 m No. 4/0 copper cable in conduit	5 mΩ
Rectifier positive to bulkhead: 30 m No. 4/0 copper cable in conduit	5 mΩ
Header cable under wharf, center connection, effective length, 155 m No. 1/0	51 mΩ
Bulkhead resistance	Negligible
Total resistance	63 mΩ

Voltage requirement. The required rectifier voltage is obtained by using Ohm's law: $E = IR$, or required voltage is equal to total current multiplied by total circuit resistance, or $E = 215 \times 63 = 13,500$ mV (13.5 V). For best results, a current and voltage requirement test, including the cables to the rectifier location, should be made after the anode installation is complete. A direct-current welding generator capable of furnishing the above current and voltage should be used. For test purposes, complete polarization is not required. However, current should be applied for several days, and potential measurements should be made from the bulkhead to a reference electrode in the water within 1.5 (5 ft) of the bulkhead (to minimum IR drop in the readings) at several locations and at several depths from surface to bottom. Readings of -0.85 V to a copper sulfate reference electrode or -0.84 V to silver–silver chloride reference electrode indicate adequate protection.

General. Woven grounding straps bolted to the sheet pile may also be used if they are protected against the environment. The point must be made that the connections must be flexible. A sheet pile bulkhead may deflect enough to break welds to a rigid 25-m (1-in) steel rod.

Some comments should be made regarding the example. A continuous 25-mm- (1-in-) diameter steel rod should be welded to each sheet pile in two places for the full length of the bulkhead. This connection rod should be well above the splash zone and should be well coated to ensure permanent connections. Electrical continuity to each and every sheet pile is essential for the success of the cathodic protection system.

Care should be taken to ensure that the steel bulkhead is electrically isolated (insulated) from all other structures or piping. Again, this is essential if the protected installation is to remain reasonably troublefree.

10.4   Cathodic Protection and Coatings in Combination

A combination of cathodic protection and coatings provides the advantages of both. Protective coatings are known to be the primary ways to protect steel, and cathodic protection is needed as a backup when the continuity of the coating is affected by application or damage. If no coating were used, the cost of cathodic protection would be greatly increased in terms of both equipment needed and current required for protection of the bare (uncoated) structure. Examples of structures for which cathodic protection is used in conjunction with coatings include sheet piling, production platforms, piles, docks, and similar structures continuously immersed in water. In these instances, the protective coating must possess:

- Good dielectric strength
- Good alkali resistance
- Good adhesion characteristics
- Low moisture absorption and transfer rates
- Good coating thickness
- Resistance to the passage of ions

Carefully conducted tests and field use show that most coatings that are designed for immersion in seawater and have the properties described above will perform satisfactorily at steel potentials ranging from -0.8 to -1.3 V with respect to a copper–copper sulfate reference cell. Above a 1.3-V potential, many coatings will show degradation such as cathodic disbondment.

If a good coating can be applied to both sides of the bulkhead pile sheets before installation, cathodic protection current requirements are decreased drastically. The coating, if applied, should be as good as the state of the art permits: white metal sandblast and inorganic zinc primer followed by two coats of coal-tar epoxy for a dry film thickness of at least 0.41 mm (16 mils). In any event, the piling, after installation, should receive such a coating from the low water line upward through the splash zone to the top of the bulkhead. Any cathodic protection is marginal above the low water line and nonexistent in and above the high water line.

10.5   Marine Exposure

The characteristics of coating systems and the structural material to be protected, as well as the specific marine exposure, will determine which coating systems can be used effectively. To achieve good structure protection, the specific marine exposures must be carefully considered when a coating system is being selected. Marine exposures are generally considered to be marine atmosphere, splash and spray zone, and submerged zone. More than one of these exposures may occur on any single structure.

For example, a marine atmosphere is one which carries airborne salt. Since only pure water evaporates from a body of salt water, this physical process does not put salt into the air. Instead, salt becomes airborne only under conditions in which finely divided salt water droplets (spray and mists) are projected into the air by wind and wave action. These fine droplets may remain as such for some time, or the water may evaporate and leave a tiny, solid particle of salt. Wind may carry the droplets or the salt particles some distance from the point of origin. It will be seen, therefore, that the term "marine atmosphere" is not a precisely definable exposure condition. The term might be applied to any situation in which the salt content of the air is great enough to exercise some effect on corrosivity and on protective coating performance.

The effects produced may range from near zero to very intense. The concentration of airborne salt, both close to the shoreline and at increasing distances from it, is difficult to predict even generally, since shoreline topography, wave heights, prevailing wind direction and velocity, and inland physical features are important factors. However, the intensity of the corrosive effect declines rapidly as the distance from the shore is increased; in most cases, supposed acceleration of corrosion many miles inland is largely imaginary. It has been reported that the effect of marine spray is negligible at distances 3 km (2 mi) inland and that analysis of iron corrosion products in seaside towns usually shows more sulfur (from industrial contamination) than chloride (from salt spray).

There is no doubt, however, that steel surfaces subjected to atmospheric exposures that are intensely marine in character present protection problems which are not solved by surface preparation and paint coatings customarily used for inland structures. The effects introduced by the salt are varied. One obvious effect is that the corrosion-accelerating influence of the salt causes even the smallest discontinuity and thin spot in the coating to become a focal point for rusting which rapidly enlarges the original corrosion site. This effect is heightened by the fact that the corrosion products (rust) formed in salt-bearing atmospheres do not have a protective influence against further corrosion to the degree that they do in inland locations; i.e., corrosion continues at a high level in marine atmospheres, whereas the rate usually drops off considerably in most inland atmospheres.

TABLE 10.19 Typical Uses of Coatings

Generic type	Use
Alkyds	On metals in mild marine atmosphere
Silicone alkyds	On metals in mild marine atmosphere
Acrylic (solvent-based)	On metals in moderately severe marine atmosphere
Acrylic (water-reducibles)	On masonry, concrete, and wood in moderately severe marine atmosphere
Chlorinated rubber	On metals in splash and submerged zones
Coal tar:	
Mastics and coal-tar cutbacks	Splash and submerged zones
Modified coal tar	Splash and submerged zones
Epoxies (catalyzed)	Splash and submerged zones; have excellent toughness
Epoxies, phenolic	Splash and submerged zones but are brittle
Phenolics	Heat-cured; high corrosion resistance but brittle; used for tank linings
Polyurethanes (or urethanes)	Usually heat-cured; splash and submerged zones; tough with high abrasion resistance
Vinyls	Marine atmosphere, splash and submerged zones; poor solvent resistance
Zinc, inorganic	Splash and submerged zones; used as single- or multiple-coat system
Zinc, organic	Splash and submerged zones; generally the higher the zinc content the better the coating
Underwater coatings and mastics	Good for in-place structures in submerged structure

The electrochemical reactions involved in the salt-accelerated corrosion processes result in alkalies and other products which may be harmful both to the paint film itself and to adhesion to the coating of the metal.

The net effect of the presence of even small amounts of deposited salt is to increase the need for more care in surface preparation and paint application; in more severe cases, it brings about a need for a more resistant coating system than is customarily used on inland, weather-exposed steel. The need for thoroughness in surface preparation and paint application cannot be overemphasized; it is increased by the fact that crevices, joints, junctions of joining members, interior angles, pockets, undersides of horizontal and inclined members, and similar surfaces that tend to be protected from the direct action of rain, which would wash away the salt, are the places of greatest corrosion and are also the places which tend to receive the poorest paint job.

Structures exposed to moderate and moderately severe marine atmospheres should receive a more advanced paint system. Thorough inspection is probably at least as important as the proper choice of coating.

10.6   Uses in Coastal Structures

Generally the generic coating systems discussed in this section have found satisfactory use in the exposures given in Table 10.19.

11

Summary

11.1 General

This chapter summarizes the principal properties and uses of materials in coastal structures, beach protection devices, and erosion control measures. Generally more than one material is used in a single coastal structure, and compatibility and effectiveness of the materials working together must be considered in each case. The selection of materials for a specific coastal structure may require consideration of the cost of labor and availability in addition to the physical properties of the materials. Such considerations influence the design of structures when more than one material can be employed to perform the same job. By considering the properties and past performances of materials, the coastal engineer can select the proper material to achieve a design objective. Materials are generally considered first for their structural properties and then their durability in coastal structures. In addition to the detailed information given in the preceding chapters, the general summary that follows may assist in the selection of materials.

Most, if not all, of the common construction materials have been used separately or in combinations of two or more in the creation of coastal structures. For example, breakwaters, both detached and shore-connected, are commonly constructed of earth and stone and in many instances are capped with concrete armor units. From time to time steel and concrete sheet piles have been added to the commonly used earth and stone to perform special functions. Also, asphalt has often been used as an earth and rock binder for capping such structures and holding the basic materials in place. Bulkheads and retain-

---

This chapter was written by Lawrence L. Whiteneck and Lester A. Hockney.

ing walls have been constructed of stone, sheet piles made of concrete or steel, mass concrete, and wood. Groins and jetties have been built of those materials as well.

Marine and harbor structures of more complex design usually require the use of a variety of materials in construction, and selection is based not only on physical properties but also on availability at the site, ease of installation, and economy of construction. When temporary structures are called for, recycled materials such as broken or crushed concrete, crushed asphalt concrete, blocks, and salvaged or scrap metals (such as ships, barges, and railroad cars) have been used. The recent development of a large variety of synthetic materials has resulted in the production of improved coating systems and synthetic films for filter cloths as well as foams for improved buoyancy. The synthetic rubbers are used as energy absorbers in fender piles, bumpers, and other protective devices.

Many materials, when used in coastal structures, require special treatment. Wood, for example, will have a substantially improved service life when it has been properly pretreated with creosote and other preservatives. Metals, and more specifically steel, will require protective coatings or cathodic protection (usually both) to be durable in the coastal environment.

11.2   Materials

11.2.1   Stone

Stone refers to individual blocks, masses, or fragments that have been broken or quarried from bedrock exposures or obtained from boulders and cobbles in alluvium. Crushed or broken stone includes all stone in which the shape is not specified. Stone for coastal structures should be free from laminations and weak cleavages and be of such character that it will not disintegrate from the action of air, seawater, or handling and placing. A stone of high specific gravity is desirable because it increases the resistance to movement by the action of waves or currents. Durability of stone can be affected by mineral composition, texture, structure, hardness, toughness, and resistance to the effects of wetting and drying and freezing and thawing. Stone is generally classified as granite, basalt and related rocks, limestone and marble, sandstone, and miscellaneous stone.

Although no standard testing procedure has yet been developed for determining the quality of stone, other than past experience with specific quarries, there are testing programs that can be used. Judgment is necessary in applying and interpreting the results of any testing program. Interpretation requires a great deal of experience and should be left to geotechnical experts. Any test program should include petrographic examination, determination of absorption and bulk specific gravity [ASTM Standard (C97-47 or C127-73), a soundness test (AASHTO T-104-66 or ASTM C38-23), and an abrasion test (Los Angeles rattler, Wetshot rattler, or ASTM 535-69 (75)]. Other tests may prove useful depending on specific project requirements. Properties contributing to the durability of stone can be both physical and chemical, and chemical changes are best evaluated by experts.

Stone size and shape. Stone size is important in coastal structures. Bedding layer material, core rock, or quarry run material is usually 10 to 20 cm (4 to 8 in) or less in size. Underlayer stone may range from a few kilograms to about 300 kg (136 lb). Armor stone is the largest size and ranges up to 22 t (24 tons). Stones larger than about 22 t (24 tons) are generally not easily handled. Although the three ranges of stone sizes are required for the different parts of a rubble mound structure, an adequate number of classes within each range is also necessary. In fitting stones into a structure, the shape as well as the size is important. Design requirements usually specify that the greatest dimension of an individual stone be no more than 3 times the least dimension.

In addition to the physical properties of stone, the method of quarrying will determine the size, range, and classes within a size range that are produced. Depending on the area topography, a quarry will generally be developed as either a side-hill or a pit-type operation. The size of quarry face developed in any given operation is usually determined by the thickness of the formation. The method of blasting and the type of explosive, as well as the geological and physical characteristics of the material, will determine the degree of fragmentation that will result from the quarry operation. Generally a high powder factor (quantity of explosive per unit volume of rock) will produce a greater degree of fragmentation than will a lower powder factor. Also, greater fragmentation will be achieved in a massive rock by using a large number of small-diameter holes at close spacing than by using large-diameter holes at greater spacing. It also appears that best fragmentation is achieved when holes are detonated individually rather than simultaneously.

Use in coastal structures. Stone has many uses in coastal structures, including offshore structures, shore-connected structures, and anchors. Breakwater, jetty, and groin designs often include several sizes of stone for use in the core and underlayers and in the covering of armor layer. Seawalls and revetments may also be constructed from stone. For protection of pier foundations a quarrystone blanket can be laid under the pier in the scour area.

11.2.2 Earth

Earth or soil is a large assortment of materials of various origins. For engineering purposes, soils are generally classified as gravel, sand, silt, clay, and organic material, but most soils are composed of a mixture of two or more of those materials. Although there are several soil classification systems, the one most widely used in engineering is the Unified Soils Classification System (USCS). Gravel is usually considered to range in size from the No. 4 sieve to 7.62 mm (3 in). Gravels are cohesionless materials. Sand is defined as a grain size between 4.76 mm and 0.075 mm (0.19 and 0.003 in) (No. 4 and 200 sieves, respectively), and it can be further classified as coarse, medium, or fine. Sand is normally a cohesionless material; but because of the surface tension effects of pore fluids, it presents an apparent cohesion when damp or moist.

Silts and clays are known as fine-grain materials. Silts also can have an apparent cohesion, but they have relatively poor strength characteristics that limit their usefulness. Clay materials are largely cohesive, have strength characteristics dependent on past stress history, and may be difficult to compact at high moisture contents. Minerals included in the clay composition influence the properties of the soil. Organic materials formed by the decay of vegetable matter can be entrained in soils; they usually have a spongy nature and a fibrous texture. Usually, organic soils have high moisture and gas contents and relatively low specific gravities.

The major significant engineering properties of soil are shear strength, compressibility, and permeability. The types of problems encountered in the design of coastal structures which utilize those characteristics are slope stability, bearing capacity, settlement, and erosion. Other useful properties of soils in the design of structures include dry density, water content, specific gravity, resistivity and corrosion potential, grain size distribution, plasticity characteristics, chemical properties, and durability.

Placement methods. Soil placement methods are usually determined by the fill location, underwater or above water, and the need for some degree of compaction. Earthfills made from land are usually truck-dumped and bulldozed into place; waterside delivery can be by barge or hydraulic pumping. Fill compaction above the water can be accomplished by using mechanical equipment. A fill placed under water will usually require some form of superimposed loading for a period of time to compact it. The loading time depends upon the depth of fill and amount of loading. It usually varies from ½ to 2 years. The compactibility of the soil will also affect the loading time.

Use in coastal structures. Earth is commonly used in virtually any port or harbor development, land reclamation, or coastal protection structure. In addition to fill of all kinds, earth is used in making soil cement as well as fill material for plastic bags and other containing units.

11.2.3 Portland cement concrete

Concrete is used as unreinforced or mass concrete, steel-reinforced concrete, or prestressed or posttensioned concrete. The latter types are usually made in the form of precast structural elements. Specific properties of concrete can be modified and improved by admixtures for special purposes such as accommodation to placing and installation requirements. The specific use of concrete in any structure will determine the mix design and curing process necessary to obtain a satisfactory result. Experience or consultation with experienced designers of concrete structures is necessary to ensure a durable concrete appropriate to the needs of the structure.

Durability is generally a requirement in coastal structures, and the designer and constructor share the responsibility for creating structures which will function as designed over the anticipated lives of the structures. Such structures have a high resistance to the disruptive attack of most environments including salt water, alkalies, most acids, corrosive atmospheres, freeze-thaw cycles, and marine flora and fauna. Good concrete is also highly resistant to abrasion.

Failures of concrete structures have been studied, and some of the more common causes of failure and methods of preventing them are discussed in Chap. 4. Determination of the cause of structural failure requires a careful analysis of the site conditions, the concrete ingredients, and the original design criteria by experienced professional engineers. Concrete failures usually result from the wrong type of cement, unsound aggregate, contaminated mixing water, improper admixtures, or inadequate curing. In addition to avoiding all those possibilities for creating poor concrete, the design engineer must have experience and good judgment in preparing plans and specifications to ensure that concrete is used within its physical capabilities.

Uses in coastal structures. Thousands of marine structures have been satisfactorily designed and constructed of concrete and have had a long history of excellent performance. Because the resources required to make good concrete are generally available in all regions of the world, concrete has wide application for use in coastal and waterfront structures. Its successful use in seawalls, bulkheads, revetments, groins, jetties, breakwaters, and a variety of other structures over many years is evidence of its excellent properties for coastal engineering use.

11.2.4 Other types of concrete and grout

Asphalt. Asphalt is a residue product of the refining of petroleum. It can be used alone as a membrane or coating, or it can be used with other materials as a binder to produce mixes for a variety of purposes. It can be combined with sand and graded aggregate to form a voidless and impermeable asphalt concrete or with an open-graded aggregate to form a stable porous mixture. A composite asphalt structure can easily be constructed with layers of different asphalt mixes each of which performs a particular function. An example might be an impermeable asphalt layer supported by an open-graded asphalt drainage layer with an asphalt mastic placed with a screen over the compacted subsurface. The drainage layer serves to prevent damage to the watertight outer layer by draining away any seepage through the outer layer or any groundwater intrusion.

Physical properties, adhesive properties as a binder, and viscous properties under service conditions are the considerations when asphalt is used in coastal structures. The manner of asphalt placement as well as the service conditions will require certain minimum and maximum viscosities.

Engineers have made considerable use of asphaltic materials in many structures for coastal protection. Asphalt concrete is used to pave or revet the slopes and tops of earth or sand seawalls. It can also be used to pave, or cap, the top surfaces of quarrystone jetties, breakwaters, groins, and cellular steel breakwaters. Asphalt mastic mixtures are used as grout to fill in the voids of quarrystone jetties and groins, the riprap facings of seawalls, and revetted slopes. In foreign countries special equipment has been designed to place a sand-asphalt mastic under water in a continuous operation. The blanket is designed to prevent scour of large areas of the seabed. As more and more emphasis is placed on pollution control, engineers are finding that asphalt offers an economical and effective means of lining dredge disposal sites and waste storage areas that are sometimes necessary in the construction of coastal structures. Asphalt has an excellent history of performance in its use in coastal structures when properly designed and used in accordance with its physical properties and capabilities.

Preplaced aggregate and grout. Pouring portland cement grout in the voids of preplaced aggregate is a specialized construction method. Generally, large stone is used, and the voids are filled with grout. The result is a type of mass concrete used as a seawall or bulkhead. The physical properties as to durability, resistance to abrasion, etc., are much the same as those of the stone and concrete components. One difference is in the cement grout mix design. Pozzolans and fluidizers are added to improve handling during placement and bonding to rock or old concrete.

Portland cement grout. Portland cement grout has the same physical properties as portland cement concrete of similar mix design. Grout, however, is usually modified in its mix design because of its intended use and placement methods. The result is usually a grout mix of cement and water with sand. Very small gravel and clay used as inert fillers, or even bentonite used as a stabilizer can be added when the grout is placed under water. Grout is easily placed by pouring, pumping, or injecting into place. To fill joints or narrow cracks, it can usually be poured into place. To fill large voids or holes, pumping is a common procedure. When ground beds for foundations or the areas behind bulkheads are to be stabilized to prevent leaching of the soil, the grout can be injected into the ground or structure. The injection procedure can be the same as pumping but at relatively high pressure.

Soil cement. Soil cement is a mixture of pulverized soil and measured amounts of portland cement and water compacted to a high density. The physical properties of soil cement are high density in comparison to uncemented soil and rigidity, which results in a structural slab-like material when small quantities of cement are used. In good soils, 7-day compressive strengths of 2027 kPa (300 lb/in^2) are obtainable. Soil cement is used primarily as a base course for stabilizing and compacting soils for foundations, bank protection, and subbase construction. It has been used for earth dam cores, reservoir linings, and slope protection.

Sulfur cement. Sulfur cement concrete and grouts are a relatively recent development and therefore do not have a long history of use in coastal structures. Recently, the availability of large quantities of sulfur has resulted in the increased use of sulfur in construction projects as a binder or admixture of aggregates. Molten sulfur mixed with sand and aggregates produces a sulfur concrete of excellent strength.

Sulfur-asphalt binder materials have higher densities than normal asphalt because sulfur is about twice as heavy as asphalt. The sulfur-asphalt binder usually results in a lower percent of voids than the asphalt cement without the sulfur addition. Sulfur does increase resistance to gasoline, diesel fuel, and similar solvents. It also improves stress fatigue characteristics. The finely dispersed sulfur particles add strength to impregnated fabrics.

Sulfur cement materials reach their full strength quickly upon cooling, but the inherent flammability and low melting point of sulfur impose some limitations on their use. Because of their quick-set characteristics, however, they may find many uses in emergency repairs that could have considerable longevity. Given more experience and additional development, sulfur cement products will probably find more uses in coastal construction.

11.2.5   Structural and sheet metals

Steel. Steel is the most utilized of all metals in marine service and coastal structures. Carbon steel is an alloy of iron and carbon in which the carbon content is less than 2 percent; in structural steel the carbon content is limited to less than 0.35 percent. Adding small amounts of alloying elements during the steel-making process can improve the mechanical properties as well as the corrosion resistance of steel. Small amounts of copper, nickel, chromium, silicon, and phosphorus have been effective in improving corrosion resistance.

In addition to strength, the mechanical properties of steel structures are ductility, brittleness, malleability, flexibility, hardness, resilience, and toughness. Ductility is the capability of a material to be drawn out without change in volume. Brittleness is lack of capability to be deformed without rupture. Malleability is the opposite of brittleness; it refers to capability to be forged or rolled into thin sheets. Flexibility is the capability of a material to bend under stress and return to its original shape when the load is removed. Hardness is a measure of capability to resist indentation when subjected to impact. Resilience is capability to absorb energy due to applied loads without breaking. Toughness is capability to absorb large amounts of energy without rupture. Structural steel has a high degree of all of these properties.

It is relatively easy to alloy other metals with iron in making steel. Low-alloy steels contain up to 1.5 percent of elements such as manganese and silicon. Medium-alloy steels contain 1.5 to 11 percent of alloy elements, and high-alloy steels, including both ferritic and austentic stainless steels, contain more than 11 percent of alloy elements.

Most coastal structures in which steel is a principal construction material use certain steel shapes as follows:

- Sheet piles for caisson walls, cutoff walls, bulkheads, and groins
- H-sections for bearing piles and beams
- Pipe or tubing for bearing piles, conduits, and handrails
- Solid rods for tiebacks or tension members
- Reinforcing bars for concrete

Aluminum. Aluminum, a light metal in its high-purity form, is soft and ductile; it does not possess sufficient strength for structural applications. The addition of alloying elements does, however, impart strength to the metal. Elements used as alloys in aluminum are copper, magnesium, zinc, silicon, and small amounts of other elements such as chromium, usually with copper to obtain high-strength structural shapes.

Copper. Copper has several unique properties that make it a very useful material. In addition to its high thermal and electrical conductivity, it has high corrosion resistance. It is readily alloyable. The copper alloys most corrosion resistant to seawater are aluminum brass, inhibited admiralty brass, and the copper-nickel alloys.

Use in coastal structures. Steel is used as structural shapes in most types of coastal structures. It is also used in composite structures, e.g., as rebar in concrete construction. Steel alloys have found many uses as bar stock, wire, and wire fabric. Many alloys of aluminum, because of their high corrosion resistance and strength-to-weight ratios, also have found applications in marine structures. Copper, in addition to direct uses as pipe and sheathing, has a high alloying capability in bronze and brass that makes it a very useful element in the marine environment.

When steel, alloys, and other metals are used in the coastal environment, care must be taken to avoid direct contact of dissimilar metals that can form galvanic couples. When dissimilar metals are in electrical contact with each other and are immersed in an electrolyte, a potential difference will be established, an electric current will flow, and rapid corrosion will take place. If two dissimilar metals must be joined, several precautions must be taken: insulating the metals, avoiding unfavorable effects by keeping the cathode area small, placing a more anodic third metal in contact with the other two metals to provide sacrificial protection, and investigating other possible ways to protect the structure.

11.2.6   Wood

As a construction material, wood is available almost everywhere and at reasonable cost. It is a cellular organic material composed principally of cellulose, which comprises the structural units, and lignin, which cements the structural units together. A tree has distinct zones: bark, sapwood, heartwood, and pith. There is no consistent difference between the weight and strength properties of heartwood and sapwood. Because wood is produced by nature under various uncontrolled environmental conditions, such as geographical location, precipitation, exposure, and elevation, the product is highly variable. Also, trees are alive, they produce wood of different properties at different ages. For a given characteristic or property of wood, such as bending strength, both the mean value and the variation about the mean should be considered.

Lumber-grading rules are, in effect, specifications of quality. The size and number of knots, slope of grain, and other strength-reducing characteristics are judged and graded according to uniform standards so that working stresses can be assigned to specified qualities. Common construction species generally available in the United States are Douglas fir, southern pine, spruce, hemlock, redwood, cedar, and other pine species such as lodgepole, ponderosa, and white.

Properties. The major mechanical properties of wood as they affect engineering design are recapped in the following paragraphs.

Tension parallel to grain creates a tendency to elongate wood fibers and cause them to slip by each other. Resistance to tension applied strictly parallel to the grain is wood's most important strength property; but if tension is applied at an angle to the grain or if the cross section of the piece is reduced by knots or holes, the strength is materially lower. Tension perpendicular to grain tends to separate the wood fibers along the grain; it is the direction in which wood has the least strength.

Compression parallel to grain has a tendency to shorten the wood fibers in the lengthwise direction. Resistance of wood to compressive force is good, but it is affected by the angle of the load to the grain and by the presence of knots and holes. Compression perpendicular to grain, such as the bearings under the ends of a beam or under a column, tends to compress the wood fibers together. The wood becomes more dense, and the action may cause slight displacement at the bearing face.

The largest stress usually occurs along the neutral axis of a beam. While lumber dries, checks and splits may occur. They reduce the area in the plane of maximum shear and therefore the shear strength for design. Shear perpendicular to grain is not a design factor because effective control is applied through limits on design stress for shear parallel to grain. Fiber stress in bending creates compression in fibers on one side and tension in fibers on the other side of a beam. The higher stresses occur in the fibers most distant from the center. Deviations in the slope of grain and the presence of knots and holes in the outside faces reduce the resistance in the extreme fibers.

Preservative treatment.   To extend its useful life long enough to make it an economical and practical material for use in the coastal zone or other marine environments, wood must be protected from its natural enemies: fungi, bacteria, insects, and marine organisms. The most effective method of treating wood with preservatives is by pressure. The pressure-treating process requires placing the wood in an airtight chamber in which either a vacuum or a pressure can be created while the preservative is introduced. The preservative generally will penetrate the wood surface from 1.5 to 4 cm (0.6 to 1.6 in) and coats the walls of the wood cells in that volume. Penetration to 10 cm (4 in) is required in some cases. Although two processes, empty cell and full cell, have been used successfully to preserve wood structures in marine environments, the full cell process is most commonly accepted as the preferable one.

The wood preservatives commonly used are grouped into two broad classes: preservative oils and waterborne preservatives. The preservative oils are considered the best wood protection in a marine environment; they include byproducts of petroleum such as creosotes, coal-tar creosotes, and mixtures of these with other oils. The mixtures may include solutions of toxic chemicals such as pentachlorophenol or copper naphthenate. Waterborne preservatives include solutions of chromated zinc chloride, fluor-chrome-arsenate-phenol, chromated copper arsenate, and other toxic chemicals.

Other protective methods.   In a marine environment, wood structures can be protected by materials which are not strictly preservatives; they include sheet metals, concrete jackets, and flexible synthetic sheets such as vinyl and polyethylene films. Because virtually all organisms that cause wood deterioration are aerobic, surrounding a wood element such as a pile with a jacket that prevents seawater containing free oxygen from coming in contact with the wood creates an environment hostile to the organisms.

Durability.   Wood, when properly treated with appropriate preservatives, has a good history of satisfactory service in marine and coastal structures. Wood piles supporting piers and wharves, when not subjected to abrasion, have lasted many years. Wood sheet piles in groins, jetties, bulkheads, and similar structures will perform satisfactorily. Care must be used in installing wood members to ensure that construction joints and connections do not damage preservative protection and that field repairs are carefully and adequately made.

11.2.7   Plastics

Chemically the term "plastics" is applied to a large group of synthetic materials, including synthetic rubber, that are processed by molding or forming into final shapes. Plastics that are soft and pliable at some stage in their production are formed into shape by the application of heat and/or pressure. They are organic compounds called monomers, that are transformed into complex molecular aggregations, called polymers, by chemical processes.

Plastics in general can be divided into two distinct groups: thermoplastics and thermosetting plastics. Thermoplastics soften repeatedly when heated and harden when cooled. Thermosetting plastics go through a soft stage only once. When they are hardened, an irreversible change takes place and they cannot be softened again. For a particular end use, plastics can also be combined to draw together the best attributes of the blended components by copolymerization. The products are called copolymers. During the production of plastics, additives such as plasticizers, fillers, colorants, stabilizers, and impact modifiers can be added.

In addition to structural qualities, plastics are easily formable, corrosion-resistant, lightweight, wear-resistant, energy-absorbent, impact-resistant, flexible, and ductile. A necessary consideration is that plastics will burn—some easily, others slowly, and others with great difficulty.

Geotextiles.   Plastics in the form of geotextiles have an important use in coastal structures; they commonly function as filters in drainage and in shore and embankment protection structures. Geotextiles are a relatively new material in the construction industry, but they have had a generally successful experience record as filters in selected coastal structures since the late 1960s. Substantial improvement in design and materials selection specifications also has occurred.

The primary function of geotextiles when used as filters is to retain the protected soil (prevent piping) and remain permeable to water without significant head loss or the development of excessive hydrostatic pressure. To function satisfactorily, the geotextile filter must have physical durability and filtering integrity throughout the design life of the structure. In the selection of a geotextile for a filter, the chosen fabric, in addition to having required physical and chemical properties, should be of a kind and finished form consistent with the specific site requirements.

Fabric construction is a predominant factor in performance. Woven fabrics are commonly manufactured by crossing the yarns at right angles and overlapping them; the yarns are monofilament, multifilament, mono-multifilament, or slit-film. Nonwoven fabrics include all materials not woven or knitted. They consist of discrete fibers, which may be random or pattern-oriented in the fabric. The bonding methods described are needle-punching, heat-bonding, resin-bonding, and combination bonding. Combination fabrics are produced by combining woven and nonwoven fabrics and using one or more bonding methods.

Use in coastal structures. Because geotextiles are relatively new as construction materials, there has not been sufficient time to develop agreed-upon standard testing techniques for the most important characteristics a fabric should have for specific applications. However, they are finding many uses in coastal structures. Different fabric specifications may be required for specific uses such as replacement of stone filters under riprap, drainage control by silt retention fabrics, and road stabilization by road or highway fabrics. Fabric users should seek the advice and recommendations of knowledgeable sources with experience in the specific use being considered, such as consultants and more than one manufacturer.

Many forms of plastics other than geotextiles also are used in coastal structures. Flexible plastics are used as mold forms for concrete and for wrapping timber pile to provide protection from marine animals and for wrapping metal piles to prevent corrosion. Molded forms have applications as rubbing strips, fenders, and bumpers. Plastic extrusions in the form of pipe and culverts are common. Pipe may be reinforced or not depending on the structural strength required.

11.2.8 Recycled and other materials

Generally, recycled materials consist of a variety of things that may be available in a given locality and are normally used in emergency situations as temporary (occasionally as permanent) protective devices against damaging waves or currents. Such materials should have specific gravities greater than 1.5 to be useful unless a floating type of structure is needed. Materials in this category include salvaged concrete, concrete rubble, crushed concrete, recycled asphalt used either crushed or as rubble, or blocks and bricks, and salvaged steel structures. Normally, because of the emergency-type use of recycled materials, little consideration is given to the properties of the materials other than the specific gravities. Also, little concern is given to the environmental impacts, although the impacts of recycled materials will generally differ from those of the same materials before recycling or reuse.

Recycled or salvaged materials have been used for many years for emergency repairs or for building temporary structures. The "temporary" structures have often remained in place for many years.

Salvaged concrete, either as rubble, crushed, or unbroken, has been used to repair revetments, groins, jetties, and breakwaters. These materials may not have a pleasing appearance, especially if they contain reinforcing steel. If located in a recreation area, the reinforcing steel may create a safety hazard. Of course, exposed reinforcing bars will corrode at a rapid rate and will accelerate concrete spalling and deterioration. Generally, the concrete materials are used as substitutes for stone in coastal structures.

Recycled asphalt can be used as an underlayer in coastal structures. Although it is relatively hard and inflexible because of its age, it will retain its broken shape for extended periods of time and further deterioration is not a problem. Recovered asphalt can be crushed and used as core or bedding material in coastal structures; but unless it is well graded, it does not make a satisfactory filter material. Crushed asphalt is also finding greater use as base material for highways, roads, streets, and parking lots.

Bricks, hollow concrete blocks, and cinder blocks have been used as temporary repair materials, but they generally break down during handling and do not have much long-term value. Also, they have no value as an underlayer or armor layer.

Salvaged ships and barges have been used as temporary breakwaters by maneuvering them into selected locations, sinking them, and then filling them with rock or gravel to provide stability. Removing these devices upon their disintegration or when a permanent structure is desired is usually a difficult problem. Other salvaged materials such as railroad cars and automobile bodies have been used for bank or shore protection, but they are not satisfactory and are usually unsightly and hazardous if located where people may visit. Used rubber tires have a variety of uses such as fenders on barges, work boats, and docks. They have also been successfully used as floating breakwaters to protect basins against short-period waves. Several different arrangements have been model-tested. Flotation has been created by filling tires with urethane foam. If they are anchored in place or on the bottom, tires have served as revetments and have been used to slow the bed movement of littoral drift.

11.2.9 Protective systems

Protective systems are applicable to steel and alloys, wood, and concrete. Application to concrete is usually for esthetic reasons, but in some few cases, a system has been used to decrease water penetration into relatively porous concrete. Protective systems are classified in two categories: coating and cathodic protection. Each system can be used separately, but in many instances cathodic protection can be successfully used to supplement coating systems.

Coatings. Protective coatings range from mere decorative paints to complex and multicoat systems requiring careful surface preparation, proper coating application techniques, and the careful selection of coating systems. In the consideration of a coating requirement, the first step is to consider the type and kind of surface to be protected, i.e., wood, concrete, or steel or other metal or alloy. Next consider the environment the structure surface is exposed to, such as a marine atmosphere, a tide or splash zone, or a submerged zone in either fresh or salt water. With this information and other specific data as set forth in Chap. 10, the coastal engineer can then consider the generic category as well as the specific type of coating within a category that is best suited for the protection of a given structure.

The types of coatings and their generic classifications are discussed in Chap. 10. To evaluate a coating's protection performance adequately, it is necessary to consider the properties of the coating material, the surface preparation requirements, and the application procedure as well as the drying or curing processes. To aid evaluation, Chap. 10 discusses the surface-preparation processes including resulting metal surface anchor patterns, the number and thickness of coats, and the drying or curing processes necessary to obtain good coating systems that can be expected to protect structures properly. Coatings are applied by brush, roller, spray (both air and airless), or dipping.

Coating repair is a common procedure, but there are precautions that must be taken to ensure successful repair. Coating compatibility is essential to provide good bonding of the repair coating and avoid the common types of coating failures: blistering, undercutting, surface cracking, delamination, alligatoring, and chalking. The coating must have strength, adhesion, resistance to the environment, and, often, a pleasing appearance if it is to function properly.

Cathodic protection. Cathodic protection is an electrical method of protecting metal structures in electrolytes. The electrolyte can vary from seawater to fresh water, saturated soil, and even relatively dry soils. Ions of acids or alkalies tend to promote metal deterioration, which can occur in localized areas or over large general areas of a metal surface. Metal corrosion is a natural process involving electrochemical reactions that result in the flow of direct current from the anodic areas (the corroding areas) to the cathodic areas of the structure through the surrounding electrolyte. The flow is caused by electrical potential difference between the two areas. In cathodic protection, an outside electric current is induced in the opposite direction and in that way stops the normal corrosion process.

The design and installation of a cathodic protection system are highly technical. To ensure an effectively operating system, field conditions of the structure must be examined to determine the total amount of electric current required to protect the structure cathodically and have proper current distribution. It must also be determined that there will be no interference with other structures in the vicinity and that potential differences within either the protected structure or adjacent structures are not created. Cathodic protection requires periodic maintenance and inspection to keep it in good working order.

In general, good protective systems, both coatings and cathodic protection, are economical, require maintenance, and will substantially extend the service lives of well-constructed structures.

11.3   Some Present Investigations of Coastal Construction Materials

11.3.1   Stone

Stone is one of the most widely used materials in coastal structures and shore protection works. There are two basic areas of research on stone: the uses of stone in shore protection structures and the characteristics of stone for use in coastal structures.

Stone is used in revetments, jetties, groins, bulkheads, seawalls, and miscellaneous types of structures. Studies are being done on new types of shore protection structures such as semisubmerged offshore ones, on new structure configurations for jetties, on different distributions or arrays of stone in armor units or layers, on the reliability of breakwater model tests, and on the effects of breakwaters on waves. Programs have been initiated to monitor and evaluate the performance of existing coastal structures in terms of effectiveness, maintenance cost, and life. This area of research also deals with development of field techniques and criteria for the functional and structural design of coastal structures. Work is continuing on the evaluation of parameters used for determining the effective elevation of structures, the slopes of revetments to reduce run-up, and the size of armor stone to dissipate energy.

The use of stone in coastal structures is based primarily on experience. Continued research is needed on the development of testing procedures, criteria, and methods of quarrying to determine and produce rock characteristics that are desirable and suitable for use in coastal structures. Current research includes various tests for shrink-and-swell behavior, wetting-and-drying effects, mineralogic composition, specific gravity, and other physical and chemical properties.

11.3.2   Earth

Current investigations and studies related to the use of earth materials in the coastal and marine environment deal primarily with the behavior of soils under various near-shore conditions and the use of soils in connection with coastal structures. The investigations include numerous programs that are in progress to develop field techniques and criteria for use in design, construction, and maintenance of effective beach and dune protection. The programs seek to describe and predict the interactions between the materials that make up the coasts and the forces that act upon them.

Studies include the development of mathematical models that designers can use to determine how much sandfill is required to adequately protect a segment of shore for a certain time span and how often additional fill will be required. Studies are in progress to determine the effective use of earth materials in low-cost shore protection. New dredge disposal techniques are being studied to aid in beach nourishment projects and sand bypassing across coastal inlets. Several field research facilities and projects have been established to study coastal processes and their long-term effects on the erosion of and protection of the natural coastline materials.

Research is also continuing on the engineering properties of the various soils. The studies include the determination of density and porosity of seafloor sediments, the grain size distribution of beach materials, and the shear strength and consolidation characteristics of estuarine deposits.

11.3.3   Portland cement concrete

Concrete structures are being increasingly utilized for a wide variety of applications in the marine environment. They are becoming more sophisticated and are being located in areas that have more severe exposure (e.g., to ice or open sea) and are subjected to cyclic and impact loads. Consequently, their performance requirements become increasingly severe and critical. There are ongoing investigations into the internal responses of structural elements, environmental conditions in which structures must serve, new materials and configurations, construction practices and repairs, and new uses in the ocean.

Existing problems related to concrete design include cracking, spalling, and corrosion of reinforcing steel and purposeful overdesign and overreinforcement of structures in an effort to cover the range of uncertainty. These problems indicate the need for additional investigations to better understand the properties of concrete:

- Corrosion of reinforcement in submerged structures with varying widths of cracks. Also, when cracks are repeatedly opened and closed under a large number of cycles, there appears to be accelerated degradation which may be due to hydraulic fracturing of the concrete by entrapped water.

- Placement of mass underwater concrete is being tested. Tremie concrete mixes and placement procedures are being investigated at the University of California, and mass concrete placement in the deep ocean is being tested by U.S. Naval Civil Engineering Laboratory at Port Hueneme. Also, investigation of thixotropic admixtures to prevent segregation of concrete when flowing through water is being considered.

- Failure of dolos units and other armor units indicates the need for additional investigation to find more stable shapes or the possible need for reinforcement to tie the member internally and improve its flexural strength. Repairs to precast concrete units of this kind, as well as other concrete structures, have been carried out with epoxy injection. Continued tests on repaired units are needed to determine fatigue and ultimate strength as well as to gain a better understanding of the impact on other properties of concrete.

11.3.4   Other types of concrete and grout

Bituminous concrete. Asphalt is made from crude oil; and because refineries are using crude from many different sources, the characteristics of presently produced asphalts differ from those of asphalts previously produced. That creates problems having to do with the performance, handling, and placing of asphalt cement. Asphalt-related problems are generally of two kinds: workability and performance. Workability problems, which make asphalt more difficult to mix and place, seem to be common. They result from mixing asphalts from several different sources of crude.

There is some evidence that equipment changes, such as using drum plant mixing, in which the aggregate and asphalt are added to a drum simultaneously, can result in a softer asphalt with a higher moisture content. The vibratory compactor, which densifies by dynamic energy, is a different compaction process than steel tandem rollers.

Many variables can affect an asphalt: examples are cement, fines aggregate, and temperature of mix and roll. More research is needed to identify and clarify the roles of variables. More investigation of the compaction process is necessary.

Preplaced-aggregate concrete and portland cement grout. The preplaced-aggregate process is essentially a special application of portland cement concrete; therefore, the investigations related to portland cement may result in better use of this type of concrete also. By investigating the performance of past, present, and future projects involving the use of preplaced-aggregate and portland cement grout, other applications should be found and techniques should be improved.

Soil cement. Soil cement also is a special use of portland cement, and additional investigation can create an understanding of how to use local soils, especially the very fine and clay-type soils, successfully.

Sulfur cement concrete and grout. With the increased production of sulfur as a result of refining more sour crude, larger quantities of sulfur have become available. That has made sulfur cement concrete and grout more economically competitive for many special uses. Ongoing research and testing are aimed at improving sulfur-asphalt materials by developing additives to improve physical properties such as those that allow mixing at temperatures above 160°C to prevent dehydrogenation of the asphalt. In sulfur concrete development, investigation of other plasticizers to improve the physical properties and upgrade the heat resistance of the sulfur concrete is an important activity.

11.3.5 Structural and sheet metals

Steel. Much of the effort in the research and development of steel and steel products for coastal construction is devoted to the metallurgy of steel making in order to develop products that are more corrosion-resistant when exposed to the marine environment. Progress has been made with the development of ASTM grades A242 and A588, in which small amounts of vanadium, zirconium, columbium, and titanium are used. These alloys exhibit improved strength and yield values, but the cost of the products is high. Because improved protective systems (both coatings and cathodic protection) are available, very little corrosion-resistant steel alloy is produced. Studies in the application and control of cathodic protection are ongoing to improve the understanding of hydrogen embrittlement and stress corrosion cracking of metals, and that may lead to greater use of steel in future coastal structures.

Aluminum and copper. The studies and investigations of aluminum and copper, as well as other metals, are similar to those of steel. Much of the research work is concerned with a better understanding of the physical chemistry of the oxide films that are responsible for the passivity and corrosion resistance of these metals. The studies may lead to new alloys and resulting allotropic modifications that might improve corrosion resistance as well as enhance some physical properties of the metals, including resistance to halogen and acidic ions.

So much research is in progress that it is difficult to estimate what new knowledge might contribute to the improved use of metals in a marine environment.

11.3.6 Wood

Because wood is produced by nature under various uncontrolled environmental conditions, it is highly variable. Also, the fact that the tree is alive and produces wood with different properties at different ages complicates the analysis of the properties of wood. Recent investigations into strength properties to determine the material variability of clear wood are based on the statistical probability of sampling.

Elastic parameters. Present wood design often involves curved members and three-dimensional stress distributions. Therefore, recent emphasis has been directed toward the determination of the strength and elastic characteristics of wood in all the principal directions. Prediction equations that have been developed allow the estimation of all the elastic parameter values of wood (Bodig and Goodman, 1973). Also, there is considerable interest in predicting the elastic and strength characteristics of wood at any arbitrary ring and grain angle. The determination of the elastic parameter values is based on the linear part of the stress-strain curve. For ultimate stress design, knowledge of the nonlinear part of the stress-strain curve is very important. Investigation is also being done on the stress interaction behavior of wood.

One of the problems associated with the theoretical prediction of the strength of wood is the lack of understanding of the mechanism of failure. Fracture mechanics of wood, as well as the concept of energy of distortion limitation, are being investigated.

Time-dependent characteristics. Because of the time-dependent stress-strain behavior of wood, a large amount of investigation has been concerned with the rheological properties of the material. Nonlinear time-dependent relations, cyclic loads, and cyclic environmental factors complicate the relations. Among the various rheological properties, creep behavior appears to be the property most often needed in designing with wood. The effect of duration of load on the strength properties also is being investigated. Dynamic forces act for a very short time, and under those conditions wood appears to be stronger and stiffer than under static loading.

Wood composites. When compared to other construction materials and on a pound-per-pound basis, wood is one of the most efficient available materials in stiffness and strength along the grain. However, its efficiency is much lower if across-the-grain direction is considered. Thus, for specific engineering purposes, it is necessary to rearrange the wood in relation to its natural form. This necessitates the manufacture of composites such as laminated beams, plywood, particleboard, hardboard, and fiberboard. Further modifications can be made by using high-density overlays, impregnations, and preservatives.

Some investigation and testing of laminated wood utilizes a proof-loading concept to establish the laminating combinations and their associated design stress. The research is intended to determine what tensile proof load should be used to justify strength levels and what percent of the tension zone laminations should be subject to the proof load level. To gain more flexibility in laminating combinations, a project which will provide criteria for combining different species was in progress at the time of writing.

11.3.7 Plastics

The number of available plastic materials and resins is so great and the variety of synthetics in each family of plastic resins is so large that it is virtually impossible to identify significant investigation and research that is important to coastal engineering materials development. Investigations and development occur in three general areas: processing and machinery, new resins, and resin modification by additives.

Processing and machinery. The development of new resins will not lead to improved products until the machinery for processing such resins or modified resins can be developed. The new machinery will control the plastic manufacturing in such a manner as to produce a given product with the required physical properties.

Machines have been developed to mold liquid polymers, and technology for injection molding of ultrahigh-molecular-weight polyethylenes is available. Systems for extruding polypropylene by using water cooling of the bubble to obtain good toughness and high clarity are being developed. A new development in injection molding is making a solid skin of one type of plastic and a foamed core of another. The manner in which plastics are made affects such physical properties of a material as impact resistance, flexural strength, and heat distortion, and so those properties can be substantially improved by improving processing capabilities. The improvement of physical properties will provide for longer service lives of both rigid and flexible plastics. As flexural strength, resistance to impact, and heat distortion are increased, plastics may find an increasing use as structural members.

New resins. Polyester resins form a large family. In the manufacture of resins, three basic controls (i.e., density or degree of crystallinity, molecular weight, and molecular weight distribution) result in a great variety of resins. The abundant glycols and dibasic acids developed from petroleum intermediates provide a wide latitude in designing polyester resins to meet specific requirements. Unsaturated polyesters can compete with epoxies, phenolics, and other plastics in electrical, physical, and mechanical properties. They predominate in applications requiring corrosion resistance. For example, non-air-inhibited types are used in boat hulls, buoys, and decks and for coating wood, concrete, metals, and other structures. High-temperature-resistant resins such as linear aromatic polyesters represent another new development. This particular polymer also has a high resistance to most organic solvents.

Polyethylene has a very simple molecular structure, but it is capable of almost infinite variation and modification. The most recent development has been in the very high density polyethylene resins that have a hard crystalline character. These developments may result in improved properties such as impact resistance, tensile strength, and abrasion resistance for use in buoys, fenders and bumpers, and unreinforced pipe.

Virtually all resins can be given different properties by the incorporation of additives. Antioxidants are used to prevent degradation of resins at high temperature. Ultraviolet stabilizers prevent deterioration in atmospheric exposures. Fillers are used for their reinforcing properties; an example is the use of chopped glass fibers to increase strength and stiffness. Air can be considered a filler when injected into a resin during processing to produce a cellular or foamed plastic.

Dispersion resins are fine-particle resins which can be dispersed in plasticizers to produce liquid systems that are essentially 100 percent solids. These systems are used in the manufacture of protective coating and paint systems. Many of these systems are used to coat, impregnate, or saturate fabrics and yarns as well as to coat paper and leather.

New developments in the use of plastics in coastal structures will be continuous for many years to come because a large variety of plastic resins is available, their molecular structure can be rearranged to form new plastics, and the physical properties of the new plastics can be changed and improved through the manufacturing process and the addition of additives and plasticizers.

Geotextile filters. The development, investigation, and testing of fabrics is fragmented, and there are many activities that overlap in effort. The manufacture of filter fabric is changing in some instances because of the ongoing development of nonwoven fabrics with controllable thickness, elongation, and filtration capabilities. Methods are being investigated to characterize fabric as to the size and shape of openings and the details of clogging of the fabric. Tear propagation in fabrics is being studied. Mechanical property analysis to indicate the amount of deformation that a fabric will undergo is being performed. Information about anchoring to indicate the required friction between the fabric and the soil is also being investigated. Filtration mechanisms (and particularly the soil structure arrangement resulting from the water flow), flow rate, permeability, and piping are being evaluated, and laboratory test methods are being recommended. The results of these investigations and others that will come along in the future will provide data for expanded and better use of fabrics.

11.3.8 Protective systems

Coatings. One of the principal means of preventing deterioration of structures is through the use of protective coatings. Coatings may be specified on a formulation basis, on a performance basis, or on a combination of the two. The formulation-type specification does not take advantage of the manufacturer's experience and formulating knowledge; the responsibility for obtaining a suitable coating belongs to the specification writer and the technical sources. The principal alleged merit of a performance-type coating is that the coating does take advantage of the manufacturer's knowledge and experience and may be a real advantage if the manufacturer is highly experienced in formulating coatings for the particular contemplated use. A principal difficulty with performance-type specifications is that acceptance tests, which purport to show that a coating is satisfactory for a specific use, must necessarily be finished in a short time and frequently have little significance in predicting actual performance. Performance tests include flexibility, hiding power, immersion resistance, gloss, resistance to weather, and salt spray tests.

The development of improved coating systems involves three areas of investigation and testing: surface preparation, coating application techniques, and improved materials. In many situations, environmental constraints have required innovations and improvements in all three areas.

Surface preparation is accomplished in many ways: solvent cleaning, hand and power-tool cleaning (wire brushing), pickling, flame cleaning, and blast cleaning (sandblasting).

Good coating performance requires good adhesion to the structure surface. Preparing the surface for coating application is critical. In solvent cleaning, mineral spirits are frequently used, but the product must be sufficiently refined as to not leave oily residues upon evaporation. Solvents that leave no residue on a surface to impede the bond of coating require additional investigation, but study in this area is limited because solvent cleaning is usually used with other surface preparation processes.

Hand and power-tool cleaning of metal surfaces is widely used; but the process usually produces some areas of polished metal, which are not conducive to good coating adhesion. Developmental studies are being done in an effort to devise means of removing loose contaminating particles from a structure surface by designing wire tips for brushes that will remove contaminants and improve brush wear.

For pickling, acids such as sulfuric, hydrochloric, phosphoric, and nitric are used. Inhibitors are added to minimize metal loss. Acids leave considerable residue on metal surfaces that can cause coating adhesion problems. The elimination of residues by hot-water rinsing helps, but redeposited salts or absorption of atomic hydrogen can cause metal embrittlement. Means to prevent these possibilities must be considered.

Flame cleaning removes only loose rust particles and grease; therefore, it must be followed by wire brushing and coating application while the surface is dry but cool. Because this process is costly, it is not used to any great extent, and little improvement in its use is being studied.

Blast cleaning is the most effective method of cleaning metal surfaces. The degree of blast cleaning can be determined by type of blasting material, the pressure used in blasting, and the time of blasting per unit area. The metal surface anchor pattern developed as a result of blast cleaning can be controlled. Certain coating systems may require a deep anchor pattern, others a shallow anchor pattern. Some blasting materials produce a rounded anchor pattern, others a sharp pattern. Because different types of coating primers require different adhesion conditions, surface preparation must be considered as a part of the coating system.

Coating application techniques are an integral part of a protective coating system. As new and better coating systems are developed, new application systems must also be created. Application systems presently in use are brush, roller, various spray methods, flow, and electrostatic processes. The electrostatic processes came about with the development of plastic resins which are applied in the dry powder form, and the application system involves a specific surface preparation process. Still in the development and improvement stage is an improvement in the process that will produce a required thickness of near-perfect coating, without pinholes or holidays, by using one of a variety of plastic resins. Present electrostatic applications are very good, but the adaptability of the procedure to a wider variety of resins and the development of a greater number of coating resins for use in the process must continue.

Asphalt coatings consisting of a dispersion of high-molecular-weight hydrocarbon compounds (asphaltenes) in heavy residual oils are made into hot-applied asphalt enamels, solvent-reduced asphalt coatings, and emulsions. Because the asphalt residue from which the coatings are made is used as a raw material for many other products, the cost of such coatings is rising and there is relatively little asphalt coating research and development activity.

Coal-tar pitch, the residue from distilled coal tar, is used to manufacture coal-tar coatings by cutting back the pitch with coal-tar solvents and usually adding mineral filler (extender pigments) such as magnesium silicate. Most recent developments of the use of coal tar in coating systems are in the coal tar–epoxy systems. These systems contain epoxy resins, pigment, solvents, curing agents, coal-tar pitch, and gelling agents. The broader use of coal-tar pitch with a greater variety of resins is continually under development at present. The development of plastic resins for use in new coating formulations is part of the research and development activities discussed in Chap. 10.

Cathodic protection.   Maximizing the efficiency of corrosion control requires a thorough understanding of the variable environment to which structural materials are exposed. That is particularly true of metals. However, environmental considerations are important in considering the durability of all construction materials.

Optimizing cathodic protection systems in the marine environment requires a detailed knowledge of the seasonal variability of dissolved oxygen and temperature and the saturation rate of the water with respect to carbonates. Although complete protection can be and is being achieved on many structures without any prior knowledge of those variables, the design of the most economical system utilizing a combination of impressed current, sacrificial anodes, and coatings is not possible without detailed knowledge of the environmental conditions.

The four variables important to corrosion for which there exists a large enough database to permit general surface water mapping of the oceans are temperature, salinity, dissolved oxygen, and pH. Other variables may also be important but not enough data are yet available for a significant evaluation of their impact on a global scale.

Premature anode material failures as a result of variation in environmental conditions are being examined in more detail. Attention is being given to electrochemical reactions and conditions at the anode-environment interface when chloride and sulfate ions are discharged in the anodic process and affect the anode material.

Although some ten cathodic protection criteria for buried steel structures have been used throughout the world, a universally acceptable criterion is still not available. Different criteria frequently give conflicting evaluations of the state of protection. This situation has been due primarily to the lack of suitable electrochemical procedures to monitor and evaluate the actual state of protection at the structure-soil interface. A great deal of investigation of this problem, in both the field and laboratory, is being done and reported regularly in the literature.

Some studies indicate that the current density for cathodic protection of embedded steel in concrete is controlled primarily by the rate of oxygen diffusion through the concrete. Measurements indicate that the resistance to oxygen diffusion may be 10 times higher through the interface between cement paste and steel than through the concrete cover.

Studies to provide a better understanding of the electrochemical and the electrode kinetics reactions at the surface of the metal-environment interface and to improve field measurement techniques are ongoing in many places in the world. This information will provide engineers with a means of determining better and more efficient cathodic protection designs required in any local conditions in the future.

Specifications of and Applications for Steel Suitable for Marine Service

ASTM designation	Title of standard	Application
A36-77	Standard specification for structural steel	Bridges, bulkheads, general structures
A131-78	Standard specification for structural steel for ships	Ship construction, tanks (shapes, plates, rivets)
A242-79	Standard specification for high-strength, low-alloy structural steel	
A252	Standard specification for welded and seamless steel pipe piles	Structures, forms for cast-in-place concrete piles
A283-79	Standard specification for low and intermediate tensile strength carbon-steel plates, shapes, and bars	General structure, tanks
A284-77	Standard specification for low and intermediate tensile strength carbon-silicon steel plates for machine parts and general construction	Machine parts, general construction
A328-75a	Standard specification for steel sheet piling	Sheet piling, dock walls, and cofferdams
A440-77	Standard specification for high-strength structural steel	Bridges
A441-79	Standard specification for high-strength low-alloy structural manganese-vanadium steel	Bridges, buildings; weight savings and added durability

ASTM designation	Title of standard	Application
A514-77	Standard specification for high-yield-strength, quenched and tampered alloy steel plate suitable for welding	Welded bridges and other structures
A529-75	Standard specification for 290 MPa (42,000 lb/in^2) minimum yield point, ½-in (12.7-mm) maximum thickness	Buildings
A572-79	Standard specification for high-strength low-alloy columbium-vanadium steels of structural quality	Bridges, building structures
A573	Standard specification for structural carbon steel plates of improved toughness	Steel plates and sheet piling
A588-80a	Standard specification for high-strength low-alloy structural steel with 345 MPa (50,000 lb/in^2) minimum yield point to 10-cm (4-in) thickness	Bridges, buildings; weight savings and added durability
A633-79a	Standard specification for normalized high-strength low-alloy structural steel	Serves at $-45°C$ ($-49°F$) and higher
A690-77	Standard specification for high-strength, low-alloy steel H-piles and sheet piling for use in marine environments	Dock wall, seawalls, bulk-heads; provides 2 to 3 times greater resistance to seawater splash zone than ordinary carbon steel
A699-77	Standard specification for low-carbon manganese-molybdenum-columbium alloy steel plates, shapes, and bars	General application; grades 3 and 4 suitable for temperature down to $-45°C$ ($-49°F$)
A709-80	Standard specification for structural steel for bridges	Carbon and high-strength low-alloy steel plates and sheets
A710-79	Standard specification for low-carbon age-hardening nickel-copper-chromium-molybdenum-columbium and nickel-copper-columbium alloy steels	Plates, shapes, and bars for general use

Geotextile Filter Tests

1. Tensile Strength and Elongation Test

Test five stronger principal direction (SPD) and five weaker principal direction (WPD) samples, unaged, in accordance with ASTM Standard D 1682-64 Breaking Load and Elongation of Textile Fabric-Grab Test Method. The jaws shall be 2.54 cm (1 in) square and the constant rate of travel 30.5 cm (12 in) per minute. Care should be exercised to make sure the fabric is properly aligned to the jaws. If not properly aligned, the results will be inaccurate. Test should be conducted at 22.8° ± 2°C (73 ± 3°F).

2. Seam Breaking Strength

Test five unaged samples in accordance with method ASTM Standard D 1638-68, using 2.54-cm (1-in) square jaws and 30.5 cm/min (12 in/mm) constant rate of travel.

3. Puncture Strength

Test five unaged samples using Standard ASTM D 751-68 and determine the puncture strength by using the tension testing machine with ring clamp, except that the steel ball should be replaced with a ⁵⁄₁₆-in (8-mm) diameter solid steel cylinder centered within the ring clamp.

4. Burst Strength

Test five unaged samples in accordance with ASTM Standard D 751-68 and determine the bursting strength by using the diaphragm test method.

5. Abrasion Resistance

Test five SPD and five WPD unaged samples in accordance with ASTM Standard D 3884-80 (formerly D 1175-71) by using the rotary platform, double-head method. The abrasive wheels must be the rubber-base type equal to the CS-17 "Calibrase" manufactured by Taber Instrument Company. The load on each wheel must be 1000 g (2.2 lb), and the test must be continued for 1000 revolutions. After abrasion, determine the residual tensile strength by the 2.54-cm (1-in) ravelled strip method.

6. Freeze-Thaw Test

Subject five SPD and five WPD samples, 10.2 ± 0.51 cm (4 ± 0.2 in), unaged, to 300 freeze-thaw cycles as described in test method CRD-C 20. Each cycle should be a duration of 2 h ± 4 min duration. Then test samples by using ASTM Standard D 1682 grab test method as described in 1 above.

7. High-Temperature Test

Place five SPD and five WPD samples, 10.2 ± 0.51 by 15.2 ± 0.51 cm (4 ± 0.2 by 6 ± 0.2 in) unaged, in a forced-draft oven at 82.2 ± 2°C (180 ± 3°F) for 48 ± 2 h. Then test each sample at the test temperature by using ASTM Standard D 1682 grab test method as described in 1 above.

8. Low-Temperature Test

Place five SPD and five WPD samples, 10.2 ± 0.51 by 15.2 ± 0.51 cm (4 ± 0.2 by 6 ± 0.2 in) unaged, in a refrigerator at − 27.7 ± 1.6°C (− 17.8 ± 3°F) for 48 ± 2 h, then test each sample at the test temperature by using ASTM Standard D 1682 grab test method as described in 1 above.

9. Acid Test

Submerge five SPD and five WPD samples, 10.2 ± 0.51 by 15.2 ± 0.51 cm (4 ± 0.2 by 6 ± 0.2 in), unaged, in a 1-L glass beaker filled to within 5.1 cm (2 in) of its top with a solution of sufficient hydrochloric acid in about a liter of distilled water to produce a pH of 2 ± 0.1. Cover the beaker with a watchglass and place in a constant-temperature bath at 62.8 ± 2°C (145 ± 5°F). By using a 0.635-cm (0.25-in) glass tube inserted into the spouted beaker to within 1.27 cm (0.5 in) of the beaker bottom, air is bubbled gently through the solution at the rate of one bubble per second continuously for 14 days. The solution should be changed every 24 h, with the new solution warmed to 65.6 ± 0.5°C (150 ± 1°F) before replacing the old solution. Test each sample then for tensile strength and elongation by using ASTM Standard D 1682 grab test method as described in 1 above.

10. Alkali Test

Submerge five SPD and five WPD samples, 10.2 ± 0.51 by 15.2 ± 0.51 cm (4 ± 0.2 by 6 ± 0.2 in), unaged, in a 1-L glass beaker filled to within 5.1 cm (2 in) of its top with a solution of equal amounts of chemically pure sodium hydroxide and potassium hydroxide dissolved in about a liter of distilled water to obtain a pH of 13 ± 0.1. Cover the beaker with a watchglass and place in a constant-temperature bath at 62.8 ± 2°C (145 ± 3.6°F). Using a 0.635-cm (0.25-in) glass tube inserted into the spouted beaker to within 1.27 cm (0.5 in) of the beaker bottom, air is bubbled gently through the solution at the rate of one bubble per second continuously for 14 days. The solution should be changed every 24 h, with the new solution warmed to 65.6 ± 0.5°C (150 ± 9°F) before replacing the old solution. Then test each sample for tensile strength and elongation by using ASTM Standard D 1682 grab test method as described in 1 above.

11. JP-4 Fuel Test

Submerge 10 SPD and 10 WPD samples, 10.2 ± 0.51 by 15.2 ± 0.51 cm (4 ± 0.2 by 6 ± 0.2 in), unaged, in JP-4 fuel at room temperature for 7 days. Then test each sample for tensile strength by using ASTM Standard D 1682 grab test method as described in 1 above.

12. Determination of Equivalent Opening Size (EOS)

Calhoun method, 1972. Based on the Calhoun (1972) method, test five unaged samples. Obtain about 150 g (5.25 oz) of each of the following fractions of a sand composed of sound rounded to subrounded particles:

U.S. Standard Sieve Number

Passing	Retained on	Passing	Retained on	Passing	Retained on
10	20	40	50	70	100
20	30	50	70	100	120
30	40				

Affix the cloth to a standard sieve having openings larger than the coarsest sand used in such a manner that no sand can pass between the cloth and the sieve wall. The sand should be oven-dried. Shaking is accomplished as described in EM 1110-2-1906, Appendix V, paragraph 2d (1) (g), except shaking is continued for 20 min. Determine by sieving (using successively coarser fractions) that fraction of sand of which 5 percent or less by weight passes the cloth; the equivalent opening size of the cloth sample is the "retained on" U.S. standard sieve number of this fraction.

Corps of Engineers 1977 guide specification method . Five unaged fabric samples are tested. Obtain 50 g (1.76 oz) of each of the following fractions of standard glass beads:

U.S. Standard Sieve Number

Designated EOS	Passing	Retained on	Designated EOS	Passing	Retained on
20	18	20	70	60	70
30	25	30	100	80	100
40	35	40	120	100	120
50	45	50			

Suitable glass beads can be obtained from:

Cataphote Division
Ferro Corporation
P.O. Box 2369
Jackson, Mississippi 39205
Telephone: (601) 939-4631

Within each size range, 98 percent of the beads should be within the specified range. The fabric should be affixed to a standard sieve 3.5 cm (8 in) in diameter having openings larger than the largest beads to be used in the test. The fabric should be attached to the sieve in such a manner that no beads can pass between the fabric and sieve wall. Shaking should be accomplished as described in paragraph 2d(1) (g), Appendix V, EM 1110-2-1906, except the times for shaking should be 20 min. Determine by sieving (using successively coarser fractions) that size of beads of which 5 percent or less by weight passes through the fabric; the equivalent opening size (EOS) of the fabric sample is the retained-on U.S. standard sieve number of this fraction.

13.   Determination of Percent of Open Area (POA)

Each of five unaged samples should be placed separately in an 80- by 80-mm (2- by 2-in) glass slide holder and the image projected with a slide projector on a screen. Select a block of 25 openings near the center of the image and measure to the nearest 2.54 μm (0.0001 in) the length and width of each of the 25 openings and the widths of the fibers adjacent to the openings. The percent open area is determined by dividing the sum of the open areas of the 25 openings by the sum of the total areas of the 25 openings and their adjacent fibers.

14. Determination of Gradient Ratio (GR)

A constant-head permeability test should be performed in a permeameter cylinder on soil specimens representative in classification and density of those materials to be protected, and in accordance with EM 1110-2-1906, Appendix VII, with the following modifications:

1. A piece of hardware cloth with 0.64-cm (¼-in) openings should be placed beneath the filter fabric specimen to support it. The fabric and the hardware cloth should be clamped between flanges so that no soil or water can pass around the edges of the cloth.

2. The soil specimen should have a length of 10.16 cm (4 in). Piezometer taps should be placed 2.54 cm (1 in) below the fabric and 2.54, 5.08, and 7.62 cm (1, 2, and 3 in) above the fabric.

3. Tap water should be permeated through the specimen under a constant head loss for a continuous period of 24 h. The tailwater level should be above the top of the soil specimen. The gradient ratio should be determined from the readings taken at the end of the 24-h period.

4. The gradient ratio is the ratio of the hydraulic gradient over the fabric and the 2.54 cm (1 in) of soil immediately next to the fabric (i_1) to the hydraulic gradient over the 5 cm (2 in) of soil between 2.5 and 7.5 cm (1 and 3 in) above the fabric (i_2).

$$\mathrm{GR} \; = \; \frac{i_1}{i_2}$$

Informational Organizations

American Association of Port
Authorities (AAPA)
1612 K Street, N.W.
Washington, DC 20006

American Concrete Institute
Box 19150, Redford Station
Detroit, MI 48219

American Institute of Timber
Construction
333 W. Hampden Rd.
Englewood, CO 80110

American Wood Preservers
Institute
1651 Old Meadow Road
McLean, VA 22101

National Association of Corrosion
Engineers
P.O. Box 218346
Houston, TX 77218

National Forest Products
Association
1619 Massachusetts Avenue, N.W.
Washington, DC 20036

Portland Cement Association
5420 Old Orchard Road
Skokie, IL 60077

The Asphalt Institute
Asphalt Institute Building
College Park, MD 20740

The Museum of Comparative
Zoology
Harvard University
Cambridge, MA 02138

National Research Council
2101 Constitution Avenue, N.W.
Washington, DC 20418

The Sulphur Institute
1725 K Street, N.W.
Washington, DC 20006

U.S. Army Coastal Engineering
Research Center
Kingman Building
Fort Belvoir, VA 22060

U.S. Department of Agriculture
Forest Service
Forest Products and Engineering
Research
P.O. Box 2417
Washington, DC 20013

U.S. Navy Bureau of Yards and
Docks
Washington, DC 20390

Literature Cited

American Concrete Institute (ACI), *Manual of Concrete Practice,* Parts 1 and 3, ACI, Detroit, 1979, 570 pp.

American Concrete Institute (ACI) Committee 201, "Guide to Durable Concrete" (ACI 201.2R-77), ACI, Detroit, 1977, 52 pp.

American Concrete Institute (ACI) Committee 302, "Recommended Practice for Concrete Floor and Slab Construction," (ACI 302-69), Detroit, 1969, 34 pp.

American Concrete Institute (ACI) Committee 347, "Recommended Practice for Concrete Formwork" (ACI 347-68), ACI, Detroit, 1968, sec. 5.2, pp. 25–26.

American Institute of Timber Construction (AITC), *Timber Construction Manual,* 2d ed., Wiley, New York, 1974.

American Society for Metals, "Properties and Selection: Irons and Steels," *Metals Handbook,* 9th ed., vol. 1, American Society for Metals, Metals Park, Ohio, 1978.

American Society for Testing Materials (ASTM), "Applicable ASTM Standards," ASTM, Philadelphia, 1971 (periodically updated).

———, "Symposium on Wood for Marine Use and Its Protection from Marine Organisms," *Special Technical Publication No. 200,* Philadelphia, 1957.

American Society of Civil Engineers (ASCE), "Wood Structures—A Design Guide and Commentary," ASCE, New York, 1975.

American Wood Preservers' Association (AWPA), *AWPA Book of Standards,* Standards C2, C3, C4, C18, and C 23, AWPA, Bethesda, Md., 1979 (periodically updated).

American Wood Preservers Institute (AWPI), C. M. Burpee (ed.), "Pressure Treated Timber Foundation Piles for Permanent Structures," McLean, Va., 1967, 79 pp.

Arber, M. G., and H. E. Vivian, "Inhibition of the Corrosion of Steel Embedded in Mortar," *Australian Journal of Applied Science,* vol. 12, no. 2, Melbourne, 1961, pp. 339–347.

Barrett, R. J., "Use of Plastic Filters in Coastal Structures," *Proceedings, 10th International Conference on Coastal Engineering,* Tokyo, ASCE, vol. 2, 1966, pp. 1048–1067.

Bell, J. R., and R. G. Hicks, "Evaluation of Test Methods and Use Criteria for Geotechnical Fabrics in Highway Applications," *Interim Report No. FHWD/RD-80/021,* Federal Highway Administration, Offices of Research and Development, Structures & Applied Mechanics Division, Washington, D.C., June 1980.

Bodig, J. and J. R. Goodman, "Prediction of Elastic Parameters of Wood," *Wood Science,* vol. 5, no. 4, 1973, pp. 249–264.

Calhoun, C. C., Jr., "Development of Design Criteria and Acceptance Specifications for Plastic Filter Cloth," *Technical Reports-72-7,* U. S. Army Engineer Waterways Experiment Station, Vicksburg, Miss., 1972.

California Department of Public Works, "Chapter V," *Bank and Shore Protection in California Highway Practice,* State of California, 1960, pp. 101–126.

Callender, G., and J. W. Eckert, "Geotechnical Engineering in the Coastal Area," *Special Report, U.S. Army, Corps of Engineers, Coastal Engineering Research Center, Fort Belvoir, Va.*

Civil Engineering Laboratory (CEL), "Dual Treated Wood Piles For Marine Use," *Technical Data Sheet 74-07,* U.S. Navy CEL, Port Hueneme, Calif., 1974.

Dallaire, G., "Filter Fabrics Can Cut Costs of River-Bank and Shore Protection Structures," *Civil Engineering,* American Society of Civil Engineers, New York, March 1977.

Davis, R. E., Jr., and C. E. Haltenhoff, "Mackinac Bridge Pier Construction," *ACI Journal Proceedings,* vol. 53, no. 6, December 1956, pp. 581–596.

Diamond, S., "A Review of Alkali-Silica Reaction and Expansion Mechanisms: Two Reactive Agents," *Cement and Concrete Research,* vol. 6, no. 4, July 1976, pp. 549–560.

Dolar-Mantuani, L., "Alkali-Silica Reactive Rocks in the Canadian Shield," *Highway Research Record, Highway Research Board, National Academy of Sciences—National Research Council (NAS-NRC) No. 268,* Washington, D.C., 1969, pp. 99–117.

Dunham, J. W., and R. J. Barrett, "Woven Plastic Cloth Filters for Stone Seawalls," *Journal of the Waterways Harbors and Coastal Engineering Division,* ASCE, New York, February 1976.

duPont de Nemours, E.I., and Lo, "Chapter 15—Quarries and Open Pit Mining," *Blaster Handbook,* 1977.

Fairley, J. G., et al., "Use of Plastic Filter Cloth in Revetment Construction," *Potamology Investigations Report 21-4,* U.S. Army Engineer District, Memphis, Tenn., June 1970.

Federal Highway Administration, "AASHTO-FHWA Special Products Evaluation List," *Report No. FHWA-RD-76-41,* Federal Highway Administration, Washington, D.C., December 1975.

Gay, T. E., "Stone Crushed and Broken," *California Division of Mines Bulletin 176,* Sacramento, 1957, pp. 565–590.

Hadley, David W., "Alkali Reactivity of Dolomitic Carbonate Rocks," *Highway Research Record, Highway Research Board, NAS-NRC No. 45,* Washington, D.C., 1964, pp. 1–20.

———, "Field and Laboratory Studies on the Reactivity of Sand-Gravel Aggregates," *Journal PCA Research and Development Laboratories,* vol. 10, no. 1, January 1968, pp. 17–33.

Haliburton, T. A., C. C. Anglin, and J. D. Lawmaster, "Selection of Geotechnical Fabrics for Embankment Reinforcement," School of Civil Engineering, Oklahoma State University, Stillwater, Okla., May 1978.

———, J. D. Lawmaster and V. C. McGuffey, "Use of Engineering Fabrics in Transportation-Related Applications," Federal Highway Administration, Office of Development, Washington, D.C., in press.

Hall, J. V., Jr., and R. A. Jachowski, "Concrete Block Revetment Near Benedict, Maryland," *Miscellaneous Paper No. 1-64,* U.S. Army Coastal Research Center, Ft. Belvoir, Va., January 1964.

Highway Research Board, "Calcium Chloride in Concrete," *Bibliography No. 13,* Highway Research Board, NAS-NRC, Washington, D.C., 1952.

———, "Chemical Reactions of Aggregates in Concrete," *Special Report No. 31,* Highway Research Board, NAS-NRC, Washington, D.C., 1958, pp. 1–12.

Hill, C. L., and C. A. Kofoid (eds.), *Marine Borers and Their Relation to Marine Construction on the Pacific Coast,* San Francisco Bay Marine Piling Committee, 1927.

International Nickel Company, Inc., "Engineering Properties and Applications of the Hi-Resists and Ductile Hi-Resists," *Publication No. A-1231,* New York, January 1976.

Jarlan, G. L. E., "A Perforated Vertical Breakwater," *The Dock and Harbor Authority,* vol. 41, no. 486, April 1961, pp. 394–398.

King, J. C., "Concrete by Intrusion Grouting," *Handbook of Heavy Construction,* McGraw-Hill, New York, chap. 12, 1959.

Kofoid, C. A., and R. C. Miller, "The Boring Habit," in C. L. Hill and C. A. Kofoid (eds.), *Marine Borers and Their Relation to Marine Construction On the Pacific Coast,* San Francisco Bay Marine Piling Committee, 1927, pp. 222–338.

Ledbetter, W. B., "Synthetic Aggregates from Clay and Shale: Recommended Criteria for Evaluation," *Highway Research Record, Highway Research Board, MAS-MRC, No. 430,* Washington, D.C., 1973, pp. 9–15.

Machemehl, J. L., T. J. French, and N. E. Huang, "New Method for Beach Erosion Control," *Proceedings of Civil Engineering in the Oceans/III,* June 9–12, 1975, University of Delaware, Newark, ASCE, vol. 1, 1975, p. 142.

MacNaughton, M. F., and J. B. Herbich, "Accidental Air in Concrete," *ACI Journal Proceedings,* vol. 51, no. 3, November 1954, pp. 273–284.

Magoon, O. T., and N. Shimizu, "Use of Dolos Armor Units in Rubble-Mound Structures, e.g., for Conditions in the Arctic," *Proceedings from the First International Conference on Port and Ocean Engineering under Arctic Conditions,* vol. II, Technical University of Norway, Trondheim, Norway, 1971, pp. 1089–1108. (Also CERC Reprint 1-73.)

McCall, J. T., and R. J. Claus, "Effects of Pellet and Flake Forms of Calcium Chloride in Concrete," *Bulletin No. 75,* Highway Research Board, 1953.

McCoy, W. J., and A. G. Caldwell, "New Approach to Inhibiting Alkali-Aggregate Expansion," *ACI Journal Proceedings,* vol. 47, no. 9, May 1951, pp. 693–708.

Mielenz, R. C., "Reactions of Aggregates Involving Solubility Oxidation, Sulfates or Sulfides," *Highway Research Record, Highway Research Board, NAS-NRC, No. 43,* Washington, D.C., 1964, pp. 8–18.

Mills, A. P., H. W. Hayward, and L. F. Rader, "Chapter 13—Building Stones and Masonry," *Materials of Construction,* sec. III, 1955, pp. 301–312.

Mitchell, J. K., "Stabilization of Soils for Foundations of Structures," *Manual for Foundation Design,* U.S. Army Corps of Engineers, Waterways Experiment Station, Vicksburg, Miss., 1976.

National Association of Corrosion Engineers (NACE), Basic Corrosion Course, Houston, October 1979.

National Forest Products Association, "National Design Specification for Wood Construction," Washington, D.C., 1977, 80 pp.

Nedeco, River Studies and Recommendations on Improvements of Niger and Benue, North Holland, Amsterdam, 1959.

Newman, E. S., et al., "Effects of Added Materials on Some Properties of Hydrating Portland Cement Clinkers," *Journal of Research, U.S. National Bureau of Standards,* vol. 30, April 1943, p. 281.

Palermo, M. T., "Damage by Western Subterranean Termites," *Marine Borer Conference,* U.S. Naval Civil Engineering Research and Evaluation Laboratory, Port Hueneme, Calif., 1951.

Portland Cement Association, "Design and Control of Concrete Mixtures," *Design Manual,* 12th ed., Chicago, 1979.

Ray, D. L., "Nutritional Physiology of Limnoria," in D. L. Ray (ed.), *Marine Boring and Fouling Organisms,* University of Washington Press, Seattle, 1959, pp. 46–60.

Savage, R. P., "Experimental Study of Dune Building with Sand Fence," *Proceedings of 8th Conference on Coastal Engineering,* November 1962, Council on Wave Research, University of California, 1963.

Sherard, J. L., et al., "Explorations for Foundations and Embankment Construction Materials," in *Earth and Earth-Rock Dams,* Wiley, New York, 1963, chap. 4.

Shideler, J. J., and A. Litvin, "Structural Applications of Pumped and Sprayed Concrete," *Development Department Bulletin D72,* Portland Cement Association, January 1964, 29 pp.

Soil Testing Services, "Evaluation of In-Place Geotextiles (Carthager Mills; POLY-FILTER X) 79th Street Causeway, Miami Beach, Florida, and Bahia Honda Bridge, Bahia Honda Key, Florida," *Report to Carthage Mills, Inc.,* STS Consultants Ltd., Northbrook, Ill., January 1980.

Steel Structures Painting Council, "Systems and Specifications," *Steel Structures Painting Manual,* vol. 2, 1975, 850 pp.

Steinour, H. H., "Concrete Mix Water—How Impure Can it Be?" *Journal, Research and Development Laboratories,* Portland Cement Association, vol. 2, no. 3, September 1960, pp. 32–50.

Stevenston, C. A., "Set a Wave to Catch a Wave," *Canadian Consulting Engineer,* June 1963.

Steward, J., R. Williamson, and J. Mahoney, "Guidelines for Use of Fabrics in Construction and Maintenance of Low-Volume Roads," *Report FHWA-T5-78-205* (Federal Highway Administration Reprint), Forest Service, Pacific Northwest Region, U.S. Department of Agriculture, Washington, D.C., June 1977.

Sulphur Institute, The, "U.S. Bureau of Mines Transfers Sulphur Concrete Technology to Industry," *Sulfur Research and Development,* vol. 2, Washington, D.C., 1979.

Tuthill, L. H., "Conventional Methods of Repairing Concrete," *ACI Journal Proceedings*, vol. 57, no. 2, August 1960, pp. 129–138.

U.S. Army, Chief of Engineers, "Civil Works Construction, Guide Specification, Plastic Filter Fabric," *No. CW-02215*, Washington, D.C., November 1977.

U.S. Army, Corps of Engineers, *Design of Breakwaters and Jetties*, EM 1110-2-2904, GPO, Washington, D.C., 1971, change 3.

———, "Embankments," in *Earth and Rock Fill Dams—General Design and Construction Considerations*, GPO, Washington, D.C., 1971a, pp. 1-5–5-3.

———, "Alkali-Silica Reactions," Appendix B, and "Alkali-Carbonate Rock Reactions," Appendix C, *Engineering Manual 1110-2-2000*, GPO, Washington, D.C., 1971b.

———, Coastal Engineering Research Center, *Shore Protection Manual*, 3d ed., vols. I, II, and III, Stock No. 008-022-00113-1, GPO, Washington, D.C., 1977.

———, North Central Division, "Guide Specifications for Rock Usage in Construction of Diked Disposal Facilities and Other Rock Harbor Structures," Chicago, Ill., 1978.

U.S. Army Engineer, Waterways Experiment Station, "Investigation of Suitability of Prepacked Concrete for Mass and Reinforced Concrete Structures," *Technical Memorandum No. 6-330*, Vicksburg, Miss., August 1954, 44 pp.

U.S. Bureau of Reclamation, "Final Report of Tests on Prepackaged Concrete—Barker Dam Materials Laboratories," *Report No. C-338*, U.S. Department of the Interior, Denver, March 1949.

———, *Concrete Manual*, U.S. Department of the Interior, Denver, 7th ed., 1963, 619 pp.

U.S. Department of Agriculture (USDA), *Preservative Treatment of Wood by Pressure Methods*, Agriculture Handbook No. 40, Washington, D.C., December 1952.

U.S. Naval Facilities Engineering Command (NAVFAC), "Specification 75M-B10a for Installation of Flexible Barriers on Marine Borer Damaged Wood Bearing Piles," Alexandria, Va., 1975.

Van Bendegon, L., and A. Zanen, "Lecture Notes on Revetments," *International Course in Hydraulic Engineering*, Delft, Holland, 1960.

Wakeman, C. M., and F. J. Steiger, "The Case of the Proliferating Punctata," *Wood Preserving News*, September, 1966.

White, A. H., "Rocks and Their Decomposition Products," *Engineering Materials*, 1948, chap. 17, pp. 384–408.

Whiteman, R. V., "Hydraulic Fills to Support Structural Loads," *Journal of Soil Mechanics and Foundations Division*, ASCE, vol. 96, no. SM1, January 1970, p. 24.

Williams, J. W., "Tremie Concrete Controlled with Admixtures," *ACI Journal Proceedings*, vol. 55, no. 8, February 1959, pp. 839–850.

Woods, Hubert, "Durability of Concrete Construction," *Monograph No. 4*, American Concrete Institute, Iowa State University Press, Detroit, 1968, pp. 57–68.

Bibliography

Aluminum Association, *Aluminum Standards and Data*, 6th ed., New York, March 1979.

American Society for Metals, "Properties and Selection: Irons and Steels," *Metals Handbook*, 9th ed., vol. 1, Metals Park, Ohio, 1978.

———, "Properties and Selection: Nonferrous Alloys and Pure Metals," *Metals Handbook*, 9th ed., vol. 2, Metals Park, Ohio, 1979.

———, "Properties and Selection: Stainless Steels, Tool Materials and Special-Purpose Metals," *Metals Handbook*, 9th ed., vol. 3, Metals Park, Ohio, 1980.

Bates, J. F., and J. M. Popplewell, "Corrosion of Condenser Tube Alloys in Sulfide Contaminated Brine," *Corrosion*, vol. 31, no. 8, August 1975.

Bottom Protection Committee, Subgroup B, "Consideration about Applications of Synthetic Cloth in Civil Engineering Erosion Control Projects," *Partial Report No. 3*, Rijkswaterstaat, Deltadienst, The Hague, Netherlands, June 1966.

Building Research Digest, "High Alumina Cements," *Building Research Digest No. 27*, Garston, England, February 1951.

Clifton, J. R., H. G. Beeghly, and R. G. Mathey, "Nonmetallic Coatings for Concrete Reinforcing Bars," Final Report No. FHWA-RD-74-18, National Bureau of Standards for Federal Highway Administration, Washington, D.C., February 1974.

Copper Development Association Inc., *Standards Handbook, Wrought Copper and Copper Alloy Mill Products*, Part 2, *Alloy Data*, New York, 1973.

———, *Standards Handbook, Cast Products*, Part 7, *Alloy Data*, New York, 1980.

Duriez, M., and R. Lezy, "New Possibilities for Insuring the Rapid Hardening of Cements, Mortars, and Concretes," *Annales, Institute Technique du Batiment et des Travaux Publics No. 9*, Paris, 1956, p. 138.

Federal Highway Administration, "Coated Reinforcing Steel," *Report No. 1, FHWA Notice N5080.33*, Washington, D.C., April 1975.

Feret, L., and M. Vepuat, "The Effect on Shrinkage and Swelling of Mixing Different Cements to Obtain Rapid Set," *Revue des Materiaux de Construction*, No. 496, Paris, 1957, pp. 1–10.

Forbrich, L. R., "The Effect of Various Agents on the Heat Liberation Characteristics of Portland Cement," *ACI Journal Proceedings*, vol. 37, no. 2, November 1940, pp. 161–185.

Giroud, J. P., J. P. Girouc, and F. Bally, "Behavior of a Nonwoven Fabric in an Earth Dam," University of Grenoble, Grenoble, France, August 1977.

Gudas, J. P., and H. P. Hack, "Sulfide Induced Corrosion of Copper Nickel Alloys," *Corrosion*, vol. 35, no. 2, February 1979.

Industrial Publishing Co., "Welding Data Book," *Welding Design and Fabrication Magazine*, West Cleveland, Ohio, 1979.

International Nickel Company, Inc., "Guidelines for Selection of Marine Materials," *Publication No. A-404*, New York, March 1975.

———, "Mechanical and Physical Properties of the Austenitic Chromium-Nickel Stainless Steels at Subzero Temperatures," *Publication A-313*, New York, April 1975a.

———, "General Seawater Corrosion," *Publication No. A-1276*, New York, March 1978.

————, "Marine Biofouling," *Publication No. A-1296,* New York, September 1979.

Laque, F. L., *Marine Corrosion Causes and Prevention,* Wiley, New York, 1975.

Moffatt and Nichol, Engineers, "Low Cost Shore Protection," *Final Report on Shoreline Erosion Control Demonstration Program (Section 54),* U.S. Army Corps of Engineers, Coastal Engineering Research Center, January 1981.

Robson, T. D., "Characteristics and Applications of Mixtures of Portland Cement and High-Alumina Cements," *Chemistry and Industry,* vol. 12, no. 12, London, 1961.

Ross, R. W., and D. B. Anderson, "Protection of Steel Piling on Marine Splash and Spray Zones—The Metallic Sheathing Concept," presented at 4th International Congress on Marine Corrosion and Fouling, Juan les Pins-Antibes, June 14–16, 1976.

Teige, N. G., and R. L. Kane, "Experience with Titanium Structures in Marine Service," *Materials Performance,* vol. 9, no. 8, August 1970.

Terzaghi, Karl, *Theoretical Soil Mechanics,* Wiley, New York, 1943.

Thomas, A., "Decay Fungi in Waterfront Structures," *Marine Borer Conference,* U.S. Naval Civil Engineering Research and Evaluation Laboratory, Port Hueneme, Calif., 1951.

Titanium Metals Corporation of America, "Marine Applications of Titanium and Its Alloys," West Caldwell, N.J., *Publication No. EP rl 8-69 3M,* November 1968.

United States Steel Corp., "Steel H-Piles," *Publication No. ADUSS 25-2700,* Pittsburgh, October 1967.

————, "Steel Piling Design Manual," *Publication No. ADUSS 25-3848-03,* Pittsburgh, April 1972.

Abbreviations

A	ampere
AASHTO	American Association of State Highway and Transportation Officials
ACA	ammoniacal copper arsenate
ACI	American Concrete Institute
ALSC	American Lumber Standards Committee
ANFO	ammonium nitrate fuel oil
ASTM	American Society for Testing Materials
AWPA	American Wood Preservers Association
BPD	both principal directions
CERC	Coastal Engineering Research Center
C	clay
C_3A	tricalcium aluminate
CCA	chromated copper arsenate
CH	clay with liquid limits greater than 50
CL	clay with liquid limits less than 50
cm	centimeter(s)
cm/s	centimeter(s) per second
CZC	chromated zinc chloride
DCPD	dicyclopentadiene
DP	dipentene
emc	equilibrium moisture content
EOS	equivalent opening size
FCAP	fluorochromearsenate phenol
FRP	fiber-reinforced plastic
ft	foot (feet)
ft/min	foot (feet) per minute
ft/(ft·°F)	feet per foot per degree Fahrenheit
ft^2	square foot (feet)
ft^2/gal	square foot (feet) per gallon
ft^3	cubic foot (feet)
ft· lb	foot-pound (foot-pounds)

ft^3/min	cubic foot (feet) per minute
g	gram(s)
g/m^2	gram(s) per square meter
gal	gallon(s)
g/L	gram(s) per liter
GR	gradient ratio
in	inch(es)
kg	kilogram(s)
kg/m^3	kilogram(s) per cubic meter
km	kilometer
kN	kilonewton(s)
kN/m^3	kilonewton(s) per cubic meter
kPa	kilopascal(s)
L	liter (s)
lb	pound(s)
lb/ft^2	pound(s) per square foot
lb/in^2	pound(s) per square inch
lb/in^3	pound(s) per cubic inch
lb/ft^3	pound(s) per cubic foot
lb/in	pound(s) per inch
lb/yd^3	pound(s) per cubic yard
m	meter(s)
mA/ft^2	milliampere(s) per square foot
mA/m^2	milliampere(s) per square meter
$m/(m\cdot {}^\circ C)$	meter(s) per meter per degree Celsius
m^2	square meter(s)
m^2/L	square meter(s) per liter
m^3	cubic meter(s)
m^3/h	cubic meter(s) per hour
m^3/min	cubic meter(s) per minute
M	silt
M_i	initial moisture content
M_f	final moisture content
MH	silt with liquid limits greater than 50
mi	mile(s)
ML	silt with liquid limit less than 50
MLW	mean low water

mm	millimeter(s)
m/min	meter(s) per minute
MPa	megapascal(s)
MS	sulfate-resistant cement
mV	millivolt(s)
mV/°C	millivolt(s) per degree Celsius
mV/°F	millivolt(s) per degree Fahrenheit
mΩ	megohm(s)
μm	micrometer
N	normal
NDT	nil ductility transformation
N • m	newton-meter(s)
N/cm	newton(s) per centimeter
N/m^3	newton(s) per cubic meter
O	organic material
OH	organic silt and clay with liquid limits greater than 50
OL	organic silt and clay with liquid limits less than 50
oz/ft^2	ounce(s) per square foot
Ω • cm	ohm-centimeter(s)
Ω/V	ohm(s) per volt
PA	preplaced aggregate
PC	polymer concrete
PCC	portland cement concrete
PIC	polymer-impregnated concrete
POA	percent of open area
PPCC	polymer–portland cement concrete
ppm	parts per million
Q	Corps of Engineers designation of unconsolidated-undrained soil
R	Corps of Engineers designation of consolidated-undrained soil
RTRP	reinforced thermosetting resin pipe
s	second
S	Corps of Engineers designation of consolidated-drained soil
SA	sulfur-asphalt
SC	sulfur concrete
SPD	stronger principal direction
SPM	*Shore Protection Manual*

t	metric ton
USCS	Unified Soil Classification System
USDAFS	United States Department of Agriculture Forest Service
UV	ultraviolet
V	volt
VMA	voids in mineral aggregate
WES	Waterways Experiment Station
WPD	weaker principal direction
yd^2	square yard(s)
yd^3	cubic yard(s)
yd^3/h	cubic yard(s) per hour

Glossary

abraded strength   The result when tested in accordance with ASTM D1682, "Breaking Load and Elongation of Textile Fabric, One-Inch Ravelled Strip Method." One-inch-square jaws at a constant rate of traverse of 12 in/min (4.75 cm/min).

abrasion resistance   The ability of a surface to resist wear by friction.

alkaline   The excess of hydroxyl ions over hydrogen ions. Seawater is usually alkaline.

alkalinity   The capacity of water to accept protons, i.e., hydrogen ions. It is usually expressed as milliequivalents per liter.

anaerobic   An oxygen-independent type of respiration.

anneal   To subject to high heat, with subsequent cooling, so as to soften thoroughly and render less brittle.

anode   The positive pole or electrode of an electrolytic cell.

aquatic   Growing or living in, or frequenting, water as opposed to terrestrial.

armor   The outer or exposed layer of material(s) (stones, blocks, etc.) in a protective structure subjected to attack by wave or scour forces.

austenitic   Having a solid solution of carbon or iron carbide in iron as a constituent of steel under certain conditions.

bank   (1) The rising ground bordering a lake, river, or sea; of a river or channel, designated as right or left as it would appear facing downstream. (2) An elevation of the seafloor of large area, located on a Continental (or island) Shelf and over which the depth is relatively shallow but sufficient for safe surface navigation; a group of shoals. (3) In its secondary sense, a shallow area consisting of shifting forms of silt, sand, mud, and gravel, but in this case it is only used with a qualifying word such as "sandbank" or "gravelbank."

basin, boat   A naturally or artificially enclosed or nearly enclosed harbor area for small craft.

bathymetry   The measurement of depths of water in oceans, seas, and lakes; also, information derived from such measurements.

bay   A recess in the shore or an inlet of a sea between two capes or headlands that is not as large as a gulf but is larger than a cove.

beach   The zone of unconsolidated material that extends landward from the low water line to the place where there is marked change in material or physiographic form or to the line of permanent vegetation (usually the effective limit of storm waves). The seaward limit of a beach—unless otherwise specified—is the mean low water line.

beach erosion   The carrying away of beach materials by wave action, tidal currents, littoral currents, or wind.

benthic   Pertaining to the subaquatic bottom.

benthos   A collective term describing (1) bottom organisms attached to or resting on or in the bottom sediments or (2) the community of animals living in or on the bottom.

bioassay   The use of living organisms as an index to determine environmental conditions.

biochemical oxygen demand (BOD)   The amount of oxygen required by the biological population of a water sample to oxidize the organic matter in that water. It is usually determined over a 5-day period under standardized laboratory conditions and hence may not represent actual field conditions.

biological resistance   Ability to resist degradation due to microorganisms.

biomass   The amount of living material in a unit area for a unit time. Also called standing crop, standing stock, live weight.

biota   The living part of a system (flora and fauna).

bottom   The ground or bed under any body of water; the bottom of the sea.

boulder   A rounded rock more than 25.4 cm (10 in) in diameter; larger than a cobblestone. See *soil classification.*

breaker   A wave breaking on a shore or over a reef or other feature.

breakwater   A structure protecting a shore area, harbor, anchorage, or basin from waves.

bulkhead   A structure or partition to retain or prevent sliding of the land. A secondary purpose is to protect the upland against damage from wave action.

buoy   A float; especially a floating object moored to the bottom to mark a channel, anchor, shoal, rock, etc.

buoyancy   The resultant of upward forces exerted by the water on a submerged or floating body and equal to the weight of the water displaced by the body.

burst strength   The resistance of a fabric to rupture due to pressure applied at right angles to the plane of the fabric under specified conditions, usually expressed as the pressure causing failure. Burst is due to tensile failure of the fabric.

cathode   The negative pole or electrode of an electrolytic cell.

causeway   A raised road across wet or marshy ground or water.

channel   (1) A natural or artificial waterway of perceptible extent which either periodically or continuously contains moving water or which forms a connecting link between two bodies of water. (2) The part of a body of water deep enough to be used for navigation through an area otherwise too shallow for navigation. (3) A large strait, as the English Channel. (4) The deepest part of a stream, bay, or strait through which the main volume or current of water flows.

clay   A fine-grained soil with cohesive strength inversely related to water content. It is plastic when moist and hardens when baked or fired. See *soil classification*.

cliff   A high, steep face of rock; a precipice.

clogging   The phenomena causing either a reduction in or the elimination of the permeability of the filter.

coast   A strip of land of indefinite width (may be several miles) that extends from the shoreline inland to the first major change in terrain features.

coastal area   The land and sea area bordering the shoreline.

cobble (cobblestone)   A naturally rounded stone larger than a pebble, especially one 15 to 30 cm (6 in to 1 ft) in diameter.

collector pipe   A pipe capable of collecting and carrying water from the soil.

colonization   A natural phenomenon whereby a species invades an area previously unoccupied by that species and becomes established. To be successful, the species must be able to reproduce in that area.

contour   A line on a map or chart representing points of equal elevation with relation to a *datum*. It is called an *isobath* when it connects points of equal depth below a datum.

coral   (1) (Biology) Marine coelenterates (Madreporia), solitary or colonial, which form hard external coverings of calcium compounds or other materials. The corals which form large reefs are limited to warm, shallow waters; those which form solitary, minute growths may be found in colder waters to great depths. (2) (Geology) The concretion of coral polyps, composed almost wholly of calcium carbonate, forming reefs and treelike and globular masses. May also include calcareous algae and other organisms producing calcareous secretions, such as bryozoans and hydrozoans.

core   A vertical cylindrical sample of the bottom sediments from which the nature and stratification of the bottom can be determined. The interior material of a breakwater or groin.

creep   To slip or become slightly displaced; specifically of metal to shift longitudinally under weight.

cure   To alter industrially, as to vulcanize (rubber) or to treat (synthetic resins) with heat or chemicals to make infusible.

current   A flow of water.

current, littoral   Any current in the littoral zone caused primarily by wave action, e.g., longshore current, rip current.

dap   A notch cut in one timber to receive another, usually permitting the two timbers to be flush.

datum, plane   The horizontal plane to which soundings, ground elevations, or water surface elevations are referred. The plane is *tidal datum* when defined by a certain phase of the tide. The following data are ordinarily used on hydrographic charts:

mean low water   Atlantic coast (U.S.), Argentina, Sweden, and Norway

mean lower low water   Pacific coast (U.S.)

low water springs   United Kingdom, Germany, Italy, Brazil, and Chile

low water datum   Great Lakes (United States and Canada)

lowest low water springs   Portugal

low water Indian springs   India and Japan

lowest low water   France, Spain, and Greece

A common datum used on topographic maps is based on *mean sea level.*

deflation   The removal of earth particles by wind action.

denier   A unit expressing the fineness of silk, rayon, nylon, or other synthetic yarn in terms of weights in grams per 9000 m of length.

depth   The vertical distance from a specified tidal datum to the sea floor.

dike (dyke)   A wall or mound built around a low-lying area to prevent flooding.

dolos (*pl:* dollosse)   A type of precast concrete armor unit for use in breakwaters and as shore protection.

dolphin   A cluster of piles.

dunes   (1) Ridges or mounds of loose, wind-blown material, usually sand. (2) Bed forms smaller than bars but larger than ripples that are out of phase with any water-surface gravity waves associated with them.

durability   A relative term for the resistance of material to loss of physical properties or appearance as a result of wear or dynamic operation.

elongation at failure   The length of a fabric test specimen when it is broken in a tensile test (ASTM D1682-64) compared to its original length, expressed as a percent.

embankment   An artificial bank such as a mound or dike, generally built to hold back water or to carry a roadway.

endemic   Native to a specific geographic area.

equivalent opening size (EOS)   The number of a U.S. standard sieve having openings closest in size to the diameter of uniform particles (which will have 95 percent by weight) retained by the fabric when shaken in a prescribed manner.

erosion   The wearing away of land by the action of natural forces. On a beach, the carrying away of beach material by wave action, tidal currents, or littoral currents or by deflation.

estuary (1) The part of a river that is affected by tides. (2) The region near a river mouth in which the fresh water of the river mixes with the salt water of the sea.

fauna Animal life as opposed to plant life (flora). Generally, the entire group of animals found in an area.

fibrillated yarn Yarns formed of fibers cut from sheet plastic film.

filament A single thread (yarn) of extreme length.

fill Fibers or yarns placed at right angles to the warp.

filter fabric A permeable fabric of synthetic fibers whose function is to retain soil and be permeable to water.

flora Plant life as opposed to animal life (fauna). The entire group of plants found in an area.

foreshore The part of the shore lying between the crest of the seaward berm (or upper limit of wave wash at high tide) and the ordinary low water mark that is ordinarily traversed by the uprush and backrush of waves as the tides rise and fall.

fouling organism An organism that attaches to the surface of submerged or introduced objects regardless of whether the objects are natural or manmade.

galvanize To subject to the action of electric currents; to coat with zinc.

geotextile Any permeable textile used with foundations, soils, rock, earth, or any other geotechnical material as an integral part of a manmade project, structure, or system.

geotextile filter A permeable fabric of synthetic fibers whose function is to retain soil and be permeable to water.

gravel A coarse-grained, cohesionless material whose size ranges between 76.2 mm (3 in) and the No. 4 sieve. See *soil classification*.

groin (British, groyne) A shore protection structure built (usually perpendicular to the shoreline) to trap littoral drift or retard erosion of the shore.

groundwater Subsurface water occupying the zone of saturation. In a strict sense, the term is applied only to water below the *water table*.

gulf A large embayment in a coast the entrance of which is generally wider than the length.

habitat The place where an organism lives.

harbor (British, harbour) Any protected water area affording a place of safety for vessels.

heat-bonded A fabric web subjected to a relatively high temperature at which the filaments weld together at their contact points.

heat of hydration The heat evolved or absorbed when hydration occurs as, specifically, when water is added to a calcium aluminate powder to produce cement.

impermeable groin   A groin through which sand cannot pass.

interlocking concrete block   A cast or machine-produced concrete block having interengaging or overlapping edges.

jetty   (1) (U.S. usage) On open seacoasts, a structure extending into a body of water and designed to prevent shoaling of a channel by littoral materials and to direct and confine the stream or tidal flow. Jetties are built at the mouths of rivers or tidal inlets to help deepen and stabilize channels. (2) (British usage) Jetty is synonymous with *wharf* or *pier*.

larva   A sexually immature form of any animal, unlike its adult form, requiring changes before reaching the adult form.

littoral   Of or pertaining to a shore, especially of the sea.

littoral drift   The sedimentary material moved in the littoral zone under the influence of waves and currents.

littoral transport   The movement of littoral drift in the littoral zone by waves and currents. Includes movement parallel (longshore transport) and perpendicular (onshore-offshore transport) to the shore.

longshore   Parallel to and near the shoreline.

monofilament   A single filament of a manmade fiber, usually of a denier higher than 15.

mud   A fluid-to-plastic mixture of finely divided particles of solid material and water.

multifilament   A yarn consisting of many continuous filaments or strands.

nonwoven fabric   A textile structure produced by bonding or interlocking of fibers, or both, accomplished by mechanical, chemical, or solvent means. Also, combinations thereof excluding woven and knitted fabrics.

nourishment   The process of replenishing a beach. It may be brought about naturally by longshore transport or artificially by the deposition of dredged materials.

nylon fiber   A manufactured fiber in which the fiber-forming substance is any long-chain synthetic polyamide having recurring amide groups (-NH-CO-) as an integral part of the polymer chain.

offshore   (1) In beach terminology, the comparatively flat zone of variable width extending from the breaker zone to the seaward edge of the Continental Shelf. (2) A direction seaward from the shore.

onshore   A direction landward from the sea.

organism   Any living individual whether plant or animal.

outfall   A structure extending into a body of water for the purpose of discharging sewage, storm runoff, or cooling water.

overtopping   Passing of water over the top of a structure as a result of wave run-up or surge action.

percent opening area (POA)   The visible net area of a fabric that is available for water to pass through the fabric. Normally it is determinable only for woven and nonwoven fabrics having distinct visible and measurable openings that continue directly through the fabric.

permeable groin A groin with openings large enough to permit passage of appreciable quantities of littoral drift.

pier A structure, usually of open construction, extending out into the water from the shore to serve as a landing place, a recreational facility, etc., rather than to afford coastal protection. In the Great Lakes, a term sometimes improperly applied to jetties.

piles A long, heavy timber or section of concrete or metal to be driven or jetted into the earth or seabed to serve as a support or protection.

piles, sheet A pile with a generally slender flat cross section to be driven into the ground or seabed and meshed or interlocked with like members to form a diaphragm, wall, or bulkhead.

piling A group of piles.

piping The process by which soil particles are washed in or through pore spaces in drains and filters.

plastic filter See *filter fabric.*

plastic filter fabric See *filter fabric.*

polyamide See *nylon fiber.*

polyester fiber A manufactured fiber the substance of which is any long-chain synthetic polymer composed of at least 85 percent by weight of an ester of dihydric alcohol and terephthalic acid (FTC).

polyethylene fiber A manufactured fabric in which the fiber-forming substance is an olefin made from polymers or copolymers of ethylene.

polymer A high-molecular-weight chainlike structure from which manmade fibers are derived. It is produced by linking together molecular units called monomers consisting predominantly of nonmetallic elements or compounds.

polypropylene fiber A manufactured fiber the substance of which is an olefin made from polymers or copolymers of propylene.

polyvinylidene chloride fiber A manufactured fiber the substance of which is a thermoplastic derived by copolymerization of two or more vinyl monomers.

port A place where vessels can transfer cargo and people. It may be the entire harbor including its approaches and anchorages, or it may be the commercial part of a harbor where the quays, wharves, and facilities for transfer of cargo and docks and repair shops are situated.

pozzolan A siliceous rock of volcanic origin, first found near Puteoli (modern Pozzuoli), Italy, used in preparing a hydraulic cement.

precast cellular block A cast or machine-produced concrete block having continuous void(s) through the vertical plane; normally with smooth vertical or near vertical sides (not interlocking). Sometimes cabled together horizontally to form a mat.

puncture resistance   Resistance to failure of a fabric due to a blunt object applying a load over a relatively small area. Failure is due to tensile failure of the fibers.

quarrystone armor units   Relatively large quarrystones that are selected to fit specified geometric characteristics, including compact dimensional proportions and a nearly uniform size, usually large enough to require individual placement. In normal cases they are placed in a layer at least two stones thick.

quay   (pronounced "key") A stretch of paved bank or a solid artificial landing place parallel to the navigable waterway for use in loading and unloading vessels.

resin-bonded   A fabric web impregnated with a resin which serves to coat and cement the fibers together.

revetment   A facing of stone, concrete, etc., built to protect a scarp, embankment, or shore structure against erosion by wave action or currents.

rheology   Science dealing with the deformation and flow of matter.

riprap   A protective layer or facing of quarrystone randomly placed to prevent erosion, scour, or sloughing of an embankment or bluff toe. Also, the stone so used, usually well graded within wide size limits. The quarrystone is placed in a layer at least twice the thickness of the 50 percent size stone or 1.25 times the thickness of the largest size stone in the size gradation.

rubble   (1) Loose angular waterworn stones along a beach. (2) Rough, irregular fragments of rock.

rubble mound structure   A mound of random-shaped and random-placed stones protected with a cover layer of selected stones or specially shaped concrete armor units. (Armor units in the primary cover layer can be placed in orderly manner or dumped at random.)

runup   The rush of water up a structure or beach on the breaking of a wave. Also *uprush*. The amount of runup is the vertical height above stillwater level that the rush of water reaches.

sand   An earthy material whose grain size is between 4.76 and 0.075 mm (0.19 and 0.003 in). Within this classification sand may vary from coarse to fine. Sand is cohesionless but exhibits the appearance of cohesion when wet. See *soil classification*.

sandcore jetty   A jetty, groin, or breakwater the core material of which consists of sand rather than stone.

saran   See *polyvinylidene chloride fiber*.

scour   Removal of underwater material by waves and currents, especially at the base or toe of a shore structure.

scour protection   The protection at the base or toe of a structure to prevent removal of underwater material by waves and currents.

screed   A strike board used to level or strike off concrete pavement slabs or to cushion courses for block pavements.

seawall   A structure separating land and water areas primarily designed to prevent erosion and other damage due to wave action. See also *bulkhead.*

seiche   (1) A standing-wave oscillation of an enclosed water body that continues, pendulum fashion, after the cessation of the originating force, which may have been either seismic or atmospheric. (2) An oscillation of a fluid body in response to a disturbing force having the same frequency as the natural frequency of the fluid system. Tides are now considered to be seiches induced primarily by the periodic forces exerted by the sun and moon. (3) In the Great Lakes area, any sudden rise in the water of a harbor or a lake whether or not it is oscillatory. Although inaccurate in a strict sense, this usage is well established in the Great Lakes area.

seismic sea wave (*tsunami*)   A long-period wave caused by an underwater seismic disturbance or volcanic eruption. Commonly misnamed tidal wave.

shear fracture   An action or stress resulting from applied forces which causes or tends to cause two contiguous parts of a body to slide relatively to each other in a direction parallel to their plane of contact.

sheet pile   See *pile, sheet.*

shoal (noun)   A detached elevation of the sea bottom comprised of any material except rock or coral which may endanger surface navigation.

shoal (verb)   (1) To become shallow gradually; (2) to cause to become shallow; (3) to proceed from a greater to a lesser depth of water.

shore   The narrow strip of land in immediate contact with the sea including the zone between high and low water lines. A shore of unconsolidated material is usually called a beach.

shoreline   The intersection of a specified plane of water with the shore or beach. (For example, the high-water shoreline would be the intersection of the plane of mean high water with the shore or beach.) The line delineating the shoreline on National Ocean Survey nautical charts and surveys approximates the mean high water line.

shotcrete   A mixture of sand, small gravel, and water applied by air hose.

silt   A fine-grained soil of low plasticity which may exhibit an apparent cohesion that is due to capillary forces. See *soil classification.*

slope   The degree of inclination to the horizontal. It is usually expressed as a ratio, such as 1:25 or 1 on 25, indicating 1 unit vertical rise in 25 units of horizontal distance. Also expressed as a decimal fraction (0.04), as degrees (2°18'), or as a percent (4%).

slump   To fall or sink suddenly.

soil classification (size)   An arbitrary division of a continuous scale of grain sizes such that each scale unit or grade may serve as a convenient class interval for conducting the analysis or for expressing the results of an analysis.

spall To break up or reduce by chipping with a hammer; to chip or crumble.

species An aggregate of interbreeding populations that under natural conditions is reproductively isolated.

splash zone The zone immediately landward of the mean higher high water level affected by the wave spray.

stone, derrick Stone, generally 1 ton and up, heavy enough to require handling individual pieces by mechanical means.

strength Load capacity at failure. Depending on the usage, load can be expressed in stress, force per unit width, or force.

surf zone The area between the outermost breaker and the limit of wave uprush.

tensile strength The strength shown by a fabric subjected to tension as distinct from torsion, compression, or shear.

terrestrial Of or relating to the earth and its inhabitants as opposed to aquatic.

tetrapod A type of precast concrete armor unit for use in breakwaters and as shore protection.

thixotropic That which becomes fluid when shaken, stirred, or otherwise disturbed and sets again to a gel when allowed to stand.

toe The lower-elevation terminus of a revetment or side slopes of a groin, breakwater, or jetty. The outer limit of a scour protection mound.

tremie An apparatus for depositing and consolidating concrete under water consisting essentially of a tube of wood or sheet metal with a top in the form of a hopper.

tribar A type of precast armor unit for use in breakwaters and as shore protection.

tsunami A long-period wave caused by an underwater disturbance such as a volcanic eruption or earthquake. Commonly miscalled a tidal wave.

turbidity A condition whereby transparency of water is reduced. It is an optical phenomenon and does not necessarily have a direct linear relation to particulate concentration.

ultraviolet (UV) resistance Ability to resist deterioration on exposure to sunlight.

vertical seawall See *bulkhead.*

viscosity (or internal friction) The molecular property of a fluid that enables it to support tangential stresses for a finite time and thus to resist deformation.

warp Fibers or yarns lengthwise in the fabric.

web The sheet or mat of fibers or filaments before bonding or needle-punching to form a nonwoven fabric.

wharf A structure built on the shore of a harbor, river, or canal so that vessels can lie alongside to receive and discharge cargo and passengers.

woven fabric A textile structure comprising two or more sets of filaments or yarns interlaced in such a way that the elements pass each other essentially at right angles and one set of elements is parallel to the fabric axis. Usually has a uniform pattern with distinct and measurable openings. Commonly referred to as cloth.

yarn A generic term for a continuous strand of textile fibers, filaments, or materials in a form suitable for weaving or otherwise intertwining to form a textile fabric.

Index

ABOUT THE AUTHORS

Lawrence L. Whiteneck and Lester A. Hockney, both professional engineers, are employed by Moffat & Nichol, Engineers, of Long Beach, California. Mr. Whiteneck, who served as Harbor Engineer of the Port of Los Angeles, is a project manager and consultant with Moffat & Nichol. Mr. Hockney is a port planner with the company. He previously served in the planning section of the Long Beach Harbor Department.